Our Unfinished Biological Revolution

Seymour W. Itzkoff

Our Unfinished Biological Revolution

Seymour W. Itzkoff

Academica Press
Washington~London

Library of Congress Cataloging-in-Publication Data

Names: Itzkoff, Seymor (author)
Title: Our unfinished biological evolution | Itzkoff, Seymour
Description: Washington : Academica Press, 2021. | Includes references.
Identifiers: LCCN 2020042299 | ISBN 9781680539226 (hardcover) | 9781680539257 (paperback) | 9781680539943 (e-book)

Dedicated to the Memory of Jacob Itzkoff, 1892-1965

Other Books by Seymour W. Itzkoff

Cultural Pluralism and American Education 1969

Ernst Cassirer: Scientific Knowledge and the Concept of Man 1971 (2nd ed.) 1997

A New Public Education 1976

Ernst Cassirer, Philosopher of Culture 1977

Emanuel Feuermann, Virtuoso 1979 (2nd ed.) 1995

The Evolution of Human Intelligence

1. *The Form of Man, The Evolutionary Origins of Human Intelligence* 1983

2. *Triumph of the Intelligent, The Creation of Homo sapiens sapiens* 1985

3. *Why Humans Vary in Intelligence* 1987

4. *The Making of the Civilized Mind* 1990

How We Learn to Read 1986

Human Intelligence and National Power 1991

The Road to Equality, Evolution and Social Reality 1992

The Decline of Intelligence in America, A Strategy for National Renewal 1994

Children Learning to Read, A Guide for Parents and Teachers 1996

The Human Prospect

The Inevitable Domination by Man, An Evolutionary Detective Story 2000

2050: The Collapse of the Global Techno-Economy 2003

Intellectual Capital in Twenty-First-Century Politics 2003

Rebuilding Western Civilization, Beyond the Twenty-First-Century Collapse 2005

Who Are the Jews?

Soul of the Israelites 2004

A Nation of Philosophers 2004

Fatal Gift: Jewish Intelligence and Western Civilization 2006

The World Energy Crisis and the Task of Retrenchment 2008

The End of Economic Growth: What Does It Mean for American Society? 2009

Judaism's Promise, Meeting the Challenge f Modernity 2014

Liberty's Dilemma, America, Two Nations, Dependent/Independent 2014

Humanity's Evolutionary Destiny: A Darwinian Perspective 2016

2284, World Society, Iaian Vernier's Memoir 2016

Contents

Introduction

All animals do what natural selection programmed their ancestors to do (...). If any species in the history of life has the possibility of breaking away from short-term selfishness and of planning for the distant future, it is our species.

Richard Dawkins[1]

[1] From R. Dawkins' acceptance speech at the 2001 Kistler Prize Banquet, accessed May 7, 2020, https://vimeo.com/43114615.

Chapter 1

Wherefrom/Whereto: Evolutionary Perspective

Awakening experience

In 1968, the wonderful president of the college at which I taught gave me a grant to participate in a European study group's investigation of European educational efforts and underlying philosophical support system. During the course of five weeks, we traveled to London, Moscow, Copenhagen, Prague, and East Berlin, with short stops in Lithuania, somewhere in Yugoslavia, and Sweden.

Growing up in New York City and educated in its public schools, including a couple of years at City College of New York, later as a faculty member at Hunter College and Lehman College, all these then free tuition institutions in New York City, I was a city boy. Given New York City's buzzing, blooming diversity of races, ethnicities, social classes, the European tour, my first, was a shock.

I expected the diversity of political ideologies and the Cold War antithesis of values. What shocked me were the faces of the citizens of these locales. For example, except for a hint of Genghis Kahn in Moscow, the Soviets had the same blondish hair, blue eyes, skin complexion as residents of Copenhagen, Berlin, and the other nations. Looking at Europeans was like looking at an extended family.

Why and what meaning could one attach to the bloody internecine past and the then possibility of a new horror in the not too distant future? Of course, I thought about the American Civil War. After the five weeks of thought awakening, I guess I was a better teacher of Comparative Educational Systems, even considering the expense of the trip for my institution. To say the least, it was an eye and mind opening experience. It set me on a course of research into the evolutionary meaning of this visual and cultural reality.

Where from these people?

Thirty years later, in the late 1990s, during my third work in the study of general evolution and human evolution especially, I came across a series of articles studying the genetic relationship of Europeans. These articles attempt to trace back Europeans' origins in terms of the genetic variation that supposedly affects individuals at a genetic and probable geographic distance. The writers called this the "genetic blockade." I believe a better description of what was going on would be "the funnel effect," as in pouring a quart of oil into the oil reservoir of the motor of your automobile. Only in this case, the action would start at the narrow opening of the funnel. Perhaps this process of Europeanization occurred over quite a long extension and then suddenly widening (into where you pour the oil) Europe.

Their timeline analysis back in evolutionary history shows a much shorter period of genetic deviation than, for example, the human residents of Africa. In Eurasia, the expansion of the to-be Europeans occured some two to four hundred thousand years in the past. Again, these writers saw this happening somewhere in Eurasia, where other long-term residents of the Eurasian land mass, such as the Neanderthals, lived. These new genes, which entered the funnel, and then probably a long time afterward, exited it with very different survival adaptations, were probably sister and brother relatives of the Neanderthals. During these two to four hundred thousand years of undergoing a funnel/genetic blockade transformation, they had undergone a radical genetic and thus a phenotypic (external body) revolution.

A different animal

The products/descendants of this genetic/phenotypic revolution were the modern Europeans. First appearing in fossil and cultural form (with tools-physical evidence of their way of life) throughout mainland Europe and west Asia, we have named these humans and their associated culture the "Cro-Magnons." This evidence for the flourishing of a people with an evolving, non-literary civilization comes into view about 40,000-45,000 years ago. This is quite a long span of time, let us say, 200,000 to 45,000 years BP (before the present), for these people, and their cultural and social systems to have been hidden.

Along with the Neanderthals in c. 600,000-40,000 (Before the Present, BP), there were other humans living in Europe; Asia; west and east Africa; but not in North and South America, and certainly not in Antarctica. The characteristic heaviness of bone and body structure, which we see in the Neanderthals, at least partially attributed to adaptations to the sometime colder climate of the Eurasian north during the "ice ages," is also seen in their contemporaries south and east, in Africa and Asia.

A characteristic of the evolutionary process is the manner in which successful mammals, as they exploit their adaptations and create an economics of success, tend to become specialized physically and behaviorally, capitalizing on their success, leaving little room for new breakouts. This seems to have become true of humans, having had many millions of years heading forward onto the pathways of adaptive success, in terms of hunting, scavenging, and scrounging for fruit and vegetables. The result of this was a comfy heaviness of the body structure, ever more deepening their dependency on the status quo of environment, ecology, animal competition.

This occurred until the coming of the revolutionary creature, hominin Cro-Magnon, who had long exited that genetic funnel. In addition to wandering and spreading their genes around in small but potent quantities, the Cro-Magnon clearly had gradually discovered the meaning of behavioral/adaptive success. By

45,000 BP, the key genetic and physical dimensions of their lives, and even the technology they wielded now set them clearly apart.

Cro-Magnon, *Homo sapiens sapiens*

It is still a mystery how early they were shaped sociologically, and physically as a fully-formed clade—a series of very similar tribal, cultural units. When we see their fossils and their technology fully-formed in West Europe, and West Asia some 40,000 years ago, they seem to have tumbled out of nowhere, probably up the Danube and then spreading about in Eurasia.

Now characterized by their Aurignacian culture, possessing a tool kit revolutionary in the fabrication of a wide variety of materials, using stone, bone, and probably wood for all kinds of uses and in various shapes. These tools were mostly utilized in hunting, but some were so delicate and refined in manufacture that they stimulate wonder. What were they for—beauty, creativity? They made jewelry for both men and women, sewed their clothing, and practiced a glorious tradition of painting and sculpting. Throughout their range, there are carved, hand-held Venuses of well-fed and voluptuous women, probably depicting those awaiting them after the hunt. These possibly pornographic symbols often are esthetic renderings—modernistic, abstract envisionments of the nude.

Figure 1: Venus of Hohle Fels, the earliest Venus figurine, ca. 35,000 BP

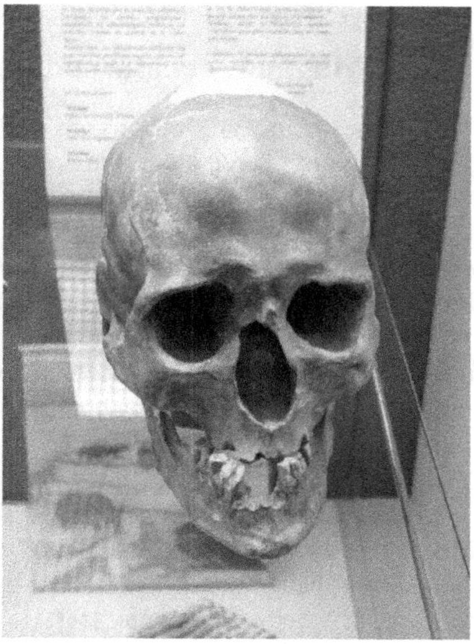

Figure 2: Cast of skull of the Chancelade man, elderly Cro-Magnon male, ca. 15,000 BP

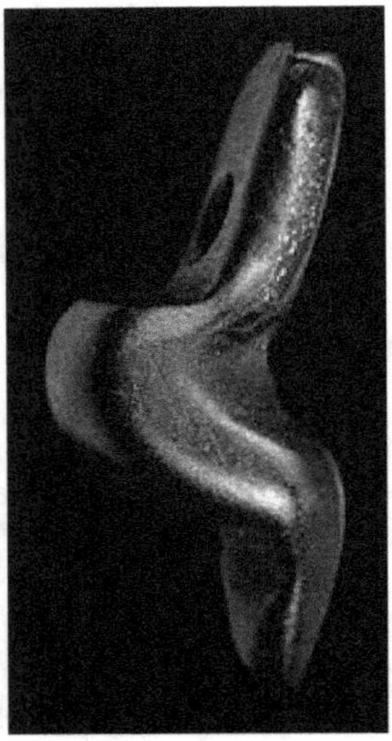

Figure 3:
Venus of Mnruz, ca 11,000 BP

The men were taller, 5'8" to 6' tall, while their Neanderthal contemporaries stood at 5'3" to 5'6" tall. The bone structure and teeth of Cro-Magnon is delicate, childlike, as compared withal their contemporaries except the mysterious Capoids (Boskopoids/ Bushman)—who had large modernistic /thin skulls, then dominating almost all of Africa, even venturing north into Spain, and themselves creating an interesting artistic culture. In the case of both sub-species/racial groups, the great explosive growth of the skull and its presumably interior neurological matter certainly created this shocking difference, at the least between Cro-Magnon and its racially similar Caucasoid progenitor, the Neanderthal. They both had white, northern skin, blondish, light colored hair, and were often blue eyed. One sub-species was to live on and dominate, and the other, the Neanderthals, would pass some its genes on into the future. Certainly, by 30,000–25,000 years ago, the Neanderthals as a unique sub-species had to have disappeared from the Earth in shock and then, sterility.

Cro-Magnon: genetic extinction/absorption

One theory about the origins of Cro-Magnon comes from Spanish researchers (1997), who discovered a grouping of fossils they called *Homo antecessor.* They trace these bones, fragments and a lower jaw (no skull found as yet) to about 780,000 years BP. The bones of the children in the group were similar to *Homo sapiens* and not proto Neanderthals. Other scientists relate these fossils to *Homo heidelbergensis*, the possible ancestors of the Neanderthals in Eur-Asia

Jumping forward in time, well out of the narrow funnel of creation to about 100,000 years ago, caches of Cro-Magnon-like tools have been discovered in the jungles of West Africa, and along the Pacific coast of China. These are the probable assigned dates. Dating back in pre-historic time is a probability, not a surety. At any rate, they tell us that the large mammal, primate propensity for *wanderlust* was and is still evident in all sub-species (races) of *Homo sapiens.* Two skulls dating to this 80–100,000 BP timeline hints at what is becoming orthodoxy in paleontological studies—the hybridization within the genus *Homo* and all its assorted species and races.

Discovered in Israel, probably the most southerly extension of Neanderthal wanderings, possibly caused by a southerly extension of the ice flows, Skhul and Qafzeh exemplars lived about ninety to a hundred thousand years ago. Their big, rounded skull profiles are not Neanderthal, and quite like subsequent Cro-Magnon fossils. There are modest brow ridges and other hints of Neanderthal heritage, and interestingly, the tools associated with this burial ground are Neanderthal (Mousterian) in heritage.

The most probable hypothesis: a wandering Cro-Magnon or band of Cro-Magnons encountered a lone female Neanderthal. The children of each separate encounter grew up in a Mousterian community, and probably contributed much to its maintenance and survival. A few miles down the road from these excavations, another historical find was uncovered. Dating to c. 60,000 BP, this was a large Neanderthal encampment. Thirty or so thousand years later in time, Neanderthal bands were again in Israel, partaking of the southern warmth, with no Cro-Magnons in sight.

The current estimate is that contemporary Europeans, inheritors of the genetic profile of the Cro-Magnons, carry within their genoplasm approximately 5% heritage of Neanderthal genetics. My suspicion is that these were mostly genes from female "guests."

"Analysis of the skeletal remains of a four-year-old child buried in a Portuguese rock-shelter 25,000 to 24,500 years ago has yielded startling evidence that early modern humans and Neanderthals may have interbred. While the boy's prominent chin, tooth size, and pelvic measurements marked him as a Cro-Magnon, or fully modern human, his stocky body and short legs indicate Neanderthal heritage," say Erik Trinkaus, a paleoanthropologist at Washington University in St. Louis, and Joao Zilhao, the Portuguese scientist who discovered the fossil. Interbreeding could answer the vexed question of the fate of the Neanderthals, the last of whom disappeared from the Iberian Peninsula 28,000 years ago.

There is another long-standing evolutionary conundrum, concerning the interbreeding of "close kin." In the natural history museum in Prague, Czech Republic, there are two skulls on exhibit that trace back c. 30,000 years, found in the soil of this nation. They are called Predmost and Mladec. Scientist have noticed that these skulls, found amidst classic Cro-Magnon tools, are very similar to the Skhul and Qafzeh skulls noted above, that appear to go back some 90,000 to 100,000 years ago, in what is now Israel. But these latter skulls are associated with Neanderthal tools.

Could it be that at the timeline for the extinction of the Neanderthals, the Neanderthals were still contributing their genetics to the Cro-Magnons, but were now participating in the Cro-Magnon civilization?

Do these skulls tell us that phenotypic (surface) appearance is not necessarily a defining characteristic of the cultural/civilizational capacities of individuals?

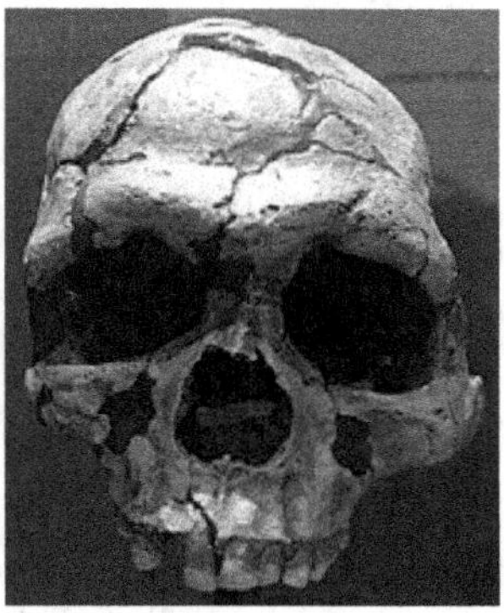

Figure 4: Cast of Qafzeh

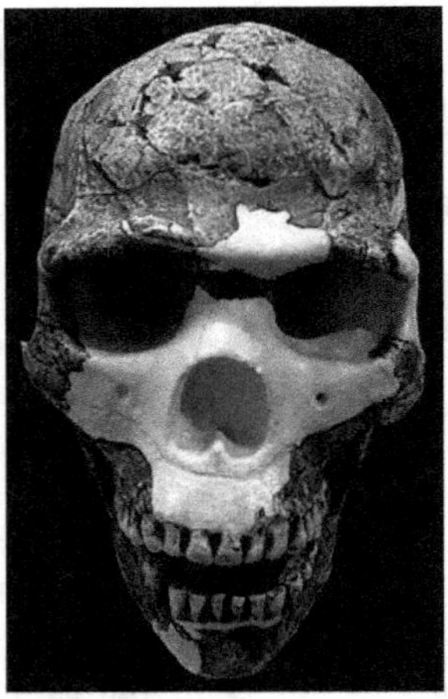

Figure 5: Předmostí skull, c.30,000 BP

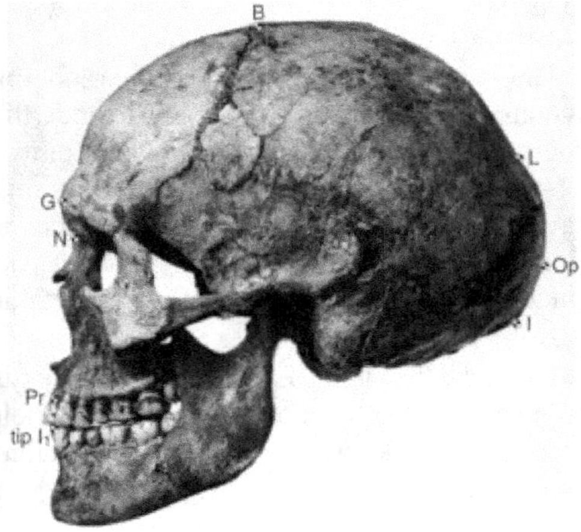

Figure 6: Cro-Magnon skull

As we will later argue, the travel of the genes of Cro-Magnon, from at least 100,000 BP, is revealed in other early modernizations of humanity—perhaps long before the extinction of the Neanderthals. The Capoid race, a sub-species, is a fascinating puzzle of the early and yet contemporary infusion of genes on the move throughout the world.

Today, the Bushman of South Africa, a remnant racial descendant of the great Boskopoid control over most of Africa, are on the brink of their puzzling extinction as a "pure blooded" exemplification of the race. We do know that their genes and "click" language have been absorbed into a variety of African ethnicities. Perhaps one of the greatest African leaders, Nelson Mandela, reveals in his phenotypic physiognomy a Capoid racial heritage along with his acknowledged Bantu affiliations. As the Capoid race continues its puzzling decline and ultimate disappearance, those ancient talents are still relevant in the modern dynamics of humanity.

A tragic historic genocide

The Tasmanian aboriginals, thought to be of Melanesian origins who, like many other prehistoric groups, entered and wandered this continent, proceeded southward. In their first encounter with the West in the mid 18th century, by Captain Cook and a number of French explorers, they did not know how to make fire, nor did they understand the causal relationship between the sexual act and the consequent pregnancy of women.

First arriving in Tasmania (then a peninsula of Australia) around 40,000 years ago, the ancestors of the aboriginal Tasmanians were cut off from the Australian mainland by rising sea levels at the end of the Wurm ice flow, c.10,00–6000 BC. They thus were entirely isolated from the rest of the human race for at least 8,000 years, until continual British contact.

Before British colonization in 1803, there were an estimated 3,000–15,000 Palawa-speaking aboriginals. The Palawa population suffered a drastic drop in numbers within three decades, so that by 1835, only some 400 full-blooded Tasmanian aborigines survived, most of whom were incarcerated in camps, where all but 47 died within the following 12 years. No consensus exists as to the cause.

The traditional view, still affirmed, held that this dramatic demographic collapse was the result of the impact of introduced diseases, rather than the consequence of policy. Some authorities claimed that by 1830 in Tasmania, disease had killed most of them, but warfare and private violence had also been devastating. Other historians attributed the depletion to losses in the "Black War" of the uprisings in 1830. The Tasmanian aboriginals were more likely to fight than their fellows on the Australian mainland.

Others point to the prostitution of women in a society already in decline. Earlier, by the 1790's, seal hunters from the United States and England were on the islands between the Australian mainland and Tasmania, engaging in this profitable trade. The seal hunters extended their efforts to the mainland, where they often killed the men and created harems of Tasmanian women, both for sex as well as for assistance in the trade.

Then came the colonists: the British Navy. The sailors' lore spoke of seducing the young but nubile Tasmanian girls with "treats," the girls responding from their tree hideouts with giggles and then acceptance. The negative impact on the native population caused the English authorities to therefore place the remaining population on a series of islands north of Tasmania, with much governmental assistance. The last pure-blooded Tasmanian died in 1876.

However, there are contemporary photographs of schoolchildren and their teachers on Cape Barren Island, that appear healthy and well-dressed, who are mostly descendants of English men and the Tasmanian natives. Many authorities view this sequence of decline and extinction as acts of genocide. Yet, there sadly was a belated attempt by the colonial rulers to rescue this ethnicity from oblivion. Today, there are approximately 20,000 inhabitants of Tasmania that claim their descent from this now-extinct aboriginal people.

Early hybridization of *Homo sapiens*

There seem to be two phases in the hybridization towards sapiency in the human clade. The first took place historically, at points still relatively unknown, before the full weight of the Cro-Magnon, and *Homo sapiens sapiens*' adaptive impact on the world began to reveal its potential potency.

This first phase is exemplified in the discovery a few decades ago of the Flores fossils, from the Indonesian Island of the same name (Brown et al., 2004). At first, the fossils were placed in the 18-13,000 BP range, which is shocking because these humans were 3.5 feet tall with a brain the size of a chimpanzee's, and possessed only the earliest *Australopithecine* (Lucy) level of hominin evolution. In general, this was an extremely primitive line of hominids, and some think they are not "worthy" of inclusion in the genus *Homo.*

What was this creature doing, all the way in East Asia, at such a primitive level of human evolution? Well, the orangutans and gibbons traveled from Africa to this same region of Asia c. 15 million years ago. One can thus accept this Australid sub species migration at the level of ca. 3–2 million years ago (see van den Bergh et al., 2016). Today, the scientific consensus is that these Flores hominid fossils date to about 50,000 BP. This means that they could have survived on these relatively hidden islands for another 10,000–15,000 years. Living nearby are groups of indigenous, pigmy Indonesians, who are fully sapient, but dwarfed by their inclusive living quarters.

By 50,000 BP, groups of humans carrying some Cro-Magnon genes could have been pushing down south worldwide, to avoid the cold of the ice-flows. Perhaps the genes of the Flores folks were shared with some of these migrants who stayed the course on other islands. A possible example could be the not-distantly separated Indonesian pygmies, or the Australian and Tasmanian folks further south.

Another even more mysterious prehistoric event involves a human group called the Denisovians. Here, we have a finger bone and some teeth that trace back to about the same period as the Flores folks. The location of this find is in the north, at the boundary between Siberia and Outer Mongolia. DNA analysis places their origin not back to the 3–2 million years ago boundary, but to perhaps 800–700,000 BP and with significant genetic association with current Melanesian/Papuan populations on the islands and lands of Southeast Asia.

Since in classic terms of the five races—Caucasoid, Negroid, Mongoloid, Capoid (Khoine-San), and Australids—the latter were probably the earliest of the migratory groups of humans to leave Africa, their genetic and phenotypic (physical) imprint being found from Pakistan, north to Siberia, and east and south into Pacific areas. The Tasmanian originals were an ethnic variant of the Australids.

In the past 200,000–40,000 years, much genetic interchange has occurred. However, with the gradual absorption of all the borderline erectine/sapiens type of humans as we approach modern history, especially with the explosion of Cro-Magnon genetic migrations from some 40,000 years ago, the modern races (sub-species) of humans, even in their merely suggestive unities of appearance, now define the historical shape of human sapiency.

In other words, we still speak of race, even though it is largely an evanescent cultural phenomenon. It is really difficult to think about any tribal or national set of humans as having a "pure" racial identification. The exception, as I noted above in my 1968 experience, were the Europeans. In the fifty plus years since that time, our lives have radically changed for all the formerly straight-line descendants of the Cro-Magnon, *Hss*—e.g. rapid hybridization.

Migrations and Hybridization: Prelude to the Revolution

The c.45k BP entrance into Europe and West Asia, in force, must have been stimulated by a climatic change, which set off a demographic bonanza for

Neanderthal or Cro-Magnon hunters. Apparently their first named cultural tradition, the Aurignacian seems already to have been formed, as it appears uncannily uniform throughout their secured economic and home site locations, from east to west. Perhaps they first moved up the Danube Valley, then separated and multiplied.

The last of the European glaciations, the Wurm, ended c.12,000 years ago. It was one of the coldest, and seems to have wreaked havoc amongst the vast array of large mammals—the prey of the Cro-Magnons over time. There is evidence that those who stayed behind were herding reindeer, as our most northern Europeans do today.

Otherwise, the ice age weather below the descending ice-caps was autumnal in summers, mildly cold in winters, with not enough snow to limit the foraging of these mammals during the usual cold periods. If there ever was a "garden of eden" for humankind, the art work, the vast productive technology, and the general cultural elan argues for happy hunting grounds, for all except the Neanderthal. Consider, that for some 30,000 years, the modern human model of social life—the relationship between the sexes, the governing of these small communities—must have made an indelible imprint even on our most "modern" human nature.

The cultural epochs during this ca. 30,000-year period—Aurignacian, Gravettian, Solutrian, and Magdelenian—all denote differing styles of art, technology, style/decoration, within a seemingly non-literate civilization. But wait—Alexander Mars hack, a magazine writer who became fascinated with this "civilization," brought to our attention markings on bone, slate, ivory, and almost any non-decayable surface. These markings are similar to those that hunters cut into the stock of their rifles, or other signs of notation. These did not appear to decorative: they seemed sequential in the alignment patterns of the cuts. To Marschack, these were 12,000-year-old symbolic references to the regularities of experiences of an as-yet naïve people—seasonal changes, menstrual and pregnancy periods, and the migration of animals.

There is more to the symbolic thought of the Cro-Magnons. In later chapters, we exemplify the heretofore unrecognized literacy of the Cro-magnons.

Thousands of years later, still naïve but better organized groups such as the Stonehenge builders in England and elsewhere would celebrate the longest and shortest days of the year with their alignment of the boulders. Going back to the latter part of the Upper Paleolithic era, the Cro-Magnons were already developing the concept of abstraction, making symbolic correlations between the events occurring outside of their bodies with notations that would remind them of these events away from the sensory experience. Examples include, the phases of the moon, and even solar or lunar eclipses.

Clearly, there had to have occurred migrations out of this "Garden of Eden," during the vast (for us humans) time span. Mostly south and east, we find many signs of the impact of the Cro-Magnons far beyond this new homeland, Europe. One of the most interesting revelations of the migration of the Cro-Magnons and other human groups during this 40,000-year period of culturalized Cro-Magnon living, is the so-called "Solutrian" problem—the seeming derivative Clovis cuture of the southwest U.S.A., with the Solutrian interregnum of the Cro-Magnon, in the latter c. 20,000 BP and the Clovis at about c. 13,000 BP.

The Amero-Indians who created this culture were part of a migration over the Bering land mass during a long cold/dry spell of the ice ages, from c. 20,000 BP. This several thousand-year trickle of many diverse groups, traveling from eastern Europe or from northeast Asia, such as the Denisovians, with their mixed Cro-Magnon, Neanderthal Australid genetic markers, was significant for humankind. It is quite likely that the Cro-Magnons brought their culture and genes east with them, hybridizing on route with other groups, thence within their newly discovered North and South American habitats.

The discovery of a very Cro-Magnon looking, c. 9,000 BP skeleton in the state of Washington, called Kennewick man, created political controversy. It supposedly was subject to native Amero-Indian burial laws, here a dispute with the scientists who wanted to study this unique specimen. Supposed DNA evidence overrode the physical evidence, rendering the skeletal remains tribal, to be buried, out of the reach of modern science.

It is clear that all the Amero-Indian migrant tribes in North and South America have a mixed racial and ethnic ancestry, with the earliest South American tribal groups being mostly Australid in origin. For example, were we to compare the facial profiles of Pharaoh Ramses II, d. 1213 BCE; Emperor Frederick the Great d.1786 CE; and Chief Sitting Bull d. 1890 CE, one would not be far wrong in calling them first cousins. The only difference here is that the ancient characters had red hair, and the chief, black. But the Sumerians, c. 3000 BCE, descendants of the Cro-Magnons, had black hair!

Ongoing natural selection of humanity

These movements, 40–10,000 BP, were now happening all over the world, almost out of sight, out of mind. Another interesting and puzzling event in nature's ever-ongoing natural selection of life, and of course humans, lies before our eyes today. This story goes back into the ice ages and is concluding in our own time. It takes place from Spain, to the southern reaches of the continent of Africa. Because it involves the disappearance of a once widely dispersed race, a sub-species of *Homo sapiens*, it is even more tragic than that which befell the Tasmanians if not the rediscovered Flores primitives and their many hominid kin throughout the ages.

Generically, this group was once called the Boskopoids. Possessing high skulls of comparable Cro-Magnon volumes, they occupied almost all of Africa, venturing as far north as Spain. Racially, they were distinct Capoids, today known as Koihsan, or the San. Their traditional colloquial names are Hottentots and Bushman. They have a brownish skin color, and have tightly curled hair. The great anthropologist Carleton Coon conjectured that the Koihsan were once related to the Mongoloid race, except for the hair. That unity would have taken place at least a million years ago, not as recently as with the Japanese and the Koreans. The colloquial joke is that the Koreans were the Japanese who stayed at home, the Koihsan were the Mongoloids who remained in Africa.

Culturally, the Bushman wall paintings, mostly illustrating these Capoids/Bokopoids at the hunt and more like stick figures than the three-dimensional perspective of the Cro-Magnons, battling with neighbors a very lively people, reaching a apogee of expansion in Spain and North Africa, c.15,000 BP. But something was amiss. In a French museum, circa 1980, this writer saw an illustration of a Cro-Magnon wall drawing from Spain. It showed a boxy, cage-like contraption hanging from the top of a cliff, alongside the cliff wall. Inside the cage was a human male. The teeth, hair and facial characteristics clearly described a male Khoisan. This was evidently an act of disdainful humor at the expense of loser Khoisans. Sure enough, these ancestors of the contemporary San people were driven from their traditional homelands not only in exploratory Spain, but throughout the northerly span of Africa.

Evidence is that the Cro-Magnons were moving across the straits of Gibraltar, the slow way around the eastern expanse of the Mediterranean Sea. Today, we call these people Berbers, now an Arabic resistant minority, living and upholding an ancient cultural/linguistic tradition in the Atlas Mountains of Morocco.

The declining prospects of the Khoisan were not yet at an end. The Bantu ethnicity of the Negroid race in West Africa began to move out of their jungle redoubt in about 1000 BC, during the reign of King David of Israel—a time of prosperity and advance in Phaoronic Egypt. Long before this, as Cro-Magnons were leaving a now Holocene climatic change, c.4000–3000 BC, the forests of the north were moving in to give animal life cover from the hunters of the Mesozoic era. An early dynastic Pharoah gave an order for "no more black slaves from south of the first cataract," the land of the Aswan Dam.

Even earlier, in 6000–5000 BCE, these refugees from the now seasonal climate extremes of the north had crossed beyond the Nile, to its sources in what is now Sudan, Somaliland, Ethiopia, and lands beyond, to exploit the riches of Africa. Their success stimulated Negroid migrants to wander east, across a less threatening Sahara, to partake of the riches that these new inhabitants of Africa were creating. As slaves or free people, they must have shared in the wealth and the genes of these northerners. And they must have taken these genes back across the Sahara to the homeland of their own ancestors.

The invigorated Bantu, in their search for the new and possible, encountered the remnants of the Khoisinoid peoples, and they too drove the San back down into the most southern parts of this continent. Come the colonialists of the 19th century, the fate of these people changed by hybridization with both the Bantu and the Caucasoid colonialists. Else, they were being pushed hither and yon from the agricultural and hunting grounds they had utilized for centuries. Thus, their contemporary, brave desperation of life in the Kalahari Desert, else protected and isolated by government, as with the Amero Indians of the United States.

Here, once more, within our sights and intellectual awareness, is a unique exemplification of a once vigorous and ancient race, a sub-species of human kind, now being pushed into extinction. Their genes will live on, as with other extinct peoples. Nature is inexorable in its positive selection, or extinction, of so many experiments in the evolution of life.

We are not immune to evolution

As I write in 2020, we are, as living things always have been, preoccupied with the dynamics of what we think to be the realities of dominance and survival. But there are processes seemingly beyond the ken of the wisest leadership groups. This is the long range nature of our existence on this planet. Oddly and mysteriously enough, we have the ability to look long into the future. Theodosius Dobzhansky, an eminent biologist and evolutionist, once alerted us to what he saw as our greatest evolutionary survivalist capacity—"the ability to bind time."

This was the mental capacity to see in contemporary events a message from nature and laws—regularities that exist over time. By seeing our momentary perceptual and social experiences as part of these regularities of nature, we can attempt to predict and anticipate the future. The entire law-making process of modern civilized peoples is an example of this attempt of our innate intelligence to predict and anticipate. In the minds of these lawmakers, no doubt there is usually a glance back in time, at the least for assurance.

In the next chapters, we will attempt to lay out some these important events and processes that created this revolutionary biological creature, *Homo sapiens sapiens*. Today, this moniker, *Hss*, is no longer the sole possession of one sub-species, once represented by the Cro-Magnon Caucasoids. This mysterious revolution continues now in the social dynamics of all the existing ethnic varieties of *Homo*.

We will take a brief look at how the evolutionary elements have led to our dominance over major portions of life. We will also give a glance into an always-precarious future.

In July 2019, anthropologists reported the discovery of 210,000-year-old remains of a *H. sapiens* and 170,000-year-old remains of a *H. neanderthalensis* in Apidima Cave, Peloponnese, Greece, more than 150,000 years older than previous *H. sapiens* finds in Europe. "Apidima 2 dates to more than 170 thousand years ago and has a Neanderthal-like morphological pattern. By contrast, Apidima 1 dates to more than 210 thousand years ago and presents a mixture of modern human and primitive features. These results suggest that two late Middle Pleistocene human groups were present at this site—an early *Homo sapiens* population, followed by a Neanderthal population." [Harvati et al., 2019: 500] Israel Hershkovitz, a paleoanthropologist at Tel Aviv University in Israel who found the fossils in Misliya Cave, thinks that because members of ***H. sapiens*** were in the Middle East about 200,000 years ago, they could have made an early excursion to southern Europe, too. *Homo sapiens* might have spread to the Arabian Peninsula around 180,000 years ago — much earlier than previously thought. The following image is of the upper jaw and teeth, found in an Israeli cave and reported in *Science* on 2018, 25 January.

On the originating hypothesis for eventual Cro-Magnons, see Itzkoff (2000: 314-316).

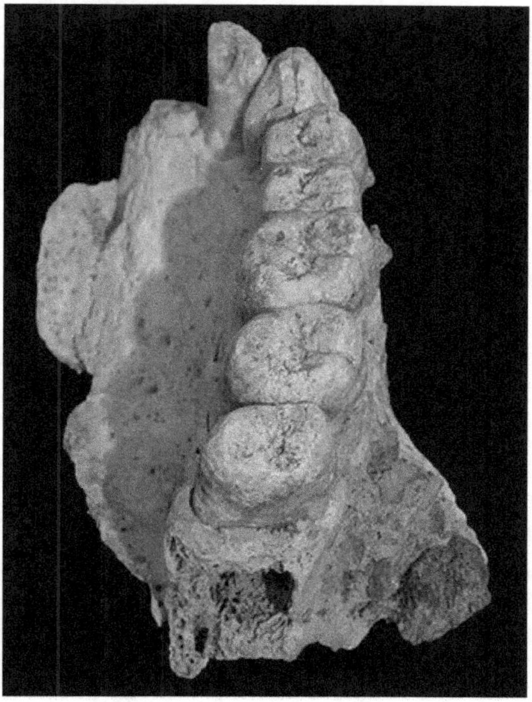

Figure 7: The upper jaw and teeth, found in an Israeli cave, are thought to be the earliest evidence of Homo sapiens outside Africa [Credit: Israel Hershkovitz, Tel Aviv Univ]

Chapter 2

Life's Struggles

Conversation

Two octogenarians discussing their own lives and family at their Florida condo, alongside of the pool, of course:

Irving: Well, Milt, we shouldn't be worrying. We'll be gone, the kids will have to solve their own problems. What can we do? I read an article which said that in a billion years, the sun is going to expand over the earth. Woof, it'll all be ashes.

Milton: What are you saying, Irv. That's terrible. Haven't we got enough to worry about? This news doesn't make *my* day.

Irving: Don't get excited, Milt. A billion years: we'll all be far gone by then.

Milton: Whew. Irv, relief, a billion? I thought you said a million.

Creationist beginnings?

Our solar system planets are thought to have formed from the solar nebula—the disc-shaped cloud of gas and metallic dust left over from the Sun's formation. The currently accepted method by which the planets formed is accretion, in which the planets began as dust grains in orbit around the central proto-star. Through direct contact, these grains formed into clumps up to 200 meters in diameter, which in turn collided to form larger bodies (planetesimals) of around ten kilometers (km) in size. These gradually increased through further collisions, growing by small leaps over the course of the next few million years.

For Earth, this all started around 4.54 billion years ago, and our planet is approximately one-third the age of the universe. Volcanic outgassing created the primordial atmosphere and then formed the ocean. The early atmosphere contained almost no oxygen.

Organic compounds (complex organic molecules) necessary for life had perhaps formed in the proto-planetary disk of surrounding cosmic dust grains, before the formation of the Earth

In the early millions of years, much of the Earth was molten because of frequent collisions with other bodies, which led to extreme volcanism. While Earth was in its earliest stages, 4.4 billion years ago, a giant impact collision with a planet-sized body, named Theia, is thought to have formed the Moon.

The Earth cooled, causing the formation of a solid crust, and this allowed liquid water to gather on the surface. Most researchers now see water on an oceanic level as taking place shortly after the impact with the moon, probably c. 4.3 billion years ago, and no later than 4 billion years ago.

The origin of the basic biomolecules, while not settled, is less controversial than other issues of origin. The basic chemicals from which life is thought to have formed are: Methane (CH_4), Ammonia (NH_3), Water (H_2O), Hydrogen sulfide (H_2S), Carbon dioxide (CO_2) or carbon monoxide (CO), and Phosphate (PO_4^{3-}). Note: Molecular oxygen at these beginnings was absent.

Origin of life

Sadi Carnot, France, 1824: the second law of thermodynamics is concerned with the direction of natural energy processes. It asserts that a natural energy exchange runs only in one direction, and is not reversible. For example, heat always flows spontaneously from hotter to colder bodies, and never the reverse.

The above is the fundamental challenge that all forms of life, even the primordial, reproducing one-celled bacterium, must face. They are the pivot for all subsequent living things.

Scientists have suggested that the chemicals in the modern cell gives clues as to what the early seas must have been like. As early as 1926, Macallum noted that the inorganic composition of the cell cytosol dramatically differs from that of modern sea water: “the cell … has endowments transmitted from a past almost as remote as the origin of life on earth.” For example: “All cells on land contain much more potassium, phosphate, and transition metals than modern cells in oceans, lakes, or rivers.” “Under the anoxic, CO_2-dominated primordial atmosphere, the chemistry of inland basins at geothermal fields would [be like the chemistry inside] modern cells.” (Macallum, 1926)

In the search for this primordial form of life, RNA (ribonucleic acid) is identified as present in all living cells. Its principal role in ongoing evolutionary processes is to act as a messenger, carrying instructions from DNA for controlling the synthesis of proteins. It is thought to work both as an enzyme and as a container of genes. Later, DNA took over this major genetic role.

The RNA world hypothesis proposes that life based on ribonucleic acid (RNA) pre-dates the current world of life based on deoxyribonucleic acid (DNA), RNA and proteins. RNA is able both to store genetic information, like DNA, and to catalyze chemical reactions, like an enzyme. RNA may have supported pre-cellular life and been a major step towards cellular life. There are some RNAs which work as enzymes. Some viruses use RNA for heredity. Many of the most fundamental parts of the cell (those that evolve the slowest) require RNA.

RNA is a complex molecule. In the 1920’s–30’s, scientists since the Soviet scientist, A. Oparin (1938), and the British scientist, J.B.S. Haldane (1929), believed that external heat sources were necessary to bring inert proto organic elements into an interactive reproductive balance with nature.

These and many other views have since been subject to critique. Even today, the intellectual motivation to discover a contemporary duplication, even hypothetical, of creating life anew from the inanimate is enormous. Some of the importance of this quest lies with the persistence in a large community of believers in *Creationism*, a theological response to the scientific challenge of life's beginnings.

Below are two recent scientific hypotheticals, placing before the reader a sense of this important ongoing search. The first (refil, Morowitz & Smith, 2009) deviates from the older idea that a great energetic input in these warms seas was the catalytic force that created a master RNA molecule, that was able to incorporate a number of different organic elements in one great moment of synthesis and sustainability.

> Rather than supposing that a few large RNA molecules control the adaptation of a passive small-molecule reaction network, Copley supposes that whole networks of intermediate molecules support each other on the path toward complexity. In this experimental setting, networks of small and randomly synthesized amino acids and single RNA units aid each others' formation, assembly into strings and evolution of catalytic capacity. Both types of molecules grow long together. Complexity, adaptation and control are distributed in such networks, rather than concentrated in one molecular species or reaction type. Distributed control is likely to be a central paradigm in the development of metabolism, first as a viable theory. We eagerly anticipate more experimental efforts like these to explore the many facets of small-molecule system organization.
>
> (Patel et al., 2015)

The second excerpt sees the beginning of sustainable life forms in more traditional, "big energic impact" terms. However, the elements in the creation of this mix of organic molecules creating the context for energetic ingestion and duplication is an external energy implosion.

"For their current study, Sutherland and his colleagues set out to work backward from those chemicals to see if they could find a route to RNA from even simpler starting materials. They succeeded. In the current issue of *Nature Chemistry*, Sutherland's team reports that it created nucleic acid precursors starting with just hydrogen cyanide (HCN), hydrogen sulfide (H_2S), and ultraviolet (UV) light. What is more, Sutherland says, the conditions that produce nucleic acid precursors also create the starting materials needed to make natural amino acids and lipids, duplicating simultaneously."

Sutherland's team argues that early Earth was a favorable setting for those reactions. HCN is abundant in comets, which rained down steadily for nearly the first several hundred million years of Earth's history. The impacts would also have produced enough energy to synthesize HCN from hydrogen, carbon, and nitrogen. Likewise, Sutherland says, "H_2S was thought to have been common on early Earth, as was the UV radiation that could drive the reactions and metal-containing minerals that could have catalyzed them." Sutherland: "In recent years many researchers have come to the view that considering the considerable heat energy available on the Earth, subterranean, volcanic, etc as well as an ongoing dynamic cooling environment, living molecules were early formed, ca. 4.3-3.9 billion years ago, struggling against entropy, Carnot's second law of thermodynamics. The race was on for cellular integrity, metabolic utilization of resources, reproduction (mitosis) the search for biological immortality."

Figure 8: The crash of meteors on early Earth likely generated hydrogen cyanide, which could have kick-started the production of biomolecules needed to make the first cells. [Credits: Joe Tucciarrone/Science]

Hadean Eon

Figure 9: Moon

4000 Mya and earlier.

Domains and Kingdoms

Today, there is thought to be three great Domains of life:

1. Bacteria—derived from the earliest, tiny duplicating entities with a cell membrane, creating a permeable boundary within which existed proteins, as well as RNA and a variety of nutrients essential for metabolism. J. Monod (1971): "A simple eubacterium could have 2500

different protein molecules helping in the metabolism of the cell. Humans have over one million different proteins in their bodies. Mitosis in bacteria occurred after the ingestion of other organic cells leading to continuing complexity and change in the bio-chemical makeup of the bacteria, whether it was an autotroph (metabolizes metals, gases) or an organic heterotroph. Most metabolism took place in anaerobic (non-oxygenating) environments."

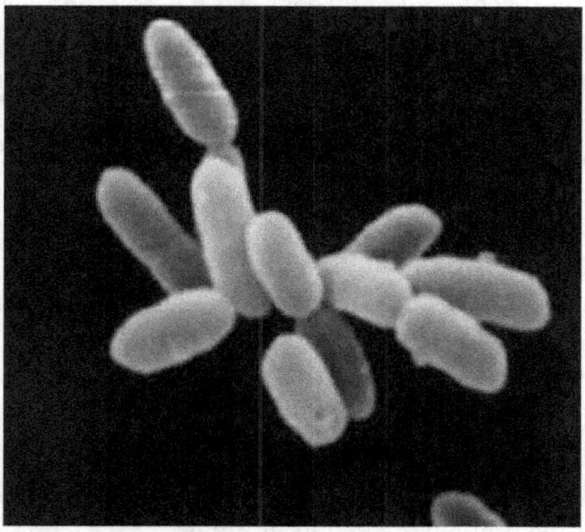

Figure 10: Halobacterium sp. strain NRC-1

2. Archaea—in 1977, American biologist Carl Woese defined this domain of prokaryotes (as with Bacteria, tiny one-celled forms of life), as without nuclei and without other forms of later advanced Eukaryote structure (Woese & Fox, 1977). Woese used his new RNA comparison method to categorize and contrast different organisms. He sequenced a variety of different species and happened upon a group of methanogens that had vastly different patterns than any known prokaryotes or eukaryotes, later evolving large complex cells, as in humans. These methanogens' anaerobic (non-oxygen), autotrophic fermentation (inorganic) metabolizers were much more similar to each other in fuction than they were to other organisms sequenced, leading Woese to propose the new domain of Archaea. His experiments showed that the Archaea were more similar to Eukaryotes than prokaryotes, even

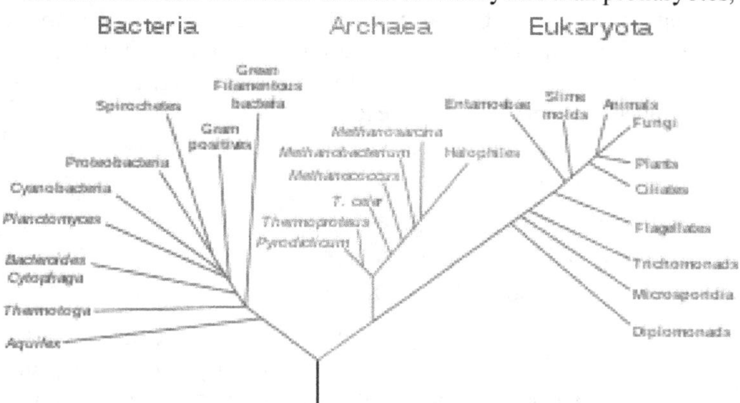

Figure 11: A speculatively rooted tree for RNA genes, showing major branches Bacteria, Archaea, and Eukaryota.

though they still were more similar to other prokaryotes in structure. This led to the conclusion that Archaea and Eukarya shared a more recent common ancestor than Eukarya and bacteria in general. The development of the nucleus occurred after the split between the two forms of prokaryotes, tiny one-celled creatures and this yet unknown common ancestor. Archaea exist in a broad range of habitats, and as a major part of global ecosystems, may represent about 20% of microbial cells in the oceans. The first-discovered archaeans were extremophiles. Indeed, some archaea survive high temperatures, often above 100 °C (212 °F), as found in geysers, black smokers, and oil wells. Other common habitats include very cold habitats and highly saline, acidic, or alkaline waters, but archaea also include mesophiles that grow in mild conditions: in swamps and marshland, sewage, the oceans, the intestinal tract of animals, and soils. Their interesting relationship to the as yet in process Eukaryotes, c. 3.5 billion years ago, is that unlike the similar, vast profusion of molecular bacteria, no archaea are known to cause human disease. Another possibility is that we have not recognized, the symptoms of archaeal diseases. The similarity of some archaea, such as Lokiarchaeota to eukaryotes, could compound this problem. I have previously speculated (to myself) that proteins from archaea, with high similarity to proteins in their eukaryotic hosts, could be the trigger for some conditions that have been labeled as "autoimmune" ("Why don't archaea cause disease?"). [Posted on August 11, 2017 by Dr James Chong, *Microbiology Society*]

3. Eukaryotes—at some point, they deviated from their Archaean origins. The guess is that it took c. 2 billion years to officially welcome the fully formed modern protest (paramecium), fungus, plant, animal cell into the kingdoms of life, they too in a metabolically determined effort to refute Sadi Carnot's law of entropy. The oldest modernist eukaryotic fossil is approximately 1.5 billion years old. The origin of the eukaryotes must have occurred long before, because the fossil is of a relatively complex single-celled organism. Biologists are almost certain that eukaryotes evolved from prokaryotes because:

- Both use RNA and DNA as their genetic material.
- Both use the same 20 amino acids.
- Both have ribosomes and DNA and RNA.
- Both have a lipid bilayer cell membrane.
- Both use L amino acids and D sugars.

One endosymbiotic model of eukaryotic origins shows a prokaryotic cell capable of engulfing other prokaryotes, such as aerobic (needing and having the use of molecular oxygen) bacteria. Rather than digesting them, the bacteria remain, as symbionts, benefitting the host cell by removing harmful O2 and helping in the production of ATP (Adenosine Triphosphate).

According to an autogenous model, the eukaryotes arose directly from a single prokaryote ancestor by compartmentalization of functions brought about by infoldings of the prokaryote plasma membrane.

The origin of eukaryotes constitutes a major evolutionary transition for which we lack much information about intermediate stages. This is particularly unfortunate, given that it is arguably the most drastic evolutionary transition that has taken place since the emergence of the last *universal common ancestor* (LUCA) of all cellular life-forms.

Eukaryotic cells are fundamentally different from those of bacteria and archaea at almost every level of organization, starting with their large physical size. These major differences in cellular architecture formed the original basis for the "prokaryotic-eukaryotic dichotomy." The defining cellular features of eukaryotes include the nucleus and other endomembrane system organelles. It is interesting to note that *Aagard* archaea today carry several genes once thought to be eukaryote specific.

[The origin of the eukaryotic cell: "A genomic investigation" Hyman Hartman and Alexei Fedorov

PNAS February 5, 2002; Communicated by Carl R. Woese, University of Illinois at Urbana–Champaign, Urbana, IL (*received for review October 25, 2001)]*

Virus as a taxonomic domain of life

Wherever there is life, there are viruses. In all probability, they have existed since living cells first evolved. The origin of viruses is unclear because they do not form fossils, so molecular techniques are used to investigate how they arose. Also, viral genetic material occasionally integrates into the germline of the host organisms, through which they can be passed on vertically to the offspring of the host, infecting many generations. Long existing, this reality provides an invaluable source of information for paleovirologists to trace back ancient viruses that have been on Earth for millions of years.

On average, the mass of viruses is only approximately 5% of the prokaryotic (bacteria and archaea) biomass. Their vast number is approximately 10 to the 30th, found in all environments, and themselves ever evolving. Genes involve approximately 2% of our genome, viruses, mostly inactive, account for 8%.

As of the present, we do not know the origins of viruses in the evolutionary history of life. Some argue that they may have evolved from plasmids—pieces of DNA that can move between cells—while others may have evolved from bacteria. In evolution, viruses are an important means of horizontal gene transfer, which increases genetic diversity in a way analogous to sexual reproduction. Viruses are considered by some biologists to be a life form, because they carry genetic material, reproduce, and evolve through natural selection, although they lack key characteristics (such as cell structure) that are generally considered necessary to count as life. Because they possess some but not all such qualities, viruses have been described as "organisms at the edge of life," and as replicators.

In sum, it is clear that viruses are ancient and have origins that pre-date the divergence of life into the three domains. This discovery has led modern virologists to reconsider and re-evaluate the nature and origins of viruses. Viruses are by far the most abundant biological entities on Earth and they outnumber all the others put together. They infect all types of cellular life including animals, plants, bacteria and fungi. Different types of viruses can infect only a limited range of hosts and many are species-specific. Smallpox viruses for example, can infect only one species—in this case humans, and are said to have a narrow host range. Other viruses, such as rabies virus, can infect different species of mammals and are said to have a broad range. The viruses that infect plants are harmless to animals, and most animal viruses harmless to humans.

Recent centuries have seen, viral infections that have been responsible for two major pandemics: the 1917-1919 "Spanish flue" epidemic that killed 20–40 million people, and the ongoing HIV/AIDS epidemic that killed an estimated 1.5 million people worldwide in 2013 alone. In 2020, the world-wide pandemic caused by the Covid-19 virus, presumably originating inWuhan, China, has caused renewed analysis of this basic and ancient form of life.

Viruses that infect archaea have complex, unrelated structures, with a wide variety of unusual shapes, ranging from spindle-shaped structures, to viruses that resemble hooked rods, teardrops or even bottles. Other archaeal viruses resemble the tailed bacteriophages, and can have multiple tail structures.

Recent research has revealed new functions for viruses, which exposes their important action in relatively benign environments as well as their typical causal actions for diseases and pandemics.

"When a virus encounters a host cell, a chain reaction of molecular events is set in motion. The virus attaches itself to the outer wall of the cell, enters inside, travels to the cell's genome, merges with its genes and then tricks the cell's genome into making copies of itself ... many millions of them ... The virus that causes Aids makes protein spheres because they protect HIV's genes as they journey from cell to cell in its hosts."

"As it turns out the same thing happens with the memory gene Arc, which also works by moving genetic information from cell to cell in humans, changes to Arc have been linked to a range of disorders, from schizophrenia to dementia. Arc is the gene that memories are made of ... sequencing the gene for Arc, decoding the string of molecules that make it up ... the memory gene was for all purposes a modified virus ... In some manner in ancient eras it put to work on an important bodily function ... the invading parasite was itself hacked to our benefit ... German scientists have discovered that as many as 85 of the genes in the human brain are derived from viruses ... Our genome is a graveyard for ancient viruses." (Shubin, 2020)

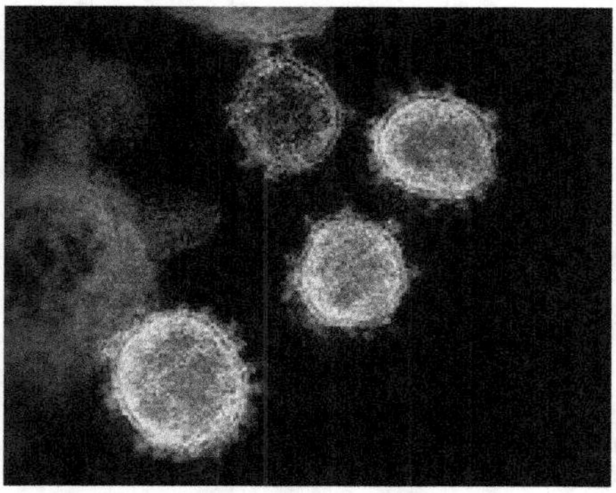

Figure 12: Electron micrograph of SARS-CoV-2 virrons with visible coronae

In short, the Virus is a very active member of life's four root Domains.

Perspective: the first 3.5 billion years

Cells are divided into two main classes, initially defined by whether they contain a nucleus. Prokaryotic cells (bacteria) lack a nuclear envelope; eukaryotic cells have a nucleus in which the genetic material is separated from the cytoplasm. Prokaryotic cells are generally smaller and simpler than eukaryotic cells; in addition to the absence of a nucleus, their genomes are less complex and they do not contain cytoplasmic organelles or a cytoskeleton.

In spite of these differences, the same basic molecular mechanisms govern the lives of both prokaryotes and eukaryotes, indicating that all present-day cells are descended from a single, primordial ancestor. How did this first cell develop? And how did it evolve to produce the complexity and diversity exhibited by present-day cells?

Bacterial cells are spherical, rod-shaped, or spiral, with diameters invisible to the naked eye. Their DNA contents range from about 0.6 million to 5 million base pairs, an amount sufficient to encode about 5000 different proteins. The largest and most complex prokaryotes are the cyanobacteria, appearing c.2.5 billion years ago, marking the modest beginnings of the evolution of photosynthesis.

Many eukaryotes are unicellular organisms that, like bacteria, consist of only single cells capable of self-replication. The simplest eukaryotes are the yeasts. Yeasts are more complex than bacteria, but much smaller and simpler than the cells of animals or plants. Other unicellular eukaryotes, however, are far more complex cells, some containing as much DNA as human cells. They include organisms specialized to perform a variety of tasks, including photosynthesis, movement, and the capture and ingestion of other organisms as food.

Amoeba proteus, for example, is a large, complex cell. Its volume is more than 100,000 times that of the bacteria, *E. coli*, and its length can exceed 1 mm when the cell is fully extended. Amoebas are highly mobile organisms that use cytoplasmic extensions, called pseudopodia, to move and to engulf other organisms, including bacteria and yeasts, as food. Other unicellular eukaryotes (green algae) contain chloroplasts and are able to carry out photosynthesis.

Multicellular organisms evolved from unicellular eukaryotes at least 1.7 billion years ago, about the time of the photosynthesizing Cyanobacteria. Some unicellular eukaryotes form multicellular aggregates that appear to represent an evolutionary transition from single cells to multicellular organisms. For instance, the cells of many algae (e.g. the green alga *Volvox*) associate with each other to form multicellular colonies, which are thought to have been the evolutionary precursors of present-day plants.

Increasing cell specialization then led to the transition from colonial aggregates to truly multicellular organisms. Continuing cell specialization and division of labor among the cells of an organism have led to the complexity and diversity observed in the many types of cells that make up present-day plants and animals, including human beings. (Thought CO)

Sexual reproduction

Sexual reproduction reflects the evolution from ancient single cell Eukayoted cells into the Kingdoms of protists, plants, fungi, and animals describes how sexually reproducing animals, plants, fungi and protists could have evolved from a common ancestor that was a single- celled eukaryotic species. There are a few species which have secondarily lost the ability to reproduce sexually, such as Bdelloidea, and some plants and animals that routinely reproduce asexually.

Ironic in its modernistic life dominance is the reality that sexual reproduction produces a single form from two originating cell units. The prokaryotes can double in numbers many times in comparison with sexually reproducing Eukaryotes. However, prokaryote asexual mitotic reproduction leads to a multitude of genetic errors, many of which can be fatal.

The possibility for movement in individual Eukaryotes allowed for the meeting and reproducing of slightly different genoplasmic types, here allowing for slower, careful reproductive but pluralistic outcomes. The secured genome, inside the nucleus with strong cytoplasmic boundaries, secured the early Eukaryotes and their experimentation with sexual reproduction from easy plundering by viruses.

Some scientists have hypothesized that in the process of sharing their DNA when creating a new individual, a process of genetic repair could take place, slowing the possibility of widespread negative mutations which might affect the success of this form of reproduction. The plant kingdom's continual flirtation with asexual reproduction, and then in environmental crises reverting to the chancier sexual reproduction, sending a newly created repertoire of genes off into the future, argues for the plant's much greater freedom of adaption, given the use of photosynthetic energy processes.

Animals have either to wait for the food to come across their bodies, else move about to hunt for their energy. Early animal forms—polyps, corals, and sponges—ordinarily do reproduce mitotically. Under certain conditions, danger or necessity will move into new physical domains, requiring change to sexual reproduction.

Sexual reproduction opened up the world of life to *death*. Death of the mother cell is important for the future of the daughter cells, especially in the dynamic physical world that allowed the eukaryotes to flourish in the first place.

One might argue that the Darwinian concept of slow evolution, given the sieve that adaptation and natural selection operate in the external environment, allows for careful continuities in the Eukaryote world. This is quite different from the two prokaryote domains, given one-celled capacities for sharp genetic changes and the engorging of very different relatives, which allows for their great variations over time—variations/mutations which often foil our human medical geniuses. Naturally, if there is a great and radical external reorganization of the climate and geology, leading to new bacteriologies of our planet, we humans may be stuck.

Timeline

Date (million years ago)	Event
4600 mya	The planet Earth forms from the accretion disc revolving around the young Sun with organic compounds (complex organic molecules) necessary for life having perhaps formed in the proto-planetary disk of cosmic dust grains surrounding it before the formation of the Earth.
4500 mya	According to the giant impact hypothesis, the Moon is formed when the planet Earth and the hypothesized planet Theia collided, sending a very large number of moonlets into orbit around the young Earth which eventually coalesce to form the Moon. The gravitational pull of the new Moon stabilises the Earth's fluctuating axis of rotation and sets up the conditions in which abiogenesis occurred.
4404 mya	First appearance of liquid water on Earth.
4280 mya	Earliest possible appearance of life on Earth. This knowledge depends mainly on dating the rocks in which living things may have left a mark.
3800 mya?	This is our current "best guess" for the beginning of life on Earth. It is distinctly possible (Le Page & Lane, 2009; Marshall, 2011). At some point far back in time, a common ancestor gave rise to two main groups of life: bacteria and archaea (Margulis & Guerrero, 1991). How this happened, when, and in what order the different groups split, is still uncertain (Coghlan, 2002; Battistuzzi, Feijao & Hedges, 2004).

3500 mya The oldest fossils of single-celled organisms date from this time (Altermann & Kazmierczak, 2003).

3460 mya Some single-celled organisms may be feeding on methane by this time (Mcleod, 2006).

3400 mya Rock formations in Western Australia; some researchers claim that fossilized microbes date from this period.

Figure 13: Fragment of the Acasta Gneis exhibited at the Museum of Natural Historyy in Vienna

3000 mya Viruses are present by this time, but they may be as old as life itself (Hamilton, 2008; Holmes, 2004).

2400 mya The "great oxidation event." Supposedly, the poisonous waste produced by photosynthetic cyanobacteria (oxygen) starts to build up in the atmosphere (New Scientist, 2001). Dissolved oxygen makes the iron in the oceans "rust" and sink to the seafloor, forming striking banded iron formations.

Recently, though, some researchers have challenged this idea. They think that cyanobacteria only evolved later, and that other bacteria oxidized the iron in the absence of oxygen. Yet others think that cyanobacteria began pumping oxygen as early as 2.1 billion years ago.

Figure 14: The cyanobacterial-algal mat, salty lake on the White Sea seaside

2300 mya **Earth freezes over in what may have been the first "snowball Earth," possibly as a result of a lack of volcanic activity (Shiga, 2009). When the ice eventually melts, it indirectly leads to more oxygen being released into the atmosphere (Hecht, 2006b).**

2150 mya **First undisputed fossil evidence of cyanobacteria and of photosynthesis, i.e. the ability to take in sunlight and carbon dioxide, and obtain energy, releasing oxygen as a by-product (Burke, 1999). There is some evidence for an earlier date for the beginning of photosynthesis, but it has been called into question (Nowak, 2008).**

2000 mya? **Eukaryotic cells, i.e. cells with "organs" (Davis, 1996) known as *organelles* (Westphal, 2003), come into being. One key organelle is the nucleus: the control centre of the cell, in which the genes are stored in the form of DNA.**

Eukaryotic cells evolved when one simple cell engulfed another (Day, 1994), and the two lived together, more or less amicably (Wakeford, 1994)—an example of "endosymbiosis." The engulfed bacteria eventually become mitochondria (Dicks, 2004), which provide eukaryotic cells with energy. The last common ancestor of all eukaryotic cells had mitochondria—and had also developed sexual reproduction (Lane, 2009).

Later, eukaryotic cells engulf photosynthetic bacteria and form a symbiotic relationship with them (New Scientist, 2005). The engulfed bacteria evolve into chloroplasts: the organelles that give green plants their color and allow them to extract energy from sunlight.

Different lineages of eukaryotic cells acquire chloroplasts in this was on at least three separate occasions, and one of the resulting cell lines goes on to evolve into all green algae and green plants.

chloroplasts 1500 mya? **The eukaryotes divide into three groups: the ancestors of modern plants, fungi and animals split into separate lineages or kingdoms (Wang, Kumar & Hedges, 1999) and evolve separately. We do not know in what order the three groups broke with each other. At this time they are probably all still single-celled organisms .**

900 mya?	**The first multicellular life develops around this time (Blackman, 1998; Le Page & Fox, 2009). It is unclear exactly how or why this happens (Singer, 2004), but one possibility is that single-celled organisms go through a stage similar to that of modern choanoflagellates, single-celled creatures that sometimes form colonies consisting of many individuals. Of all the single-celled organisms known to exist, choanoflagellates are the most closely related to multicellular animals, lending support to this theory (Holmes, 2009).**
800 mya	**The early multicellular animals undergo their first splits. First they divide into, essentially, the sponges (Ghiold, 1991) and everything else—the latter being more formally known as the Eumetazoa.** **Around 20 million years later, a small group called the Placozoa (New Scientist, 2009) breaks away from the rest of the Eumetazoa. Placozoa are thin, plate-like creatures about 1 millimeter across, and consist of only three layers of cells. It has been suggested that they may actually be the last common ancestor of all animals (Dellaporta et al., 2006).**
770 mya	**The planet freezes over again in another "snowball Earth." (Walker, 1999)**
730 mya	**The comb jellies/ctenophores (Hecht, 2006a) split from the other multicellular animals. Like the cnidarians that will soon follow, they rely on water through their body cavities to acquire oxygen and food.**
680 mya	**The ancestor of cnidarians (jellyfish and their relatives) breaks away from the other animals—though there is as yet no fossil evidence of what it looks like.**
630 mya	**Around this time, some animals evolve bilateral symmetry for the first time, that is, they now have a defined top and bottom, as well as a front and a back (Hecht, 2004). Little is known about how this happened. However, small worms called Acoela may be the closest surviving relatives of the first ever bilateral animal (Hecht, 1999). It seems likely that the first bilateral animal was a kind of worm. *Vernanimaculca guizhouena*, which dates from around 600**

million years ago, may be the earliest bilateral animal found in the fossil record (Chen et al., 2004).

590 mya **The Bilateria, those animals with bilateral symmetry, undergo a profound evolutionary split. They divide into the protostomes and deuterostomes.**

The deuterostomes eventually include all the vertebrates, plus an outlier group called the Ambulacraria. The protostomes become all the arthropods (insects, spiders, crabs, shrimp and so forth), various types of worms, and the microscopic rotifers.

Neither may seel like an obvious "group," but in fact, the two can be distinguished by the way their embryos develop (Palmer, 1995). The first hole that the embryo acquires, the blastophore, forms the anus in deuterostomes, but in protostomes, it forms the mouth.

580 mya **The earliest known fossils of cnidarians, the group that includes jellyfish, sea anemones and corals, date to around this time (Mason, 2002), though the fossil evidence has been disputed.**

575 mya **Strange life forms known as the Ediacarans appear around that time and persist for about 33 million years (O'Donoghue, 2007).**

570 mya **A small group breaks away from the main group of deuterosomes, known as the Ambulacraria. This group eventually becomes the echinoderms (starfish, brittle stars and their relatives) and two worm-like families called the hemichortates and Xenoturbellida.**

Another echinoderm, the sea lily, is thought to be the "missing link" between vertebrates (animals with backbone) and invertebrates (animals without backbone), a split that occurred around this time (Williams, 2003).

565 mya **Fossilized animal trails suggest that some animals are moving under their own power (O'Donoghue, 2010).**

540 mya **As the first chordates—animals that have a backbone or at least a primitive version of it—emerge among the deuterostomes, a surprising cousin branches off.**

The sea squirts or tunicates (Morris, 2003) begin their history as tadpole-like chordates, but metamorphose partway through their lives into bottom-dwelling filter feeders that look rather like a bag of seawater anchored to a rock (Hooper, 2007).

535 mya **The Cambrian explosion begins, with many new body layouts appearing on the scene (Holmes, 1997; Rudder, 1991), though the seeming rapidity of the appearance of new life forms may simply be an illusion caused by a lack of older fossils (Jones, 2001; Wray et al., 1996).**

Looking Back and Ahead

"G.L. Stebbins believed that in this dynamic pre-biotic soup of 4.3 bya to 3.8 bya, possibly millions of pathways of metabolic activity existed as these molecules maintained themselves for a moment, then dissolved. Gradually, the evolving Earth began to produce less and less of the naturally occurring organic molecules that were excreted from the interaction of hot lava, geothermal eruptions, and the flux of many molecular minerals. As N.W. Pirie puts it, in the evolution of living forms, the transition over time is from chemical and metabolic diversity to increasing bio-chemical simplicity, specialization, and regularity. Then, morphological diversity." (Itzkoff, 2000; Stebbins, 1982; Pirie, 1957) We might add that at the apex of this multi-billion-year process came a creature whose genetic structure allowed it to evade the regularity of behavior that the eukaryote cell had given to living things. This genetic/behavioral freedom had unwittingly and perhaps momentarily created something called culture. Here was an innovation that seemed to evade that ancient and basic message of life, and allowed it to adapt and be selected by nature to ensure the survival of the species. War, genocides, the often enslavements and tyrannies by humans over humans seem to belie the messages of adaptation and the natural selection of the fit over the less lucky.

Chapter 3

How We Vertebrates Arrived

Ancient panorama

Sometime between the origins of our planet, ca. 4.5 bya, down the line some 500 mya, reproducing molecules were dividing and gobbling up adjacent molecules in the quest to avoid energy degradation. Between 4 bya and another 500 million years, ca. 3.5 bya, the two one-celled domains, Eubacteria and Archaea, separated, perhaps because of geographical distance or sharp divergence of molecular pathways of energy metabolization.

A view gaining much agreement is that sexual reproduction in the Archaea domain may have begun early. The majority settled on the concept of symbiotic ingestion of metabolically agreeable molecules. As these symbiotic absorptions were subject to natural selection, i.e. battles between adjoining molecules based on longer term compatibilities, only a few survived the eons.

One line of dissident Archaea seemed to have survived and endured over time. They were also adding more molecular elements that worked together, mostly to resist aggressive molecular viruses, reproduction mutations all of which degraded adaptability, and always a variety of environmental challenges. Early on, the symbionts that worked together biochemically said to each other, "why not," creating sexual reproduction in which the parental genotypes (DNA) now intermixed genetically, creating offspring of the future, to be or not to be accepted by natural selection. It worked, over the eons.

From an evolutionary point of view, sexual reproduction presented two major advantages: *recombinational DNA repair* (Bernstein and Bernstein, 2010) and *complementation* or *heterosis*, which mitigates the impact of harmful mutations (Bernstein et al., 1984; Bernstein et al., 1985; Bernstein, Hopf & Michod, 1987).

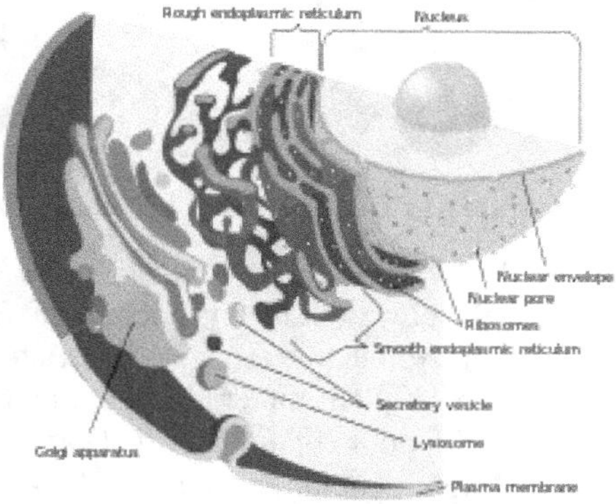

Figure15: Detail of the eukaryote endomembrane system and its components

Some celled protists reproduce sexually, as do the multicellular plants, animals, and fungi. In the eukaryotic fossil record, long evolving sexual reproduction first appeared 1.2 billion years ago, in the Proterozoic Eon. All sexually reproducing eukaryotic organisms likely derive from a single-celled common ancestor. It is probable that the evolution of sex was an integral part of the evolution of the first eukaryotic cell, as long ago as 3bya (Ploompu, 1999; Hörandl & Speijer, 2018).

The estimate is that evolution over this time frame, some one to two billion years, produced an ever larger, more complex cell that resisted the varied challenges of natural selection. But as noted in the previous chapter, all mother cells themselves were subject to personal death. Their progeny was to continue on in the relentless search for both resistant novelty and consequent reproductive success.

Eukaryote kingdoms

When did the multicellular life—fungi, plants, animals—originate? In or about 2.4 to 2.1 billion years before the present, some bacteria had ingested mineral or metallic elements that made them sensitive to light, subsequently producing a powerful chemical residue—oxygen. These "cyanobacteria" created masses of green "mats" and gave a great boost to the earth's photosynthetic capabilities.

The continual extrusion of oxygen had its impact on the warm, CO_2-heavy atmosphere and waters. Climate change—but with a difference. Warm was out, cool was in. Scientists call this hypothetical event the "great oxygenation event" (GOE), mostly impacting the anaerobic bacterial world. Conversely, it could have given a great boost to the energy production and availability of food resources for diverse forms of life, including the now comparatively gigantic eukaryote cell.

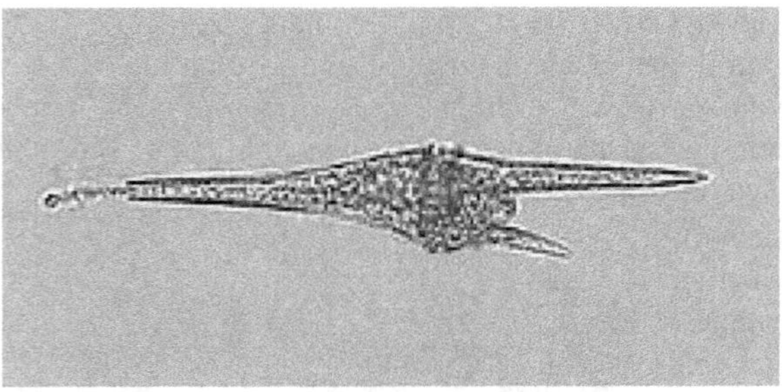

Figure 16: Dinoflagellate Ceratium furca

During the next several hundred million years, by 1.5 bya, the separation of structure and function of the four kingdoms of the Eukarya Domain could have occurred to form protists (e.g. Paramecium), fungi, plants and animals. Naturally, of those cells which became plants, some eventually followed the fortuitous biochemical pathway of the cyanobacteria. But now, being eukaryotes, they were able to mobilize much greater energies, and extrude more oxygen per cell. The multiplication of plant material (algae) had to have had a nourishing impact on the evolving kingdom of animal life. Current estimates view the origin of fungi as c.1.5bya (Wang, Kumar & Hedges, 1999). Other kingdoms and their metazoan multi-cellular organisms were certainly not far behind in the making.

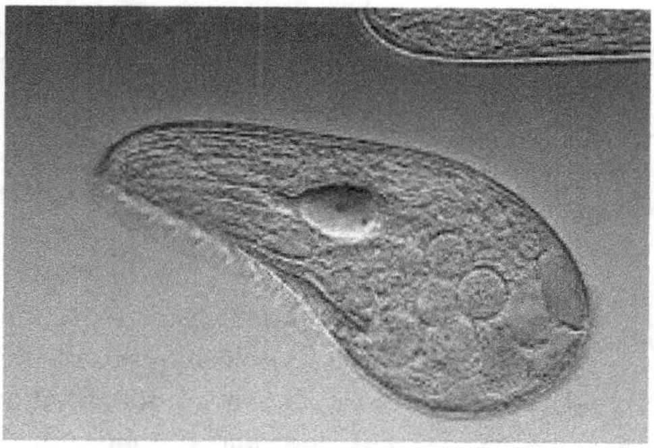

Figure 17: Blepharisma japonicum, a free-living ciliated protozoan

There are many hypotheses making the rounds in the various scientific communities concerning ancient climate change at this point, to attempt to explain the extraordinary explosion of descendants of the primal Eukaryote cell. Here, it is the kingdom of animal life that we are concerned with. The destiny of the diverging plant kingdom was now closely determined by the sun, water availability, and the temperature range of the seas.

Life choices: offense/defense

About one billion years ago, the so-called Varangian glaciations occurred. It might have been the second massive glaciation/frost event, the first coinciding with the cyanobacteria's appearance, c.2.5 bya. The view is that this latter glaciation coincided roughly with the breaking up of the Rodinia super continent.

For many billions of years, living cells, prokaryote and eukaryote, had been launched into reality as aggressive predators searching for immortality, ingesting their own kind and other molecular forms that could be absorbed for another moment of this presumed immortality. This challenge was the rock of a billions-of-years requirement by nature, to exist at the edge of plenitude and/or extinction. When we first observe the fossil hints of metazoan animal life forms, some 800 mya in the Ediacaren period, we see hints of the first animal colonies, sponges, jelly fish, and the Echinoderm phyla (starfish, sea lilies, sea urchin, sea cucumbers).

Most of these animals were soft, two layered bodies ingesting energy—i.e. filter feeding themselves, protists, plants, even bacteria as they came with water. Mostly anchored, what movement they could generate in addition to the tides, and waves that passed them by also brought the organic substances that fed them. They were also radial in design: starfish, fivefold radiation, jelly fish, four fold, comb jellies (cilia movement) plus the later animal colonies, the corals.

Still mysterious are the sea bed tracing of bi-lateral worm-like creatures. Not only possessing two of everything as in eyes, arms, etc. but also frontal and back distinction, these tracing are yet to reveal either a hard external skeleton—a notochord prelude to the vertebrates or even the probable descendant, an invertebrate modern worm.

The Ediacaran animals—many phyla, so much like anchored plants—seemingly disappeared as truer animals began to expand in numbers from the orgy of organisms available in the seas. From the mats of cyanobacteria two billion years ago to the radial filter feeders of c. 800mya, and more recent organisms, e.g.

corals, the orgy of metabolic dining courses almost shouted: “enjoy!” Out of the life dynamics of this newer contra 2nd Law of Thermodynamics environment, into the seas poured a five-star menu for the plankton, algae, jellyfish, comb jelly masses.

Figure 18: Dickinsonia costata an iconic Ediacaran organism, displays the characteristic quilted appearance of Ediacaran enigmata

Sometime, somewhere, out of this effluence of subsistence and growth possibilities, the vertebrates were formed along with their close cousins, the chordates. The chordates, along with the Echinoderms (sea lilies, sea urchins and star fishes, etc.), are three germ layered animals, along with the various invertebrate forms of animal life that are more pancaked in structure.

With these mostly filter feeding Echinoderms, there is as yet a question about a later evolutionary affinity, this with the Chordates and Echinoderm. This is even more puzzling in the fact that Echinoderms and Chordates, including the Vertebrates are what is called “deuterstomes.” All other animals are “protostomes.” Protostomes in the embryological stage develop the mouth of the growing creature before the anus. In the Deuterstomes, the opposite is the case. Why this association with such a primitive phylum, only the star fish deviates from rooted sessile feeding Echinoderms. It specializes in moving to eat filter feeding and sedentary hard shell animals, clams etc.

The larvae of both phyla, the Chordates and the Echinoderm, all reveal a free-swimming stage with a hint of a notochord, seemingly as an aid to stability in the moving waters. Later in their ontogenetic development, exemplars of the three phyla go their differing evolutionary ways. Remember, the Echinoderms show many characteristics believed central to the physical character and behavioral characteristics of Ediacaran phyla of c. 800–700 million years ago.

Part of the mystery is contained in another developmental affinity that the chordates have with the echinoderms, and that seemingly alone. In the earliest stages of development, becoming freed from the mother, the larvas of both phyla are free swimming. Later in their development, they take on their adult destinies. The Echinoderms become sessile filter feeders, rooted to the bottoms of seas; most of the chordate sub-phyla also revert to the usually safe world of passive, defensive filtering, as with the completely invertebrate Echinoderms. Only the true vertebrates, when they appear on the scene, have retained this free-swimming survival behavior.

Over the eons, the Eukaryote cells moved through the seas, pursuing their meals—other organic cells. Thus, free *swimmingness* is really the ancient form of earning a living. It was only when nature created the enormous food resources centered on plant life, this coming from photosynthetic metabolism and thus expiring oxygen. Consequently, it led to the 21% oxygen ratio we have today, accomplished sometime between 1 billion and 600 million years ago. These conditions became the adaptive and then the selective magnet for animal life, taking the easier defensive tack of letting the food come to them, rather than aggressively moving about to hunt it down.

The reason why the Echinoderms and the Chordates are the only animal phyla in their embryonic development to create the anus first, the mouth second, deuterstomes, and survive might be due to special selective circumstances (ancient), weather or geological changes, which allowed a variety of protostome animals (mouth first, anus second) to avoid this possible extinction crisis. The dim mists of time do not allow us to know for sure, and we can only wonder about these unknown, yet radical circumstances in the evolution of life.

A possible explanation for the deuterstome/protostome separation of the Echinoderms and the Chordates from all the other protostome animal phyla is the relationship enigma of the sponges, possibly the earliest existing metazoan phylum. Researchers are probing their unique morphology, molecular/DNA and reproductive patterns to see if an outsider relationship to the two deuterstome phyla can be established. At the same time, we may even find out why there are no Echinoderms species flourishing in fresh waters, and how the star fish wandered off the sessile plantation!

Chordates: prelude to a revolution

It is noteworthy to recall that the invertebrate echinoderm phyla also reveals tadpole-like, free-swimming larva in their earliest developmental stage. Keep in mind that most frogs, air breathing vertebrate amphibians who evolved much later in evolutionary history, c.350 million years later than Pikaia, have a watery tadpole stage in ontogenetic development. Here is another example of the genetic imprint of ancient adaptations on the developmental form of the "modern" individual.

The cephalochordates, or "lancelets" comprise one of the three subphyla in the phylum Chordata (the other two being the vertebrates and tunicates). Lancelets are also small, fishlike animals that are generally less than five centimeters in length, although some reach 6 to 9 centimeters. Lancelets are found most commonly in shallow, subtidal tropical, subtropical, and temperate sand flats, where they burrow in clean gravel or sand (not silty sediments) with just the head exposed.

A group of uncertain phylogeny, small eel-like conodonts, are known from microfossils of their paired tooth segments from the late Cambrian to the end of the Triassic (485–200mya). As of 2012, scientists classify the conodonts in the phylum Chordata, on the basis of their fins with fin rays, chevron-shaped muscles and notochord. Milsom and Rigby envision them as vertebrates, similar in appearance to modern hagfish and lampreys (**Milsom and Rigby, 2004**). Some analysts do not regard conodonts as vertebrates, because they lack the main characteristics of these groups.

Chordate/vertebrate adaptations

G. Ledyard Stebbins noted that the split between the defensive, perhaps even defeatist adaptive tack taken by fixed sessile feeders, such as the acorn worm colonies, and the lancelet mud feeders; with the contrasting free-swimming opportunists who would become the true vertebrates, is reflected in the reproductive patterns taken by the two lines.

With regard to the proto-vertebrates, the egg cell of the fertilized neonates, while growing in the mother's ovary, would have acquired a large amount of fatty yolk which the embryo would digest before becoming a free-swimming larva. This shift of nutrition from filter-feeding to dependence on nutrition from mother and egg, is apparently one of many random genetic changes, and is an alternative subsequently often proffered in nature (sharks are an example).

This change tended to shift animals from more stable, environmental nutritive options, to mother/daughter binding patterns, thus eventually making it possible for some lines to move to a more adventurous and opportune group of adaptive shifts. Here, we may find one more adaptive key to the development of higher, behaviorally malleable life styles: the closer nurturing of the young by parents, and thus a lengthened state of infantile dependency, offset by the correlative maturational delay, ultimately leading to increased behavioral adaptiveness.

The fertilization of the female egg cell allowed for the internal development of the young metazoan animal in the first place. The moment in time that the fertilized egg, or immature larva is expelled or born, is disciplined by the particular selective conditions of survival. Clearly, just as plants and some simple animals can go back and forth from asexual stability of reproduction to sexual, exploratory reproduction in times of environmental stress, a parallel exists in the animals. Here, the progressive alternative lies in the longer retention of the immature within the body of the mother's egg.

At this point in evolution, in the early Cambrian era, the earliest vertebrates lacked hard parts. The pharynx at the head acted as a food strainer, and the mouth sucked mud and organic debris. Then, they too, eventually found the morphological means to exploit the relative plenty at the bottoms. There existed a long row of vertical slits through which water was expelled, and food was passed into a thin tube for digestion. Next was the gut, for absorption of nutrients. Muscles in the body wall between the slits created suction to operate the pharynx, to suck in water, then expel it through gill slits. Later, thin sheets of folded blood vessels attached to walls of pharynx between the gill slits created a larger area of gas exchange, so as to provide greater oxygenating power for a larger animal.

The gill arches in the wall of the pharynx between the slits provided the muscular support for the process. Later, a heart pump would develop to move the nutrients through the blood fluid that carried greater amounts of molecular oxygen to the gut, passing it back to the gills, where it was re-oxygenated as circulation. The heart developed strong muscular tissue to maintain the contractile process throughout the life of the growing vertebrate.

The vertebrate adaptive plan, like the vertebrate body plan, needed much time for its eventual perfection and proliferation in the Ordovician era, some 450mya. As in so many subsequent evolutionary transitions, the late success of one line should not necessarily suggest or imply its sudden revolutionary coming into being. Rather, the success of the vertebrates, given the embryological evidence of evolutionary affinities and origins, argues for an ancient and steady, if precarious, adaptive tack. The survival of the line, and its gradual perfection of body plan, amidst the quick and distinct successes of other animal lines and their own particular adaptive pathways, tells us about the very variable and relatively optimal ecological possibilities existing on our planet during the metazoan animal expansion, after 1bya.

The first sign of head development, cephalization, is seen in chordates such as *Pikaia* and *Branchiostoma* (lancelets). It is thought that development of a head structure resulted from a long body shape, a swimming habit, and a mouth at the end that came into contact with the environment first, as the animal swam forward. The search for food required ways of continually testing what lay ahead, so it is believed that anatomical structures for seeing, feeling, and smelling developed around the mouth. The information these structures gathered was processed by a swelling of the nerve cord (*efflorescence*)—the precursor of the brain. Altogether, these front-end structures formed the distinct part of the vertebrate body, known as the head.

True vertebrates

While the chordates, along with all the modern phyla, appear in the Cambrian era, c.550mya, this innovation, head first, underlines a seemingly great breakthrough in plenitude and in the expanding modern phyletic pathways toward energy stabilization. As noted below, even admission to the sub-phylum *Vertebrates* does not tell us about the full realization of that other basic and highly selective adaptation, mid-water exploration of the environment. This was the change of these vertebrates from filter feeders at the bottom, still defensive in nature into that more aggressive, even predatory journey in the open water of the seas.

From the as yet hidden research papers of early animal evolution, they appeared at the end of the Cambrian/beginning Ordovician era, c. 500–450mya, and they blasted forward. Where did they come from? That, we still don't know.

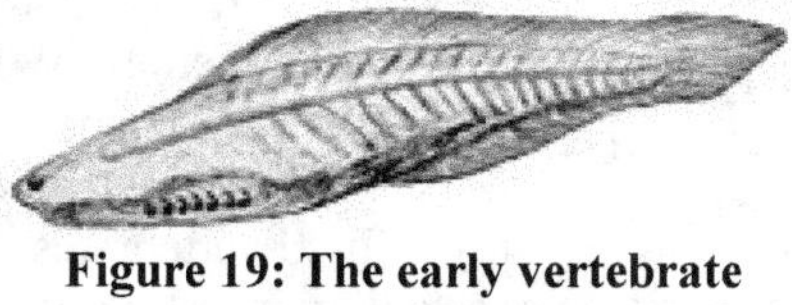

Figure 19: The early vertebrate Haikouichthys

Vertebrates originated about 525 million years ago, during the Cambrian explosion, which reveals a great rise in organism diversity. The earliest known vertebrate is believed to be the *Myllokunmingia*. Another early vertebrate is *Haikouichthys ercaicunensis*. Unlike the other fauna that dominated the Cambrian, these groups had the basic vertebrate body plan: a notochord, rudimentary vertebrae, and a well-defined head and tail. All of these early vertebrates lacked jaws in the common sense and relied on filter feeding close to the seabed.

The vertebrates are the only chordate group to exhibit cephalization—the concentration of brain functions in the head. A slight swelling of the anterior end of the nerve cord is found in the lancelet, a chordate, though it lacks the eyes and other complex sense organs comparable to those of vertebrates. Other chordates do not show any trends towards cephalisation.

Today, the first known vertebrate fossils, found at the Chengjiang locality in China, date back to the early Cambrian era. These early vertebrates, such as *Haikouichthys* noted above, are small, tapered, streamlined animals, such as modern hagfish, showing eyes, a brain, pharyngeal arches, a notochord, and rudimentary vertebrae. Vertebrates surely appeared in the early Cambrian era, c.550mya, and seem to have radiated in the late Ordovician era, about 450 million years ago—giving them one hundred million years to flex their adaptations.

The oldest classified vertebrate fish—jawless agnathans—appeared in the Cambrian era, and two groups still survive today: the lampreys and the hagfish, comprising about 120 species in total. Hagfish are considered members of the subphylum *Vertebrata*, only because they lost vertebrae secondarily. This event was inferred from molecular and developmental data. They both have skulls, but possess no vertebrae.

Summary of fish evolution

Armored fish (Placodermi)

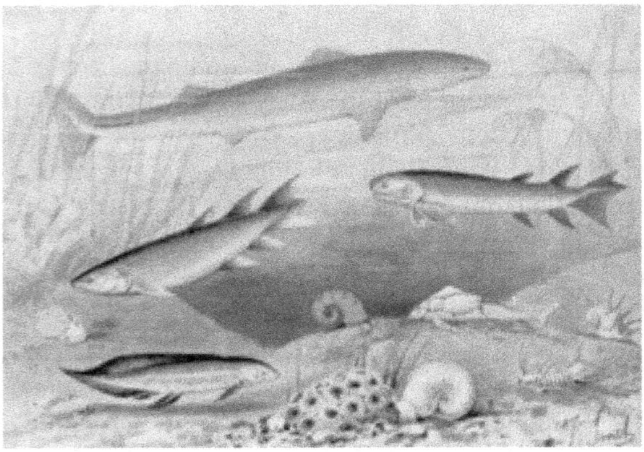

Figure 20: The Devonian period 419–359 mya (Age of Fishes) saw the development of early sharks, armoured placoderms and various lobe-finned fishes including the tetrapod transitional species.

The armored fish evolved during the Silurian period (420 mya). Like their predecessors, Ostracoderms Cambrian), and the spiny sharks, they too lacked jaw bones, but possessed paired fins. The armored fish diversified during the Devonian period, but declined and fell into extinction by the end of the Permian period. Most Ordovician (c. 450 mya) fossil vertebrates are rare and fragmentary, although available material suggests that ancestors of the sharks and jawed fish were present. By the middle Silurian, about 420 million years ago, the picture is clearer: the armored, jawless fish were quite diverse, and the first definite jawed fish had appeared—the Silurian is sometimes called the "Age of Fishes."

Cartilaginous fish (Chondrichthyes)

Cartilaginous fish, which include sharks, skates, and rays, evolved during the Silurian period, c.420mya. Cartilaginous fish have skeletons composed of cartilage, not bone. They also differ from other fish, in that they lack swim bladders and lungs. "By the late Devonian, 350 million years ago, early cartilaginous fish, sharks and bony fish were diversifying" (Shu, et al., 2003).

Bony fish (Osteichthyes)

Bony fish arose during the late Silurian, 400 mya. The majority of modern fish belong to this group (note that some classification schemes recognize the Class Actnopterygii instead of Osteichthyes).

Bony fish first diverged into two groups: one that evolved into modern fish, and the other that evolved into lungfish, lobe-finned fish, and fleshy-finned fish. The fleshy-finned fish gave rise to the amphibians. Thus, bony fish ultimately evolved into two separate groups: the Actinopterygii (or ray-finned fish) and Sarcopterygii (which includes the lobe-finned fish).

Sharks: a window for humanity

Sharks were once kings of the seas—that is, until humans decided to contest this achievement. As the primary predator of the oceans, they also developed a high intelligence, equal to the best of the non-hominid mammals. Below are some of the similarities to humans, and contrasts, as sharks fight for their existence.

Sharks and other cartilaginous fish (skates and rays) have skeletons made of cartilage and connective tissue. Cartilage is flexible and durable, yet is about half the normal density of bone. This reduces the skeleton's weight, saving energy. Because sharks do not have rib cages, they can easily be crushed under their own weight on land. Shark skeletons are very different from those of bony fish and terrestrial vertebrates.

Some sharks, if inverted or stroked on the nose, enter a natural state of tonic immobility. Researchers use this condition to handle sharks safely.

Unlike bony fish, sharks do not have gas-filled swim bladders for buoyancy. Instead, sharks rely on a large liver filled with oil that contains squalene, and their cartilage, which is about half the normal density of bone.

Members of the family Lamnidae (such as the shortfin mako shark and the great white shark) are homeothermic, as are mammals (humans), and maintain a higher body temperature than the surrounding water.

In contrast to bony fish, with the exception of the coelacanth, the blood and other tissue of sharks and Chondrichthyes are generally isotonic to their marine environments because of the high concentration of urea (up to 2.5%) and trimethylamine N-oxide (TMAO), allowing them to be in osmotic balance with the seawater, and to balance internal and external liquid pressures on the fish.

Sharks have keen olfactory senses, located in the short duct (which is not fused, unlike bony fish) between the anterior and posterior nasal openings, with some species able to detect as little as one part per million of blood in sea water.

Many sharks can contract and dilate their pupils, like humans, something no teleost fish can do.

Sharks practice internal sexual fertilization, as do many other fish and all mammals. Most sharks are ovoviviparous, meaning that the eggs hatch in the oviduct within the mother's body, and that the egg's yolk and fluids secreted by glands in the walls of the oviduct nourish the embryos. The young continue to be nourished by the remnants of the yolk and the oviduct's fluids.

As in viviparity, the young are born alive and fully functional. Viviparity is the gestation of young without the use of a traditional egg, and results in live birth. Viviparity in sharks can be placental or aplacental. Young are born fully formed and self-sufficient. Hammerheads, requiem sharks (such as the bull and blue sharks), and smooth hounds are viviparous, as are humans. Shark migration patterns may be even more complex than in birds, with many sharks covering entire ocean basins.

Sharks possess brain-to-body mass ratios that are similar to mammals and birds, and have exhibited apparent curiosity and behavior resembling play in the wild. There is evidence that juvenile lemon sharks can use observational learning in their investigation of novel objects in their environment.

Perspective

We now understand that the vertebrates were survivors. Long hidden on the peripheries, they eventually burst into a form of dominance that has lasted for some 500 million years. It is a dominance characterized by the occupation of an adaptive ecology heretofore uncontested. The innovative character of the vertebrate design is its long-run ability to broaden and expand the environmental limits of its life ecology.

There does not here exist an augur of selective dominance into a limitless future for any line of livingness. The early history of animal and chordate life tells us much about the ephemeral nature of biological success—one that can even last several hundred million years. Ultimately, the conditions and possibilities for adaptive and selective perdurance are determined by the Earth’s geological and climatic conditions. We should not diminish the ongoing struggle with other phyla, insects, and bacteria, etc. There are many evolutionary caveats that even the most dominant forms of life have to face by the various actors on this still evolving planetary stage.

Chapter 4

Land and the Brain

Revolutions

The first great breakaway revolution in the form of life from the heterogeneous masses of Eubacteria and Archae was the Eukaryote cell. It departed the Archae sometime between four and three billion years ago. It probably took another two billion years to form itself into a complex, sexually reproducing cell and then expanded in size and complexity to create metazoan organisms.

By 1.5 billion years ago, there were three diversified kingdoms of Eukaryotic cells: Fungi, plants, and animals, the latter gradually in motion, attempting to defy the second law of energy and its inevitable degradation of high energy organization in a closed environment. For animals, there were two adaptational strategies—floating, passive imbibing of organic materials, predation, and fixed (rooted) filtration of oncoming food sources, plants, fungi or animal. The Ediacaren fauna, c.800mya, already represent the latter.

The fossil, or any other evidence for the predatory floaters is scant until the Cambrian era,_590–510 mya. The early chordates and proto-vertebrates, Pikaia, begin to show up, 550mya. This was not a sudden miracle of creation. They probably were around for a long time before we found evidence of their existence—three to four hundred million years? Again, the Ostracoderms, jawless fish with real vertebrates, show up at the end of the Cambrian into the Ordovician era, 510–430 mya. Then, armored jawed fish, called Placoderms, appeared in the late Ordovician era, c.430 mya.

Finally, in the Silurian era, 430–395 mya, we find the fossil appearance of an explosion of many new types of fish—ray finned fish, lobe fin fish, cartilaginous fish, fish with lungs, fish with bladders. Where did they all come from? Most explanations for such changes point to climate/geologic changes, and point to competition. Even within the great climatic changes which might have led to enormous taxonomic (orders of animals) casualties, explaining this sudden progression of animal, even plant life is difficult. For now, we must back off.

The first known mass extinction in earth's history was the Great Oxygenation Event, 2.4 billion years ago. That event led to the loss of most of the planet's anaerobes. Researchers have identified five major and later extinction events in earth's history:

- End of the Ordovician: 440 million years ago, 86% of all species lost, including graptolites

- Late Devonian: 375 million years ago, 75% of species lost, including most trilobites
- End of the Permian, "The Great Dying": 251 million years ago, 96% of species lost, including tabulate corals, and most extant trees and synapsids
- End of the Triassic: 200 million years ago, 80% of species lost, including all of the conodonts
- End of the Cretaceous: 66 million years ago, 76% of species lost, including all of the ammonites, mosasaurs, ichthyosaurs, plesiosaurs, pterosaurs, and non avian dinosaurs

The late Devonian extinction event, c.375mya, and its climatic/geological causes was epochal. Heretofore, most of the older vertebrate and chordate forms were still pulsing around in their safe haven, committed to living. In terms of its impact on the evolution of life forms, it compares in impact with the creation and development of the Eukaryote cell. We now find the full vertebrate predatory solution to energy retentions, here in the Devonian (395–345 mya). Oncoming is a great and parallel revolution in life: vertebrates on land, breathing the air.

Genetic dimension

New environments by themselves do not create the new. When one group of living things bite the dust, we invariably see newcomers entering the scene, through the study of fossils. Some of these transitions seem sudden, but others are more gradual, and thus explainable.

Always, it depends on the availability of the genes. In order to appear in the fossil and thus evolutionary record, they have to be there. They are not created by the environment. The environment either facilitates their expression in the competition for survival, else it hammers down into extinctions those genes which cannot create a reproductively functioning phenotype (surface configuration of the individual). S.J Gould of Harvard University and Niles Eldridge (1972) coined a term, "punctuated evolution," to explain such sudden appearances of a radically deviant form in an apparently sonambulent clade (cousins) of plants or animals. This hypothesis was based on evolutionist Richard Goldschmidst's earlier writings on "hopeful monsters."

In these seemingly improbable jumps, the phenotypic changes over time between (a) a line of animals or plants undergoing slow evolution, and (b) and then the appearance of very different, but still related exemplars. These appear to be unexplainable using Darwinian propositions. On paper, the time difference may seem to be negligible in the creation of a sport, 50 to 150 million years. But we should have found transitional forms.

The real answer, as Ernst Mayr (1958) commented "there is no evidence in science of such great and sudden genetic recombination. We humans might not have as yet found the fossils of the rarer outlier intermediate forms External changes in the environment, ecology, could have changed the selective equation and the once rare exemplars could have burst forth in a frenzy of opportunity, *orthselection*, positive selection of hidden features in a straight line of change."

Throughout the evolutionary process, what we discover in the fossil record is just a tiny sampling of the genetic variation around any species or genus theme. The massive build of fish, or reptiles, and even mammals that we discover at one historic geological point in time is surely only a hint of the massive pre-adaptive potential of related but as yet undiscovered forms.

When the mini-revolution—climate, geology, or competitive pressures—occurs to push one or a group of living forms aside and into extinction, around the corner there will always be others with a slight or greatly different slant (here, reptiles/mammals) in their genetic profiling. Usually, they are selectively hanging on at the edge. Suddenly, with external changes, their selective possibilities will become positive in the oncoming time frame. And so, it has gone on for a few billion years.

Towards the land

The evolution of lungs played a crucial role in the development of complex organisms. In single-celled bacteria, gas exchange can take place entirely by simple diffusion. In larger organisms, however, only a small proportion of cells are close enough to the surface for oxygen from the atmosphere to enter through diffusion. Thus, two major adaptations made it possible for organisms to attain great multi-cellularity: efficient circulatory systems that conveyed gases to and from the deepest tissues in the body; and a large, internalized respiratory system that centralized the task of obtaining oxygen from the atmosphere and bringing it into the body, whence it could rapidly be distributed to any part of the circulatory system.

In consideration of the evolutionary history of the respiratory system of the proto-vertebrates and early vertebrates, the anatomy and physiology of lung breathing fishes indicates that lungs are an adaptation for providing the heart with oxygen (Farmer, 1997, 1998, 1999). Thus, lungs may have allowed early fishes to become large and pre-adapted to a variable marine environment.

The lungs of today's terrestrial vertebrates and the gas bladders of today's fish are believed to have evolved from simple sacs, as out pocketings of the esophagus, that allowed early fish to gulp air under oxygen-poor conditions. These out pocketings first arose in the bony fish. In most of the ray-finned fish, the sacs evolved into closed-off gas bladders, while a number of carp, trout, herring, catfish, and eels have retained the condition of the sack being open to the esophagus.

In more basal bony fish such as the gar, bichir, bowfin and the lobe-finned fish, the bladders have evolved to primarily function as lungs. The lobe-finned fish gave rise to the land-based tetrapods. Thus, the lungs of vertebrates are homologous to the gas bladders of fish (but not to their gills).

In embryonic development, both lung and swim bladder originate as an out pocketing from the gut; in the case of swim bladders, this connection to the gut continues to exist as the pneumatic duct in the more "primitive" ray-finned fish, and is lost in some of the more derived teleost orders (modern fish). There are no animals which have both lungs and a swim bladder.

Most fishes (the Actinopterygii, or teleost/ray-finned fishes) have several bones at the bases of their pectoral fins, and their fins are composed of a set of webbed rays, not muscle- and skin-covered bone.

Lobe fins are rare among living fish and are only possessed by the coelacanth and lungfish. However, lobe *limbs* are possessed by many living organisms — including humans. That's because we, and in fact all tetrapods (four-limbed vertebrates, many of which live on land), share a more recent common ancestor with the coelacanth and lungfish than we do with ray-finned fishes. We tetrapods are sarcopterygians (lobe finned types) too!

Tetrapods evolved from a group of organisms that, if they were alive today, we would call fish. They were aquatic and had scales and fleshy fins. However, they also had lungs that they used to breathe oxygen. Between 390 and 360 million years ago, the descendents of these organisms began to live in shallower waters, and eventually moved to land. As they did, they experienced natural selection that shaped many adaptations for a terrestrial way of life. Like other terrestrial sarcopterygians, modern humans still carry the evidence of our aquatic past in the way our arms and legs attach to our bodies, as well as in the many other features that link us to our fishy origins.

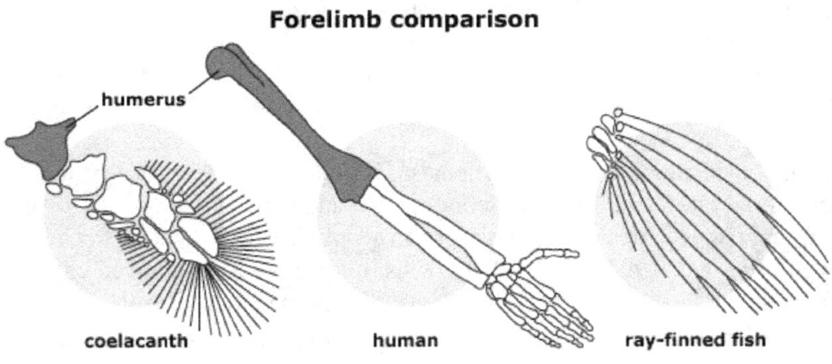

Figure 21

Which came first, the gill or the lung?

Since fish appear in the fossil record earlier than the clade we call tetrapods, it's tempting to assume that modern fishes bear the same traits that their and our common ancestor did. This line of reasoning is intuitive, but it is not correct. Though it is true that both modern ray-finned fishes and the ancestor we tetrapods have in common with them are finned and aquatic, the same pattern of reasoning does not hold water when it comes to lungs.

The available evidence suggests that gills were present in the very earliest fishes—the common ancestor of hagfish and ray-finned fishes. However, lungs—

gas-filled organs that serve the function of respiration—also evolved very early on. The common ancestor of the lobe- and ray-finned fishes had lungs *as well as* gills. So what happened to these lungs and gills? In the lobefins, lungs stuck around, and tetrapods, coelacanths, and lungfish all inherited them and use them to obtain oxygen. Coelacanths and lungfish also retained their gills.

Modern tetrapods, on the other hand, bear evidence indicating that we once had gills but that these were lost in the course of our early evolution. The ray-finned fishes retained gills, and some of them (e.g. the bichirs, *BYK-heerz*) also retained lungs for the long haul. In the lineage that wound up spawning most ray-fins (and in at least one other lineage), lungs evolved into the swim bladder—a gas-filled organ that helps the fish control its buoyancy.

Coelacanth: didn't make it

Figure 22: A coelacanth swimming near Sulawesi, Indonesia Preserved specimen of Latimeria chalumnae, [USNM 205871, from the Fish Division specimen collection. Photo by Sandra J. Raredon].

In 1938, a South African fisherman pulled a strange catch from the waters of the Indian Ocean. The iridescent blue animal had oddly fleshy fins that looked something like limbs. Scientists had seen fish like it before—but only preserved as fossils in ancient rocks. The fish was a coelacanth (*SEE-luh-kanth*), a member of the clade Sarcopterygii (*sar-KOP-tuh-RIJ-ee-eye*), and was thought to have gone extinct 80 million years ago!

Like other sarcopterygians, the coelacanth has "lobe-fins"—muscular fins with a single bone that articulates with the rest of the body.

What do coelacanths eat?

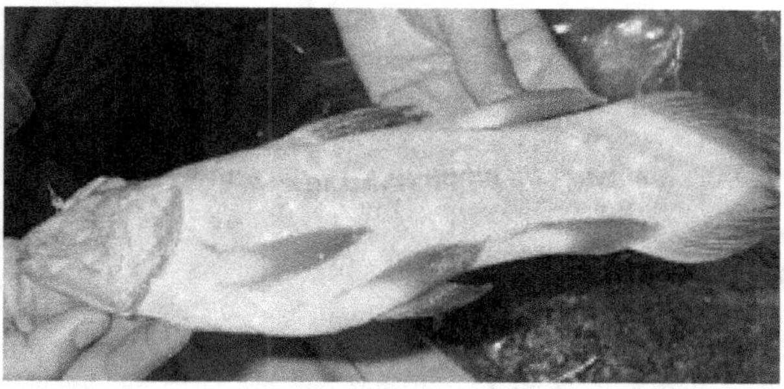

(Left) Figure 23: *Latimeria chalumnae* pup, SAIAB 37324. This specimen is on loan to the National Museum of Natural History, from the South African Institute of Aquatic Biodiversity (SAIAB) and is currently on display at the NMNH Ocean Hall Exhibit.

The coelacanth is a "passive drift feeder," moving slowly and passively near the substrate, where it feeds primarily on cephalopods (cuttlefish, squid, and octopus) and fishes. It is capable of moving quickly and may do so when capturing prey or avoiding danger.

How do coelacanths reproduce?

The mode of reproduction is ovoviviparity. This involves internal fertilization of eggs, followed by a gestation period thought to be about a year in duration, during which time the embryos feed off the yolk sac of the egg, culminating in the live births of fully formed young. Only two females carrying young, or "pups," have been captured. One female contained five full-term pups, each approximately 14" long, and the other had twenty-six pups of approximately the same size.

Is the coelacanth a missing link to tetrapods?

The coelacanth's evolutionary relationships are a matter of controversy. There are several competing hypotheses and many unresolved questions, in large part owing to the many unusual characteristics found in coelacanths. Experts largely agree that coelacanths are primitive osteichthyans or bony fishes (as opposed to a cartilaginous fishes, such as sharks and rays), all living (c. 360mya) close to Africa, Australia and South America, but they disagree on the exact placement of the coelacanth in the evolutionary history of vertebrates. Coelacanths might best be described as occupying a side branch in the basal portion of the vertebrate lineage, closely related to, but distinct from, the ancestor of tetrapods (four-legged vertebrates).

According to genetic analysis of current species, the divergence of coelacanths, lungfish and tetrapods is thought to have occurred about 390 million years ago. Coelacanths were once thought to have undergone extinction latest, 66 million years ago during the Cretaceous-Paleogene extinction event.

As noted, Coelacanths are ovoviviparous, meaning that the female retains the fertilized eggs within her body while the embryos develop during a gestation period of over a year. One male is used to fertilize all the eggs, potentially through female sperm choice or last-male sperm precedence.

"The Coelacanth behaves more like a gymnast than a fish, with unusual twists and turns. Its ears are similar to ours, it now bears live young." (Nisbet, 1991, p.22) This, in a creature who has chosen the depths rather than the shallow waters and the land. In the depths, it grew even larger, now as an ancient isolate.

On land

The first fully terrestrial vertebrates were amniotes, small lizard-like creatures. Like their amphibious tetrapod predecessors, they had lungs and limbs. Amniotic eggs, however, have internal membranes that allow the developing embryo to breathe but keep water in. Hence, amniotes can lay eggs on dry land, c. 325–295 mya, while amphibians generally need to lay their eggs in water.

The first amniotes apparently arose in the Pennsylvanian sub-period of the Carboniferous (c. 325–295mya). They descended from earlier amphibious tetrapods, which lived on land that was already inhabited by insects and other invertebrates, as well as ferns, mosses and other plants. Synapsids have a single hole (temporal fenestra) low on each side of the skull. One synapsid group, the pelycosaurs, included the largest and fiercest animals of the early Permian (c. 275mya). Non-mammalian synapsids are sometimes called "mammal-like reptiles."

The final mammal-like reptiles, the therapsids, evolved from pelycosaurs (specifically sphenacodonts) 275 mya. They replaced the pelycosaurs as the dominant large land animals in the Middle Permian , c. 250 mya. The therapsids included the cynodonts, the group that many think gave rise to mammals in the Late Triassic era, around 225 million years ago. Of the non-mammalian therapsids, only cynodonts and dicynodonts survived the Triassic-Jurassic extinction event (c. 200 mya). The last of the non-mammalian therapsids, the tritylodontid cynodonts, is thought to have become extinct in the Early Cretaceous, approximately 150–100 million years ago.

Therapsids became the dominant land vertebrates, c. 250 mya. They differ from basal polycosaurs in several features of the skull and jaws, including: larger skulls and incisors which are equal in size in therapsids, but not for eupelycosaurs. The therapsid lineage leading to mammals went through a series of stages, beginning with animals that were very similar to their pelycosaur ancestors and ending with probainognathian cynodonts, some of which could easily be mistaken for mammals. Those stages were fulfilled by:

The gradual development of a bony secondary palate which reorganized the amplification of sound for the developing hearing structures, as well as the olfactory (smell) sensibilities. It also allowed for the simultaneous eating while breathing. In the reptiles archosaurs, only the crocodiles developed this ability.

This jaw bone reconstruction had the added benefit of now wrapping the skull bones totally around the brain. Finally, throughout the evolution of the therapsids, the teeth were increasingly realigned in a vertical manner, increasing the quick mastication and concomitant digestion of food, quite different from the evolving dinosaurs that had to wait to the calm of night to put all their daylong feasting into their digestive innards.

There was also progression towards an erect limb posture, which would increase the animals' stamina. But this process was slow and erratic: for example, all herbivorous non mammaliaform therapsids retained sprawling limbs (some late forms may have had semi-erect hind limbs); Permian carnivorous therapsids had sprawling forelimbs, and some late Permian ones also had semi sprawling hindlimbs. In fact, modern monotremes still have semi sprawling limbs.

The dentary gradually became the main bone of the lower jaw which, by the Triassic (245mya) progressed towards the fully mammalian jaw (the lower consisting only of the dentary) and middle ear (which is constructed by the bones that were previously used to construct the jaws of reptiles). The result throughout the evolution of the therapsids saw the teeth increasingly realigned in a vertical manner, here increasing the quick mastication and concomitant digestion of food, quite different from the evolving dinosaurs who had to wait to the calm of night to put all their daylong feasting into their digestive innards.

Summary

Therapsida are a group of synapsids that includes mammals and their ancestors. Many of the traits today seen as unique to mammals, had their origin within early therapsids, c. 275 mya, including having their four limbs extend vertically beneath the body, as opposed to the sprawling posture of reptiles.

Fifty million years of dominance leads us to the late Triassic, c. 225 mya. Here, we note the evolving dominance of the dinosaurs amongst the true reptiles, which include the alligators and crocodiles of our own time. This dominance is viewed as a "clearing the land," product of the devastating extinctions of the Permian (c.250–245 mya)—extinctions which probably impacted the Therapsid/Cynodonts, which shrank from dog-sized creatures to rabbit size.

The relative emptiness, propelled by the Permian extinction (c. 245 mya) and the rise of the dinosaurs, likedly opened the door to true mammals in the Late Triassic, around 225 mya. Of the non-mammalian therapsids, only cynodonts and dicynodonts survived the next Triassic-Jurassic extinction event, c. 200 mya. Most of the non-mammalian therapsids, and cynodonts became extinct, most scientists now believe, c. 200–180mya.

The tiny, mouse-like true mammals, with the advantage of more efficient homeothermy (internal regulation of body temperature) and possible viviparity (live birth), may have made the crucial difference in the future of terrestrial life. Still, many puzzles persist.

Figure 24: Illustration of Pristerognathus, a cat-sized therocephalian therapsid

Mammals: brain patterns

Summary Perspective: *Mammals* are the vertebrates within the class *Mammalia* (/məˈmeɪliə/ from Latin *mamma* "breast"), a clade of endothermic amniotes distinguished from reptiles (including birds) by the possession of a neocortex (a region of the brain), hair, three middle ear bones, and mammary glands. Females of all mammal species nurse their young with milk, secreted from the mammary glands. Living mammals are divided into the Yinotheria (platypus and echidnas) and Theriiformes (all other mammals).

The early synapsid mammalian ancestors were sphenacodont pelycosaurs, a group that produced the non-mammalian *Dimetrodon*. At the end of the Carboniferous (c. 300 mya) period, this group diverged from the sauropsid line that led to today's reptiles and birds. The line following the stem group, Sphenacodontia, split off several diverse groups of non-mammalian synapsids—sometimes referred to as mammal-like reptiles—before giving rise to the proto-mammals (Therapsida) in the early Permian (c. 245 mya) era.

Early, true mammals began to appear between 225–200 mya. The modern mammalian orders arose in the Paleogene and Neogene periods of the Cenozoic (65 mya) era, after the extinction of non-avian dinosaurs, and have been among the dominant terrestrial animal groups from 65 mya to the present. Arguably, placental groups diverged from each other between 120–105 or 70–60 mya.

Origin of major adaptations

The plant invasion of the land began c. 470 mya, while the first true vertebrate fish were beginning to make their adaptive play in the waters. By the time that the first transitional crossopterygian fish made their move from snout above the muddy waters to taking a deep breath and lumbering across the land surface now covered with plants, c.360 mya, the great atmospheric revolution was completed—oxygen.

Oxygen dissolved in water equals 0.0451 grams per liter. In air, oxygen equals 6.70 grams per liter of air at room temperatures. This gives a comparison of some 150 times more oxygen in air than in water. (P.60 TIDBM) When we today use the term "brain dead," it usually denotes the lack of oxygen feeding the blood flow of an ill human. The heart needs oxygen and the brain even more so. Frail human beings extend their lives by taking it in from portable storage tanks.

Another comparison affecting the now-evolving amphibians, 340mya. Support for body of fish/amphibians: Density of air is >001 g/cm^3; density of water is $1g/cm^3$, the density of muscle is $1.05g/cm^3$; and density of bone is $3g/cm^3$. These numbers imply the necessity for land-dwelling animals to translate previous greater water support in the now thin but oxygenated air, into denser muscle and bone development on land. Obviously, with the high level of oxygen, and the existence of a wide variety of food sources on a land surface beginning to teem with life, large animals were now an early ongoing possibility. Soon, genetic innovations for on-land reproduction would open the way for a "reptilian" model of land survivors.

Reptilian leathery egg layers probably derive from the Crossopterygian fish who made the transition to the land and evolved into amphibians, next the amniotic/cleidoic egg. A Scotch fossil, c. 315 mya, in the Carboniferous, shows signs of eggs with leathery coating. This appears after the amphibian invasion of land, c. 360 mya.

The split between the ancestors of mammals and dinosaurs, and their descendants—birds—occurred some 275 mya. Birds slowly evolved from dinosaur origins. They developed homeothermy, the self regulation of internal heat, after their adaptive separation from the mainline dinosaurs, c. 150 mya. Yet, with their typically reptilian egg-laying method of reproduction, they still show great brain/body advances over all their cousins. The first mammals, by contrast, reveal themselves some 225 mya, this possibly with the evolutionary beginning of the important homeothermic adaptation.

Mammal adaptations

1. Internal temperature control, Endothermy (Homeothermy).
2. Chemical controls; importance of kidney function in cleaning out the system, concentrated urine, maintaining balance of internal bodily liquids; relative strong blood pressure is given by the heart and circulatory system to assist kidney function.
3. Importance of viviparity and endothermy for the protection of vulnerable, undeveloped neonates in a stable environment, provided by parental protection in early ontogenetic stages of development.
4. Development and integration of all the unique functions of mammalian structure after much time and evolutionary "space," hidden evolution, small size from 225 mya, and then in diversification and specialization.
5. Specialization in nocturnal visual defensive behavior (rod system) allowed for the further diversification of adaptive behavior, and gave the growing brain independent pathways to stay out of the way of the dinosaurs, before the great breakout, 90–65 mya (TIDBM P.86ff footnotes).

In the 1960s, Herndon Dowling noted that the Burmese python, Python molurus bivittatus, (a reptile) at the Bronx Zoo shivered as she remained wrapped around her clutch of newly laid eggs. Use of the thermistor showed that the shivering was thermogenic and maintained the clutch at an appropriate temperature for development

---

Placentals

They reflect a widespread reproductive pattern found in ancient shark lines and rays, as well as a wide variety of snakes (some also exhibiting homeothermic tendencies), all reflecting probable bio-chemical/genetic carryovers from more ancient times. The time of the split in the reproductive patterns of the mammals shows itself c. 175 mya, but probably began earlier.

The embryo attaches itself to the uterus via a large placenta, by which the mother supplies food and oxygen, and removes waste products. The mother, by having the growing embryo within her body for the full gestation period, can move about and into safety if threatened. Marsupials and Monotremes need more protection before and after birth. Larger placental neonates at birth, and fewer numbers at birth, also helped in the selective advance of this line of mammals.

Pregnancy is relatively long and the young are fairly well-developed at birth. In some species (especially herbivores living on plains), the young can walk and even run within an hour of birth. It is thought by some researchers that the tiny multituberculates, extinct c. 35 mya (mouse-like creatures), gave birth to tiny live and active young whom they must have nurtured with milk glands. mammary or monotreme-like clinging and lapping strategy (echidna, platypus.)

The brain is the slowest-growing organ in the fetal mammal; it is also the pacemaker for growth of other tissues, which are held to its rate of growth. Brain growth proceeds at the maximum rate allowed by its intrinsic growth law, so that the growth of all mammalian brains is governed in great degree by a single functional relation of attained size and time. The invariant rate constant for mammalian brain growth could be due to nutritional limitations (Sacher and Staffeldt,1974).

In general, the argument is that the length of embryonic development in placental reproduction governs the possibilities for brain size, with the accompanied requirement of sufficient nutrition for the mother, so as to provide for this brain growth of the fetus.

Endothermy

Here is a succinct overview of a part of this crucial evolutionary bodily adaptation's success:

> Recent palaeontological data and novel physiological hypotheses now allow a time scaled reconstruction of the evolution of endothermy in birds and mammals. A three-phase iterative model describing how endothermy evolved from the Permian (c.250mya). Ectothermic ancestors are presented in Phase One. I propose that the elevation of endothermy—increased metabolism and body in response to the fitness benefits enhanced embryo development (parental care) and the activity demands of conquering dry land. I propose that Phase Two commenced in the Late Triassic and Jurassic (200–190 mya) and was marked by extreme body-size miniaturization, the evolution of enhanced body insulation (fur and feathers), increased brain size, thermoregulatory control, and increased ecomorphological diversity. I suggest that Phase Three occurred during the Cretaceous and Cenozoic (120–90 mya) and involved endothermic pulses associated with the evolution of muscle-powered flapping flight in birds, terrestrial *cursoriality* in mammals, and climate adaptation in response to Late Cenozoic (Pleistocene, c. 2 mya) cooling in both birds and mammals.
>
> (Barry Lovegrove, 2016)

Naturally, the fact that endothermy is found in fish, sharks, and some reptiles, requires that this theory needs amplification.

Mammal brain

Within bony fishes, non-neopterygians (bichirs, paddlefish, and sturgeons) have relatively smaller brains (in terms of brain-body weight ratio) than those in the vast majority of teleosts (van Dongen, 1998). This is also true of the living lobe-finned fishes (lungfishes and coelacanths), which have relatively smaller brains than most amphibians and reptiles. Among amphibians, frogs generally have relatively larger brains than salamanders, and the brains of reptiles are generally two to three times larger than the brains of most amphibians of the same body size.

Both birds and mammals have brains that are six to ten times larger than the brains of reptiles of the same body size. Among birds, the largest brains for body size are seen in perching birds, woodpeckers, and parrots, while the relatively smallest brains are found in pigeons and chicken-like birds (van Dongen, 1998).

Surprisingly, many cartilaginous fishes have brains as large for their body size as those of birds and mammals (Bauchot ***et al.***, 1976; Northcutt, 1977; Demski and Northcutt, 1996). Among cartilaginous fishes, carcharinid sharks and myliobatid rays have the relatively largest brains, while squalomorph sharks and skates have the smallest brains for a given body size (R.G Northcutt, 2002).

Harry Jerison, in his various writings, developed a theory of the growth of the vertebrate brain. Brain/body relationship in vertebrates, fish, amphibians, reptiles remains in stasis from c. 470 mya in jawed fish to about 220 mya, this in spite of body size differences (Jerison—P.72-73 TIDBM). Leonard Radinsky sees the mammal brain as also being in body-brain stasis from c. 220 mya to 120 mya, when things begin to change towards the modern so-called “crown founders” of the various lines of contemporary mammals (Radinsky TIDBM).

Jerison’s conclusion was that from the Ordoviscan period, c. 475 mya, a time when we can obtain data to measure the brains of fishes during this period, and for 250 million subsequent years, until c.225mya, the brain/body relationship of fish, amphibians, reptiles, mammal-like reptiles was equable when one considers that bigger bodies=bigger brains (Jerison and Butler [2009] confirms Jerison, general view).

The apparent reason for this stasis was the relative non-competitive nature of the expansion of life in the seas and on land. These vertebrates were intelligent enough in the integration of their sensory-neurological apparatus, such that they could function and prosper equally well with this homologous brain/body equipment in an environmental-ecological context that was not selectively punishing.

L.Radinsky and T.Rowe postulate a 5X–10X expansion of the early mammals (225–190 mya) in terms of the relationship of body/brain size ratios, with these early mammals showing a large olfactory growth (“that is where the sensory information came from”), and an expanded cerebebellum (far greater than that of the late therapsid/cynodonts). Also, fossils reveal the beginning of a neocortex, 225–190 mya. “We found that relative brain size expanded to later mammalian levels, with enlarged olfactory bulbs, neocortex, olfactory (pyriform) cortex, and cerebellum, in two evolutionary pulses. The initial pulse (225 mya) was probably driven by increased resolution in olfaction and improvements in tactile sensitivity (from body hair) and neuromuscular coordination.” (T. Rowe, 2011)

Jerison did an analysis of the relative encephalization of the mammals, postulating an average fossil and living mammal brain to body quotient of 1.00:

a) Upper Jurassic mammals, 140mya = .28
b) Multituberculate (extinct c. 35 mya), from 60 mya = .26
c) Contemporary Virginia Opposum (Marsupial) = .22
d) Contemporary wild rat = .45

Note: The Upper Jurassic Triciniodon, 140 mya mammal, which was a down-the-line relative of early mammal Morganucodin, c. 225 mya, had more brain power than that a 60 mya multituberculate and the contemporary Virginia oppossum, a Marsupial. The contemporary placental wild rat doubled the contemporary marsupial and the tragic multiberculate placental, thought to have been made extinct by aggressive Oligocene (40–35 mya) rats.

Final thoughts

It took three hundred million years (360–60 mya) for the potential genetic expression of their reifications, to give the placental mammals metazoic eukaryote dominance on land. It will take another 60 or so million years for one super species to create conditions for selective evolutionary criteria to be subject to radical alteration. The trick now is the gradual substitution of the billions of years of sub-conscious genetic determinations to be supplanted by conscious, generation to generation decision-making by the DNA of a *hominin*.

Chapter 5

Overture to the Creation: Our Primate Heritage

Animal intelligence

1. To remember the past
2. To choose in the present
3. To plan for the future=time binding

Great Extinction

Many theories have been suggested for the extinction of the dinosaurs. Some of these have dealt with:

1. Toxic volcanic gases and dust
2. Asteroid impact
3. Food shortage
4. Climate change

Some scientists have suggested a combination of these factors. An element may also have been the growing size and appetite of predatory mammals and birds on the lookout for the delicacy of scattered reptile eggs.

A wide range of species perished almost overnight in the K-Pg extinction of 56 mya, the best-known of which being the non-avian dinosaurs. The event also destroyed a plethora of other terrestrial organisms, including certain mammals, pterosaurs, lizards, insects and plants. In the oceans, the K-Pg extinction killed off plesiosaurs and the giant marine lizards. It also devastated fish, sharks, mollusks (especially ammonites which became extinct), and many species of plankton. It is estimated that 75% or more of all species on Earth vanished. Yet, the extinction also provided evolutionary opportunities: in its wake, many groups underwent remarkable adaptive radiation—sudden and prolific divergence into new forms and species within the disrupted and emptied ecological niches.

Mammals, in particular, diversified in the Paleocene, evolving new forms such as horses, whales, bats, and primates. Birds, the last memory of the dinosaurs, survived and multiplied. Fish, and perhaps reptilean lizards, also radiated.

Mammals and primates

While we have no primate fossil material prior to the Eocene epoch, by molecular analysis, the first primates are thought to have evolved prior to the Paleocenee era (66–56 mya), and possibly as far back as 90 mya, during the late

Cretaceous era. With the extinction of the dinosaurs at the end of the Cretaceous era, many terrestrial niches became available and predation pressures were somewhat relaxed. In addition, temperatures were higher than in the recent past, and the angiosperms (flowering plants) were undergoing an adaptive radiation, i.e. a relatively rapid speciation, and spreading globally. The spread of flowering plants resulted in an adaptive radiation of insect pollinators and herbivores (plant-eaters), as well as insectivorous and herbivorous arboreal vertebrates.

"The origins and the divergence times of the most basal lineages within primates have been difficult to resolve mainly due to the incomplete sampling of early fossil taxa. The main source of contention is related to the discordance between molecular and fossil estimates:.while there are no crown primate fossils older than 56 Ma, most molecule-based estimates extend the origins of crown primates into the Cretaceous [c. 130–65 mya]." (**Mol Phylogenet Evol. 2014, pg. 165–183)**

Primate phylogenetic relationships and divergence dates are inferred from complete mitochondrial genomes (Luca Pozzi, Jason A. Hodgson, Andrew S. Burrell, Kirstin N. Sterner, Ryan L. Raaum, and Todd R. Disotell).

Figure 25: Tree shrew. "Tupaia cf javanica 050917 manc" by W. Djatmiko

Early placental mammals had a corpus callosum, connecting the neocortex of the two hemispheres, a primary motor area, M1, and perhaps one or more premotor areas. One line of evolution led to present-day primates, tree shrews, flying lemurs, also rodents and rabbits. Early primates evolved from small-brained, nocturnal, insect-eating mammals with an expanded region of the temporal visual cortex. These early nocturnal primates were adapted to the fine branch niche of the tropical rainforest, by having an even more expanded visual system that mediated visually guided reaching and grasping of insects, small vertebrates, and fruits. The neocortex was greatly expanded, and included an array of cortical areas that characterize the neocortex of all living primates. Recent discoveries in China of early primates indicate a frontal vision complex not suggestive of primeval nocturnal adaptations.

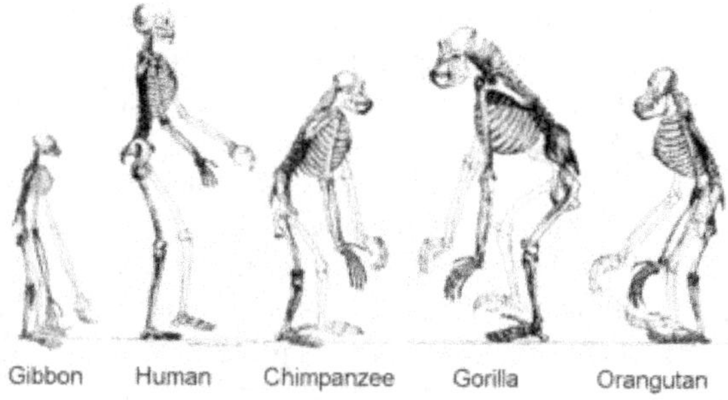

Figure 26: These hominoids are presumed to be descendants of a common ancestor

Specializations of the visual system included new visual areas that contributed to a dorsal stream of visuomotor processing in a greatly enlarged region of the posterior parietal cortex and an expanded motor system, with the addition of a ventral premotor area. Higher visual areas in a large temporal lobe facilitated object recognition, along with the developing frontal cortex, now including the granular prefrontal cortex. The auditory cortex included the primary and secondary auditory areas that characterize prosimian (lemur-like) and anthropoid primates today. As anthropoids emerged as diurnal primates, the visual system specialized for detailed line of sight-central (foveal) vision. While primates are thought to have evolved in Asia, the majority of the early primate fossils have turned up in Africa.

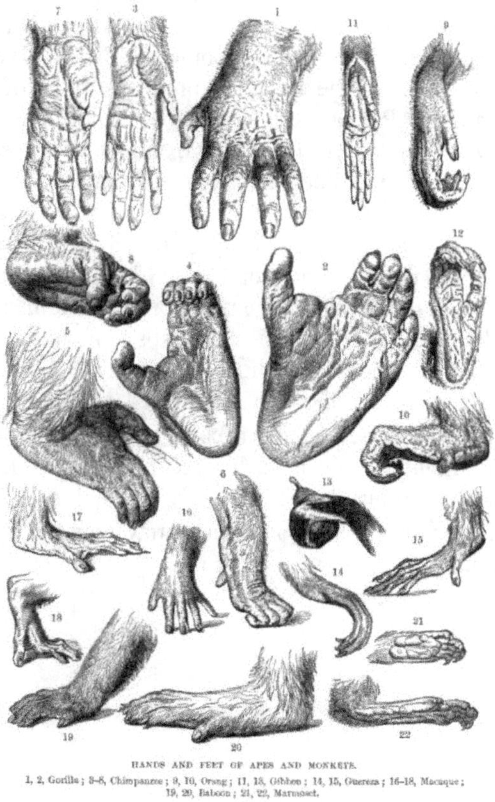

(Left) Figure 27: "Hands and Feet of Apes and Monkeys" by Richard Lydekker

Considered generalist mammals, primates exhibit a wide range of characteristics. Some primates (including some great apes and baboons) are primarily terrestrial, rather than arboreal, but all species possess adaptations for climbing trees. Locomotion techniques used include leaping from tree to tree, walking on two or four limbs, knuckle walking, and swinging between branches of trees (brachiation).

Primates are characterized by large brains relative to other mammals, as well as a highly developed sensory system as compared with most mammals. These features are more developed in monkeys and apes and noticeably less so in lorises and lemurs. Except for apes, they all have tails. Most primates have developed opposable thumbs. Many species are sexually dimorphic (differences between the sexes); differences include body mass, canine tooth size, and coloration. Primates have slower rates of development than other similarly sized mammals and reach maturity later, but have longer life spans. Depending on the species, adults may live in solitude, in mated pairs, or in groups of up to hundreds of members.

Humans and some other mammals, including most primates, have evolved trichromacy (three color perception channels in the brain) based partly on pigments inherited from early vertebrates. In fish and birds, for example, four pigments are used for vision. These extra cone receptor visual pigments detect energy of other wavelengths, including sometimes ultraviolet. Eventually, two of these pigments were lost (in placental mammals) and another was gained, resulting in trichromacy among some primates. Humans and closely related primates are usually trichomats, as are some of the females of most species of New World Monkeys, and both male and female howler monkeys.

Transitions

Given the constant discovery of new primate fossils, ancient pro-simians, as well as early monkey and ape specimens, it pays to give some priority in this case to the molecular time line—that most of the great mammal divisions of the Cenozoic, from c. 65 mya, during the Paleocene era, became all but modernly diversified during this period, 90–65 mya. An interesting example of this early diversification of the primates is the New World Monkey, that have a unique, prehensile (grabbing) tail.

Many paleontologists suspect an African origin, based on skeletal resemblances to ancient African monkeys. Now, Bond *et al.* describe three 36 million-year-old fossil teeth found in the Peruvian Amazon that support this idea. The shape of the teeth and phylogenetic analyses link the fossils to monkeys that inhabited Africa during the late Eocene, about 38 million years ago. The discovery also pushes back the monkeys' arrival date in the New World by 10 million years.

Considering that the Americas separated from Africa c. 140–130 mya, how the migration occured across the Atlantic Ocean at such a late, primate-developed stage, c. 50–40mya, is a mystery (Gramling, 2015).

Where did all these monkeys come from? Fossil teeth may now hint at an Asian origin for anthropoid primates

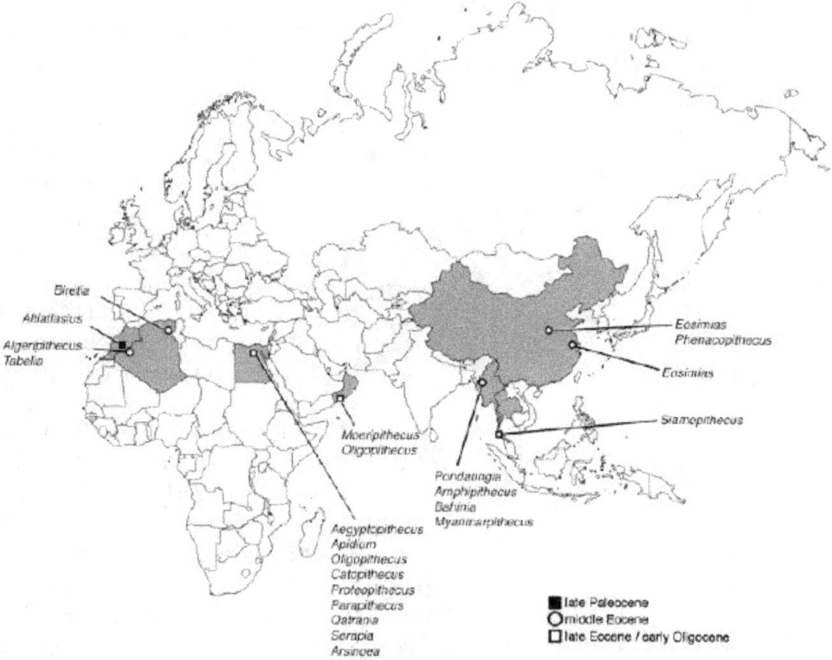

**Figure 28: Paleocene = 65—52 mya;
Eocene = 52–35 mya; Oligocene = 35–22 mya**

Interdisciplinary studies have confirmed that early anthropoids (monkeys and apes) were most closely related to small, tarsier-like primates and not the lemur-like adapiforms (which had been traditional favorites for the role). This recognition informs hypotheses about how early anthropoids lived and what their ancestors were like—in order to understand how early anthropoids evolved, we need to know what they originated from.

Another spot of contention has been where the first anthropoids evolved. Given that some of the earliest known anthropoids—such as the 37 million year old ***Biretia***—have been found in Egypt and Algeria, northern Africa seemed like the logical choice for the place of the group's origins. This hypothesis has not gone uncontested. In 1994, K. Christopher Beard and colleagues described the 45 million year old primate ***Eosimias*** from fissure-fill sites in southeastern China. Among the earliest of anthropoid primates, ***Eosimias*** was very close to the stem from which the first anthropoids radiated, hinting that the origins of the group were to be found in prehistoric Asia.

Evidence for the origin of anthropoids in Asia has continued to accumulate. In 1999, a team of paleontologists led by J.J. Jaeger described a relative of ***Eosimias*** from Myanmar, which they named "Bahinia." It lived alongside early anthropoids such as ***Amphipithecus***, and while it was not an anthropoid ancestor, "Bahinia" did possess a number of archaic traits which indicated that it was a member of a long-surviving lineage, that had branched off near the base of the anthropoid radiation. Its anatomy was like a time-capsule which recorded changes that had occurred millions of years earlier.

"Sunil Bajpai and co-authors described the 54 million-year-old relative of ***Eosimias*** **(from China, 45 mya),** named ***Anthrasimias*** from India. This extended the record of anthropoid primates back much further than the fossils from Africa, and last year's description of the geologically younger (about 38 million-years-old) anthropoid, ***Ganlea*** from Myanmar by Beard and colleagues, was another instance of an early anthropoid lineage branching off and persisting in Asia. The record of early anthropoids in Africa is nowhere near as rich. Prior to 40 million years ago, there is virtually nothing except a controversial taxon from Morocco, called ***Altiatlasius,*** which may not be an anthropoid at all. Taken all together, the fossils fall into a pattern consistent with the idea that anthropoids emerged in Asia, underwent an evolutionary radiation, colonized Africa, and then underwent a subsequent radiation." (Switek, 2010)

Figure 29: The Eocene-age fossil site located in Libya at c. 39 mya, then, contemporary with early elephants

Shapings

The separation of the monkeys from the apes, either before the migration out of Asia or subsequent, is considered to be due to climate or geological changes, at c. 50–25 mya. Monkeys in general were shaped selectively in their genetic response to a more tropical, forested, competitive environment. The apes seem to have responded to a more open country environment and more of a life on the ground. Most monkeys, with the special example of the baboons, who have come down to the ground late in their evolution, are arboreal, living their lives in the trees, similar to the modern gibbon, an Asiatic ape. It is quite possible that the gorilla and the two chimpanzee species were pushed into the arboreal of West Africa at a later stage of the Miocene, c. 10 mya.

Another interesting ape/monkey variation is the thick/thin enamel in the teeth of these primates. Humans have thick-enameled teeth, as do many apes and some Old World monkeys. Some of our reputed hominid ancestors have thick enamel, others thin. It has become an important evolutionary marker, since gorillas and chimpanzees have thin enamel teeth, whereas Orangutans have thick-enameled teeth. The teeth of primates would seem to be seemingly intransigent genetically and morphologically, i.e. immune selectively to momentary adaptive/survival needs.

It is likely that apes early in the Paleocene-Oligocene, 65–30 mya, occupied some ecological niches that would later be filled by monkeys. One of the earliest discoveries of the monkey to ape transitional primates was *Proconsul*. It lived in Africa in the eaerly Miocene, 23-17 mya.

Among the other numerous Miocene primate species presumably were the ancestors of all modern apes and humans. By 14 million years ago, the group of apes that may have included our ancestors was apparently also in the process of adapting to life on the edges of the expanding savannas in Southern Europe. They were very likely members of the genus *Dryopithecus*, which were generally similar in appearance to modern African apes. These apes evolved mostly during a relatively short global heat wave that began around 15 million years ago, and also made a new connection between Africa, Europe and Asia. This caused enough polar ice to melt, so that sea levels once again rose 80–130 feet.

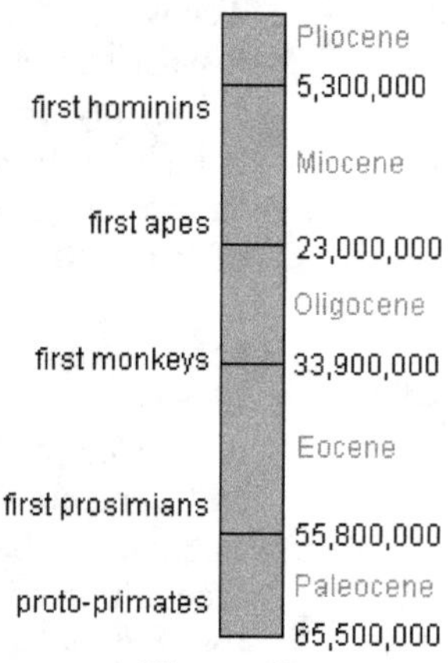

Figure 30

Toward the end of the Miocene, less hospitable cooler conditions in the northern hemisphere once again caused many primate species to become extinct, such as *Oreopithecus bamboloi*, from the swamps of northern Italy. This European ape had small, human-like teeth, and its spine entered the skull (foramen magnum) much further forward than contemporary apes. Some survived by migrating south into Africa or east (becoming ancestors of the orangutans and gibbons) into South Asia, where it remained relatively warm. About 8–9 million years ago, the descendants of the *Dryopithecines* in Africa may have diverged into two lines—one that led to gorillas and another to chimpanzees, and bonobos.

Extending our timeline

Darwin believed in the slow adaptive and selective workings of nature to produce evermore modern forms of life. Surely, once grooved into a successful adaptive niche, subsequent mutations were usually maladaptive and repressed in the geno and phenoplasm. However, we are now cognizant of outliers lurking in every species—that term, *species*, itself a broad taxonomic definition created not by nature but by humans trying to organize information.

Nature is obstinate. It doesn't create convenient gene mutations when there is sudden need. Nor do long-evolved animals have the ability as some plants do, to radically shift from vegetative reproduction to the sexual and then back again, in order to find quick solutions to danger. Within our genome, we animals better have some recessive DNA lurking within our genetic reservoir that will show up in times of opportunity or crisis.

We should always be looking for what is as yet unseen. Remember how the erratic movement of planet Uranus disturbed scientists in the early 19th century, because its movements violated Newtonian principles. Finally, a newly invented telescope discovered a new planet, Neptune, whose mass exerted gravitational influence on the orbits of Uranus. Motto: *Keep looking.*

The fact that along with a variety of transitional primates in the Eocene, c. 40 mya (see illustration above), we can recognize an ancestor of the elephant-to-be, tells us that the form of most modern mammals and primates was being shaped by nature many million years before we discovered their fossils, say up to c. 60 mya. The British Museum has an illustration of the 40 mya evolution of the horse, from the Eocene. There are differences in size and structure, but for the basic recognition of the form of horse, it was by then, already long established. (British Museum, 2000, pp. 139–167, 161)

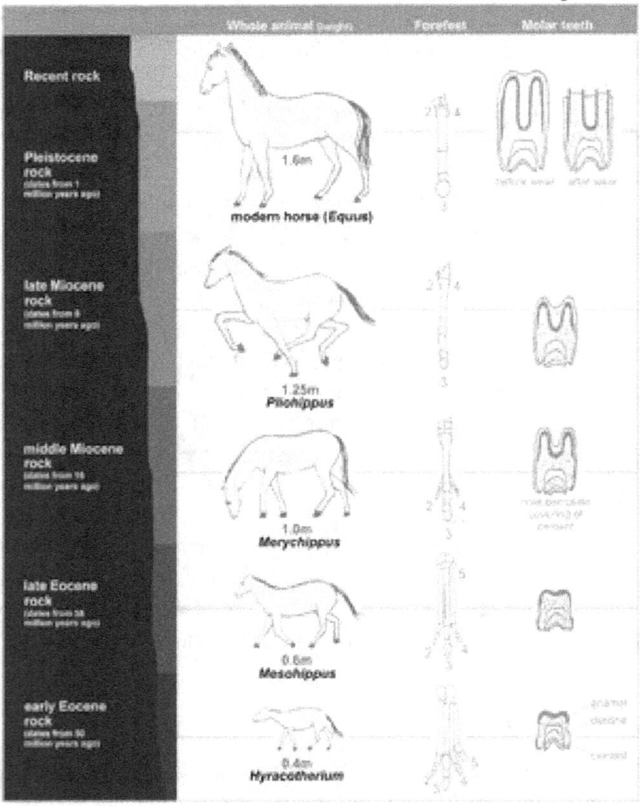

Figure 31

Then, there is the question of the New World monkeys—where

did they come from and when? The fossils found in Peru date to c. 40 mya. They seem to reflect a well-developed New World version of an older monkey line, which itself must have shared an ancestor, probably in Africa, with the Old World monkeys. Thus, one must go back another 20–30 million years. As in the traditional time line illustrated above, we don't see many Old World monkeys until quite a bit later. Certainly, they were on the move in that distant past.

Then there is the case of the multituburculates, one of the four main lines of (perhaps) marsupial mammals, possibly evolving from way back in the Mesozoic era, c. 225 mya. They disappeared in the Eocene era, 40–35mya. Possibly, they could not compete with the domineering placentals—the multituberculate neonates, who gave live birth to helpless offspring, who were quickly gobbled up by early placental rats that were already long in the developmental process of intelligent predation. Along with the multitiburculates, masses of other marsupials and monotremes hit the dust at that time, except for those in distant geographies such as Australia, where the recent introduction of placental rabbits amongst others will eventually eliminate the last of these mammal lines in the wild.

The return of animals, reptiles and mammals to the waters, just as the opportunist placental bats return the favor of an old reptile commitment to the air waves, also challenges us in our time-line decisions. For the mammals, this return by whales and dolphins, precipitated by an ancestor of the Hippos, at about 55 mya, currently raises little controversy of this orthodoxy.

Hippos have long flirted with the opportunity to explore a new or empty niche waterside where there is much nutrition. They still copulate and give birth to young under water. Sea lions and manatees still are in their own flirting stage. We ask, when did this trend to take advantage of the water's edge and then the seas themselves, begin as part of the placental mammalian radiation? How long before the hypothesized fossil date of c. 55 mya? A few tens of millions of years?

In general, the molecular people are probably correct. The evolution of the primates from a generalized mammalian class and their sub and infra class members, suddenly shifted to the wildly varietal placental mammal moment of opportunity. Here too, primates do probably stretch back beyond the Paleocene into the Cretaceous, some 90–100 million years ago.

Ancestors of the Hominids?

Again, we confront a story of intellectual scientific conflict between the molecular people and those who analyze fossils. In the old days, the search for the ancestors of *Homo sapiens* included much comparative analysis of monkeys, apes and humans fossils, as well as modern humans. An example from the "old days" perspective is the writing of Bernard Campbell in 1966, where he comes to the conclusion that the relatively unspecialized structure of modern humans is closer to some of the Old World monkeys, especially in the level of brain/body weight and the shape and structure of the hand and male penis, than to any of the great apes, ancient and modern. (TIDBM, P.142-43)

Soon however, the new techniques for molecular analysis and the projection of genetic DNA changes over time became a fascination for the laboratory, over and beyond what the "field" had and was uncovering. The revolutionary result of their work showed a closer molecular affinity of the chimpanzee to humans than even the chimpanzee to the gorilla. Moreover, they extrapolated rates of molecular change into a timeline of modern human/chimpanzee separation from a common ancestor. It was calculated to be 5–6 mya. (TIDM in P. 148 ff)

There were two morphological (structural) facts that had to be disturbed, given the molecular analysis. Chimps have thin enamel, while humans have thick-enameled teeth. Further, chimp locomotor patterns use their knuckles, while humans, even babies, navigate with flat hands on the ground. Where is the bi-pedality in chimps, which many see as the key to Homo's great explosive conquest of the animal world, the return argument is that in the millions of years following the original separation of future chimp and future Homo, both would go their own ways, facing very different adaptive challenges.

This writer did not believe in the applicability of the physicist approach to the evolution of the hominids. It just did not seem possible that such a frenzy of mutations could have precipitated such a timeline and radical changes—in the enamel of teeth as well as the back and spine, and the four limbs, these among a number of other structures and behaviors. Not only was it anti-Darwinian, but laboratory analysis could not possibly take scientific precedence over the "facts on the ground." (SWI: *TIDBM* 2000, PP161–184) Another skeptic was Jeffrey Schwartz of the University of Pittsburgh.

A number of fossil apes of the middle and late Miocene had dental patterns featuring low-cusped, grinding molars with relatively thick enamel. In females of some species, such as *Ouranopithecus*, *Kenyapithecus wickeri*, and *Gigantopithecus*, the canine teeth were small in size compared to living apes like chimpanzees. Living chimpanzees, bonobos, and gorillas differ from the pattern of these fossil apes, as they all share molar teeth with relatively thin enamel and high crowns, and large canines that project well beyond the incisors and premolars even in females. These substantial differences between living African apes and fossil Miocene apes make it unclear which pattern may be the ancestral condition for early hominids (Hawks, 2005). Furthermore, robust and gracile "Lucy"-like australopithecines, which date far back, show a more bipedal a hint towards later humans.

Martin (1983, 1985) reviewed the significance of enamel thickness in hominoid evolution. He concluded that thick enamel formed at a fast rate, and represented the ancestral condition in the human and great ape clade. Thin enamel in African apes reflected a secondary reduction in secretion rates, with outer enamel being formed at slow rate.

Jeffrey Schwartz fought the molecular battles along with others as to who is more related to whom, and then to the end product, *Homo sapiens sapiens*. He also emphasized the morphological distances between chimpanzees and orangutans and the closeness of orangutan morphology (physical characteristics) as precursor to the human level, highlighting issues of teeth, partial bipedality, locomotor structure and behavior, as well as reproductive patterns. The response did remind one of the hopes that David Pilbeam of Harvard had with *Ramapithecus wickerii*, a *Sivapithecine* ape related to the orangutans, possessing very human-like parabolic dental/skull structure, but a very ape-like post cranial structure (below the neck).

One of the puzzling aspects of the Asiatic great apes—the orangutan and the gibbon—is their probable origination in Africa and then their fossil appearance on the road to Southeast Asia. We have found sivapithecine fossils from Turkey, Pakistan, and China, and then witness their contemporary presence in Indonesia. A comparison of chimpanzees with the nine million-year-old sivapethcine from Pakistan, and contemporary orangutan skulls show that the nine million-year-old sivapithecine is uncannily similar to the contemporary orangutan (SWI, TIDBM, 2000, p.158). No modernization is apparent to confirm the modernizing chimpanzee hypothesis. Along with the probable migration of sivapithecines some 12–10 mya, is the puzzling simultaneous disappearance of the unspecialized Miocene apes from Africa, and the descent of the baboons from the trees apparently to fill the vacuum. The chimpanzees and the gorilla ancestors, in the view of some, presumably migrated west from the savannas into the jungle.

The challenge to the molecular hypothesis, which has gained ever more loyalty over the decades as written in all the contemporary textbooks, is the issue of ancestors to the human and chimpanzee lines. Hint, hint, hominin fragments dated up to c. eight mya have been periodically announced.

These fragments have now been filled in. Over the last decade, there have been a number of important fossil discoveries in Africa of what may be very early transitional ape/hominins, or proto-hominins. These creatures lived just after the presumed divergence from our common hominid ancestor with chimpanzees and bonobos, during the late Miocene and early Pliocene Epochs. The fossils have been tentatively classified as members of three distinct genera—*Sahelanthropus*, *Orrorin* and *Ardipithecus*. *Sahelanthropus* was the earliest, dating 7–6 mya. *Orrorin* lived about 6 mya. *Ardipithecus* remains have been dated to 5.8–4.4 mya. At present, the vote is still out as to whether any of these three primates were in fact true hominins, and if they are presumed to be ancestral to *Homo*. The classification of *Sahelanthropus* has been the most in question.

Ardipithecus ramidus, is seen as living 5.8–4.4 mya, had a small brain, measuring between 300 and 350 c3. This is slightly smaller than a modern bonobo or female common chimpanzee brain, but much smaller than the brain of australopithecines like Lucy (~400 to 550 cm, female height, 3'11"). Tim White of University of California, Berkeley and C.O. Lovejoy of Kent State University, who have done the most research on the fossils, believe that *Ardipithecus ramidus* is not related to the chimp-human divergence. They see this divergence as a more distant 8–11mya.

It has been noted, however, that Oreopithecus bambolii, a late Miocene ape from Italy, shows many of the characteristics that are used to denote the hominin status of *Ardipithecus ramidus*. These include relatively small canines, reduction of C/P3 honing complex (tooth alignment structures), anteriorly placed foramen magnum, and well-developed anterior inferior iliac spine (eminence on the hip bone). While other morphological features of oreopithecus leave no doubt that it is an ape, the presence of the above characters suggests that interpretations of the earliest hominins may be confounded by homoplasies (independent separate lineage). (Wood & Harrison 2011; Su, 2011)

Given the extensive reconstruction of some of the fossils, some scientists are not convinced that the evidence definitely demonstrates *Ardipithecus ramidus* walked upright. The researchers point to features of the pelvis—such as hip muscles that are positioned on the side of the body, as they are in modern humans, to maintain balance while walking—as proof that *Ardipithecus ramidus* was bipedal. But the one known *Ardipithecus ramidus* pelvis had been crushed, so the researchers had to reconstruct it to learn anything. That, says Carol Ward, an anatomist at the University of Missouri in Columbia, makes any interpretation "equivocal."

As for the environment, there's "no doubt that [*Ardipithecus*] was adapted for an arboreal lifestyle," Ward says. But where Lovejoy, White and their colleagues see primitive, monkey-like climbing, Richmond still sees typical chimpanzee- and gorilla-like features. "The hands and feet look like an ape," he says. "They would have been capable of climbing the way chimps and gorillas climb."

Still, with all of this new evidence, "the idea that we came from a chimp-like ancestor is highly unlikely," Ward says. And therefore, given a different starting point, researchers will need to reframe many of the questions about how and why hominids started down the human path.

Orrorin tugenensis

Living around 6 million years ago, O. tug. is one of the oldest early humans of our family tree. Individuals of this species were approximately the size of a chimpanzee and had small teeth with thick enamel, similar to modern humans. The most important fossil of this species is an upper femur, showing evidence of bone buildup typical of a biped—indicating *Orrorin tugenensis* individuals climbed trees but also probably walked upright with two legs on the ground. (Smithsonian Museum)

Its discoverers believe this species belongs on the human family tree. They claim that it is dissimilar to the genus *Australopithecus*, and that this genus (*Australopithecus*) should be moved to a side branch on the human tree, leaving *Orrorin tugenensis* as the earliest direct ancestor of humans. However, most authorities say there is not enough evidence to support this argument, due to the fragmentary nature of the remains. Other experts think this species may have lived before the ape and human line split so it could be an ancestor of both lines or that it was a basal member of the hominin clade.

The species lived during a critical period in the human evolutionary timeline. It is still widely believed humans and chimps diverged from a common ancestor that lived between five and eight mya. Its brain is probably similar in size to a modern chimpanzee's, but the lack of cranial material makes this difficult to determine. The teeth of an ***Orrorin tugenensis* were** large pointed canines, with thick tooth enamel. (Australian Museum)

Sahelanthropis tchadensis

The most recent discovery involves the above fossil remains, found not in East Africa, but in **the desert of what is** now Chad. *Sahelanthropus*, c. 7 mya, is viewed by many to be of the hominin clade. **Evidence** includes small canines worn at the tip, implicative of a reduced or absent C/P3 honing complex (tooth alignment and structure), and a short cranial base with a foramen magnum that is positioned anterior and orthogonal to the orbital plane, suggestive of an upright posture and habitual bipedality. **Its brain size is c.360cc, similar to that of** a chimp. (Brunet et al. 2002; Guy et al. 2005; Zollikofer et al. 2005; Su, 2011)

Brigitte Senut and Martin Pickford, the discoverers of Orrorin tugenensis, suggested that the features of *S. tchadensis* are consistent with a female proto-gorilla. Even if this claim is upheld, the find would lose none of its significance, because at present, very few chimpanzee or gorilla ancestors have been found anywhere in Africa. Thus, if *S. tchadensis* is an ancestral relative of the chimpanzees or gorillas, then it represents the earliest known member of *their* lineage. *S. tchadensis* does indicate that the last common ancestor of humans and chimpanzees is unlikely to closely resemble extant chimpanzees, as had been previously supposed by some paleontologists.

A further possibility, highlighted by research published in 2012, is that the human-chimpanzee split is earlier than previously thought, with a possible range of 7 to 13 mya (with the more recent end of this range being favored by most researchers), based on slower than previously thought changes between generations in human DNA. Indeed, some researchers (such as Tim D. White, University of California) consider suggestions that *Sahelanthropus* is too early to be a human ancestor to have evaporated.

Thus, White argues that we should push back the separation of the hominid line by many more millions of years. But where are the proto-chimpanzees and gorillas?

A possible answer follows: the fossil, named *Anoiapithecus brevirostris*, found in Spain by Salvador Moyà-Solà of the Catalan Institute of Palaeontology in Barcelona, Spain, and his colleagues, dates from a period of human evolution for which the record is very thin. While only the animal's face, jaw and teeth survive, their shape places it within the African hominid lineage that gave rise to gorillas, chimps and humans. However, it also has features of a related group called *kenyapithecins*. (Holmes, 2009)

The problem with Moya-Sola's discovery, is that often, the sub-cortical morphology of these seeming hominoids when found is typically ape or monkey-like. The hominid hints of the skull, however, do reveal possible selective pressures for higher intelligence far back into the Miocene, 12.5 mya.

The work of David Begun of the University of Toronto

Dryopithecus lived about 12.5 mya. “A new analysis of an ape that lived 12.5 million years ago suggests it is a type of gorilla. If that’s true, it means gorillas evolved much earlier than thought, and also pushes back the time when humans split from chimps by about 2 million years.”

David Begun of the University of Toronto in Canada reanalyzed fossils of *Dryopithecus* apes, which lived in what is now Europe, as early as about 12.5 mya. He says that the characteristics of the skull suggest that rather than evolving earlier than the great apes, as was previously thought, *Dryopithecus* is a proto-gorilla. (Taylor, 2015)

Final thoughts

We have still not found the missing link between us and apes.

Colin Barras

Barras argues that the entire chimp-hominids relationship—the concept of the last common ancestor—has fallen apart. Barras mentions the older work of Straus, and also Schwartz’s orangutan thesis. (Colin Barras, 2017)

Why humans are important to studies of primate diversity

It is not in the least bit controversial to picture humans, within the context of the placental mammal group, that we belong to the primates. Naish argues that many human attributes relate to **O**ld **W**orld monkeys, orangutans, australopithecines **and** not other early hominines. Our bodies are structurally monkey-like. Bipedalism relates more to the gibbon, orangutan **and** lorises—all unlike great apes gorilla, chimps. **(Naish, 2014)**

Where should we look?

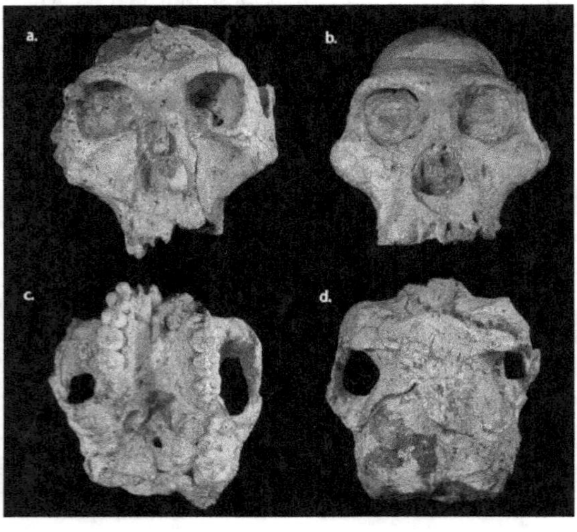

(Left) Figure 32: Paranthropus robustus and Australopithecus africanus from southern Africa. P. robustus is represented by the SK 48 cranium from Swartkrans and is shown in anterior (a), inferior (c), and superior (d) views. A. africanus is represented by Sts 5 from Sterkfontein and is shown in anterior view only (b). SK 48 is distinguished from Sts 5 by many traits.

Twenty years ago, this writer hypothesized a borderline Oiligocene/Miocene break off from the versatile proto hominid line, i.e 20 mya. At this level of time, it seemed likely that an animal with so many shared physical and behavioral attributes might finally be shearing off from the mysterious primate founder. This animal would be somewhat bipdeal, 24–36 inches in height and able to peer through the savannah grasses for threats, else climb a convenient tree; definitely a defensive creature at that point, and thus highly sensitized perceptually and in brain capacity. [TIDBM P.170-171]

Go back to the Eocene some forty mya, and we might be suspicious about a tree-inhabiting creature, not prognathus, beginning to use their frontally aligned eyes for food gathering in the daylight, without a tail; this animal could be a developed Tarsier, maybe an early monkey, even an ape-like critter. Certainly, our earliest ancestor would not predict the morphology of a five million-year-old Chimpanzee, if we could ever unearth one.

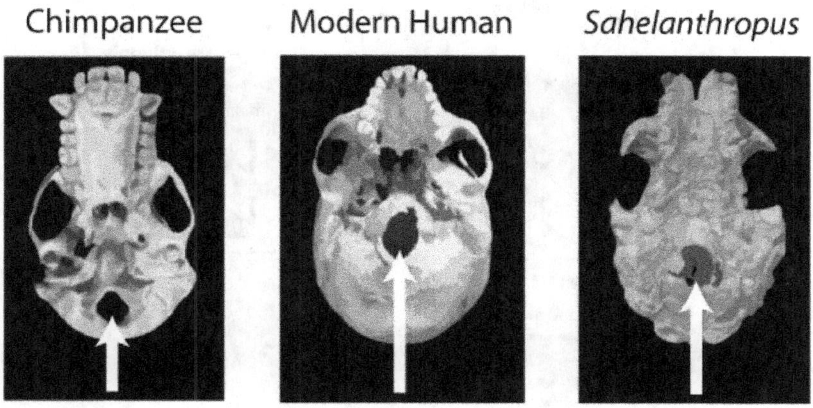

Figure 33: A comparison of the position of the foramen magnum (marked by white arrows) in chimpanzee, modern, human, sahelanthropus

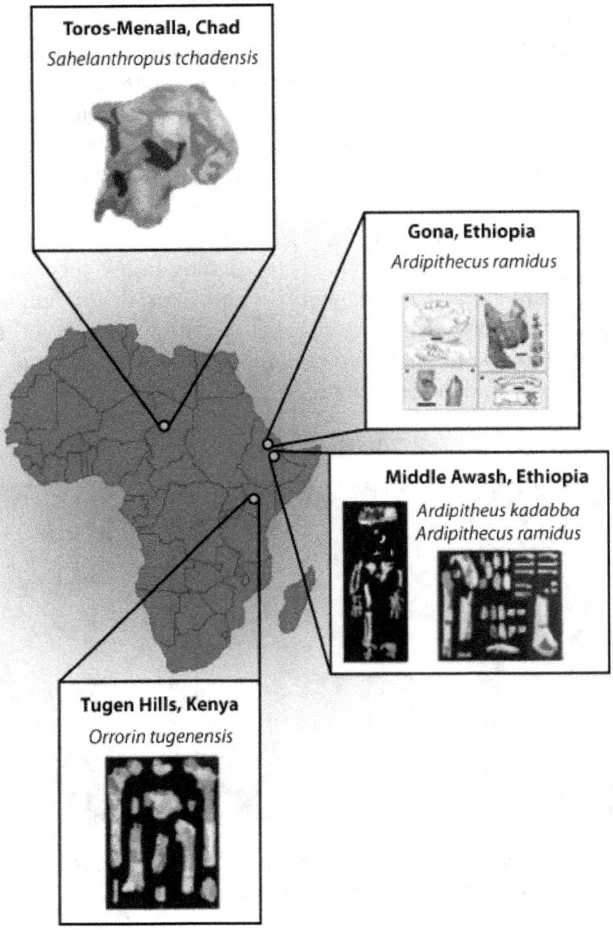

Figure 34: Map showing the placement of the earliest hominin localities and some of the specimens recovered.

[The Earliest Hominins: *Sahelanthropus*, *Orrorin*, and *Ardipithecus*

By: Denise F. Su (Curator and Head, Department of Paleobotany and Paleoecology, Cleveland Museum of Natural History), 2013 Nature Education]

Chapter 6

Hominin Patterns and Pathways

Hominins on the march

The gradual disappearance of the undifferentiated apes of the early to middle Miocene, c. 23–12 mya, is still a puzzle. Some went north, including a number of Dryopithecines (see Chapter 5), especially Oreopithecus, but the climate of Europe was not especially conducive to their long term flourishing. Others took the migratory pathway to Asia (gibbons and ancestral orangutans), which had opened up c. 15 mya.

The usual explanation is the change in climate conditions which gave an advantage to the Old World monkeys, re-forestration of the mixed plains, and forested areas that had nourished the progression of ape forms, as well as different nutritional opportunities which subsequently favored the monkeys. This could explain the shift of the gorilla and chimpanzee ancestors into the West African jungles.

But there is another hypothetical explanation. This is the continuing evolution of a partially bipedal defensive ape of somewhat higher brain size and power. The transition from the defensive mode of survival and selection could well have come about by the relative success of brain power defensiveness. Concomitantly, growth from being a one-footer to a two-and-half-footer could have made all the difference in pushing the similar evolutionary trajectory of these Miocene apes out of the way.

All animals take the adaptive path of least resistance. Usually, the decision is taken to get out of the way of a fight that might be lost, given that the Massachusetts General Hospital is not on the premises to treat the wounded. When the Autralopiethecines show up a few million years later, and in numbers and variety, the hesitant assumption is that they have been around for a while, but we haven't as yet discovered many of these early proto-hominin roamers around Africa.

The earliest australopithecines very likely did not evolve powerfully enough until 5 million years ago, or shortly thereafter (during the beginning of the Pliocene Epoch, 5 mya), in East Africa. The primate fossil record for this crucial transitional period, leading to the australopithecines, is still scanty and somewhat confusing. However, by about 4.2 mya, unquestionable australopithecines were present. By 3 mya, they were common in both East and South Africa. Some have been found dated to this period in North Central Africa also. As the australopithecines evolved, they exploited more types of environments. Their early proto-hominin ancestors had been predominantly tropical forest animals. However, as argued above, African forests might again have progressively given way to sparse woodlands and dry grasslands, or savannas The australopithecines took advantage of these new conditions. In the more open environments, bipedalism would very likely have been an advantage.

An interesting perspective on the early African dominance of the australpithecine and the selective dominance of the evolution of brain size is suggested in the following analysis:

> The brain size increase from ***A. afarensis*** to ***A. africanus*** to the "robust" australopithecines does not appear to be an artifact of body size increase but reflects progressive encephalization. The expansion of absolute brain size with the appearance of ***Homo*** is beyond what would be expected from body size increase alone. These working hypotheses have implications for how members of early hominid species behaved to enhance their chances of survival and reproduction within the constraints and requirements of their environments. For example, the relatively high level of body size sexual dimorphism in the earliest species (australopithecines) implies a polygynous mating system and a ranging pattern in which females foraged in smaller territories than the males.
>
> (Mc Henry, 1994, p. 77–97)

Our guy or gal, true proto *Homo* was probably around the savannas of Africa at this early stage, working out his/her more bipedal and brain armaments for positive selection and diversification. Omnivorous, given, as with the australopithecines their thick-enameled teeth, these smarter "apes" would eventually profit by their opportunistic predatory instincts, and then diversified and multiplied.

(Left) Figure 35: "Taung child" reconstruction (Australopithecus africanus) 3–2.2 mya

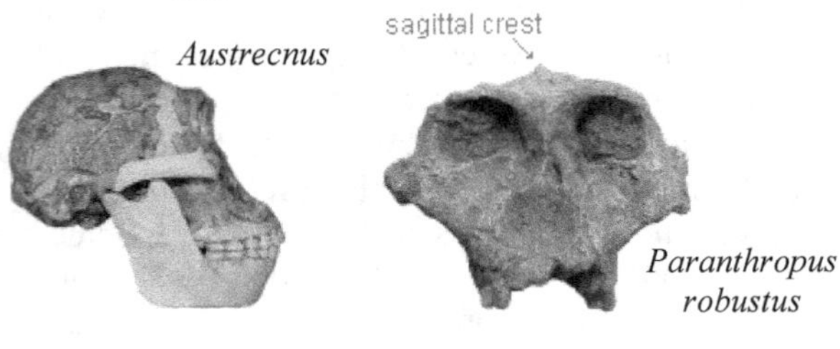

(Left) Figure 36: Paranthropus robustus, ***c.2.4–1.4 mya and*** **Australopithecus Africanus,** ***3–2.2 mya***

They say pictures are worth thousands of words. From the above, one can see some of the early actors on this later stage. Hominims' genetics in place were obviously undergoing a process which we call *orthoselection,* selection favoring phenotypic success and tending to eliminate other adjacent genetic possibilities, such that it resulted in the steady ascent of larger brains, taller and stronger animals, *viva la DNA*. The genes had to be there, a heritage now of many millions of years of Paleocene/Eocene Miocene (50–20 my), previous evolutionary success. Who could not expect that the genus, *Homo* would eventually be subject to a process of robusticity typical of the scenario of australopithecine evolution, contra *Africanus*—thicker bones, a skull with deep brows and thus hints that *Homo* too was settling in for security and domesticity? The australopithecines got hit, and became extinct by *Homo,* c. 1 mya. Is *Homo* immune from such a fate?

Evidence from the fossil record

Figure 37: Replica of fossil skull of Homo habilis. Fossil number KNM ER 1813, found at Koobi Fora, Kenya, 2.9–2 mya

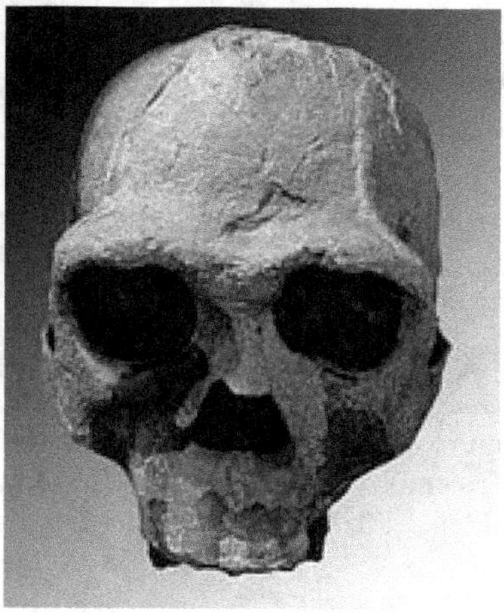

Figure 38: Replica of fossil skull of Homo ergaster (possibly African Homo erectus). Fossil number Khm-Heu 3733 discovered in 1975 in Kenya. c. 1.9–1.4 mya

Figure 39*: (Left to right)* Australopithecus africanuss*;* Homo habilis *(*ergaster*?);* Homo rudolfus *(*erectus*);* Homo sapiens *(Qafzah—92,000 BP (possible* sapiens*/Neanderthal hybrid);* Homo sapiens sapiens *(Cro-Magnon).*

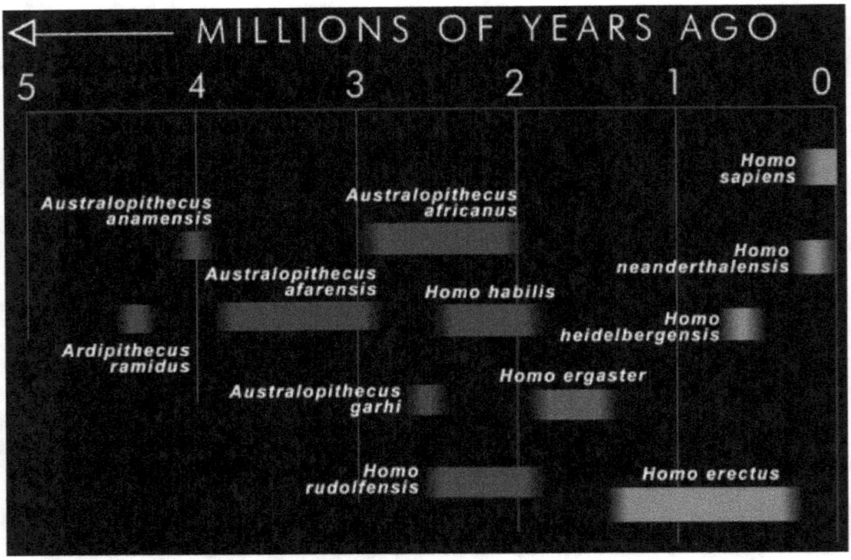

Figure 40: (Above)

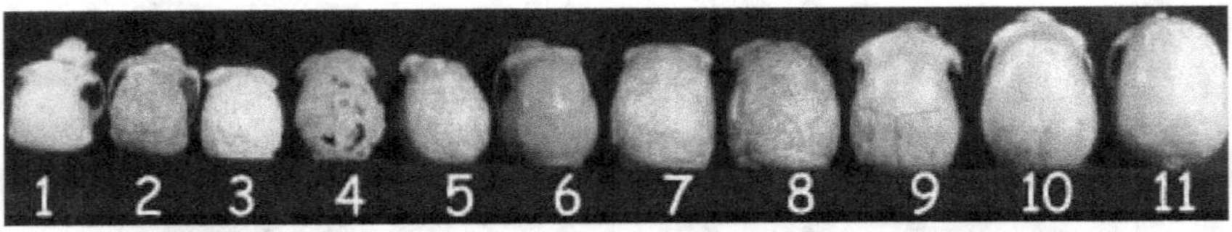

Figure 41: Skulls in primate/hominid/hominin evolution. 1. Chimpanzee; 2. A. africanus; 3. H. habilis; 4. KNM-ER 1470; 5. Java Man; 6. Peking Man; 7. H. saldensis; 8. H. saldensis; 9. "Broken Hill" Man; 10. Neanderthal Man; 11. H. sapiens sapiens

Bigger brains: complex brains for a complex world

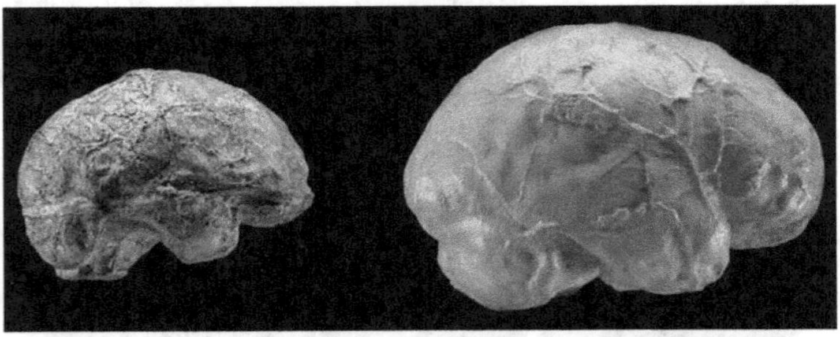

Figure 42: Left: Homo erectus brain, c. 1.8mya–c. 500,00 BP; Right: Homo sapiens sapiens brain, c. 20,000 BP

Endocasts of *Homo erectus* (left) and *Homo sapiens* (right) illustrate rapid increase in brain size. As early Homo erectus humans faced new environmental challenges and evolved bigger bodies, they evolved larger and more complex brains.

Large, complex brains can process and store a lot of information. That was a big advantage to early humans, in their social interactions and encounters with unfamiliar habitats.

Figure 43: Dmanisi, Republic of Georgia, c. 1.8 mya, Homo erectines. Five specimens in one cave at one general time position, showing enormous variability

The braincase of Skull 5 is only about 33.3 cubic inches (546 cubic cm) in volume, however, which suggests that this early ***Homo*** had a small brain despite his modern human-like limb proportions and body size: large face, small brain (see below).

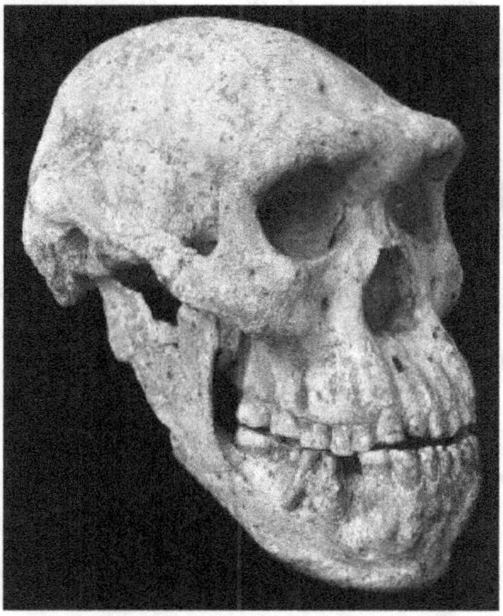

Figure 44: Skull 5 from Dmanisi, Georgia. Image credit: Guram Bumbiashvili/Georgian National Museum

At 1.8 million years ago, the dating for this primitive West Asiatic *Homo erectus* skull, there probably were australopithecines still wandering the world, just as *Homo florensiensis* **was** (see discussion on the next page).

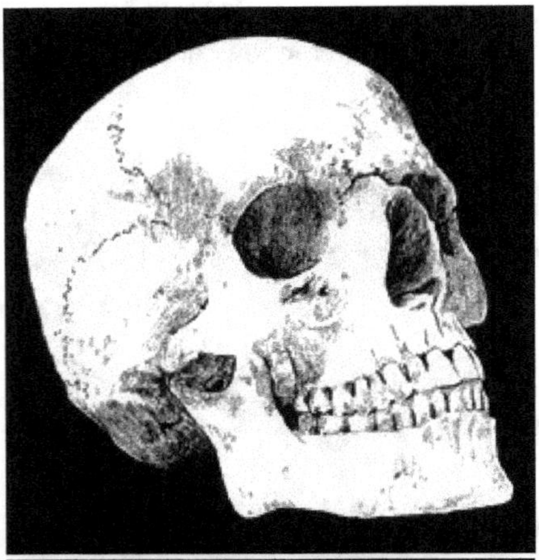

Figure 45:
Homo sapiens sapiens

Worldwide: possibly 400,000–200,000 years ago to present. Our own subspecies. Has a lighter build than earlier *Homo* forms and may have interbred or displaced other *Homo* species.

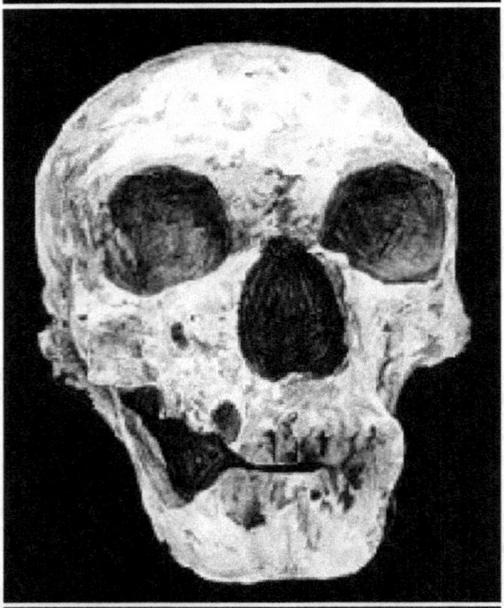

Figure 46:
Homo neanderthalensis

Europe, Western Asia: 500,000 to about 30,000 years ago. Stocky; adapted to the cold climate of their time. Our closest extinct relative; possibly a subset of our own species. Racially Caucasoid.

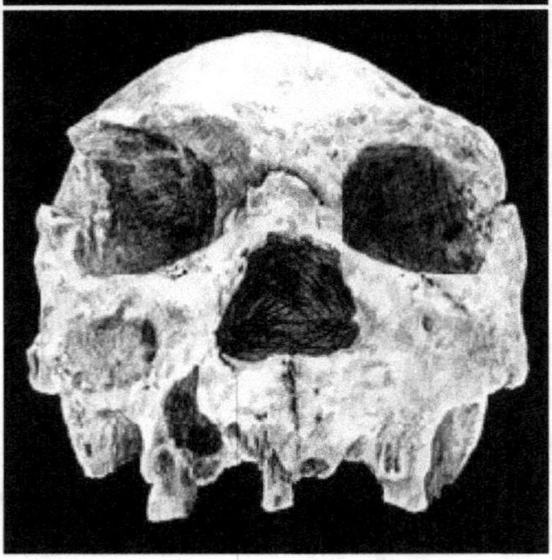

Figure 47:
Homo heidelbergensis

Europe: 800,000 to 250,000 years ago. Muscular and tall, with a sloping chin and forehead. May be a direct ancestor of *Homo neanderthalensis*..

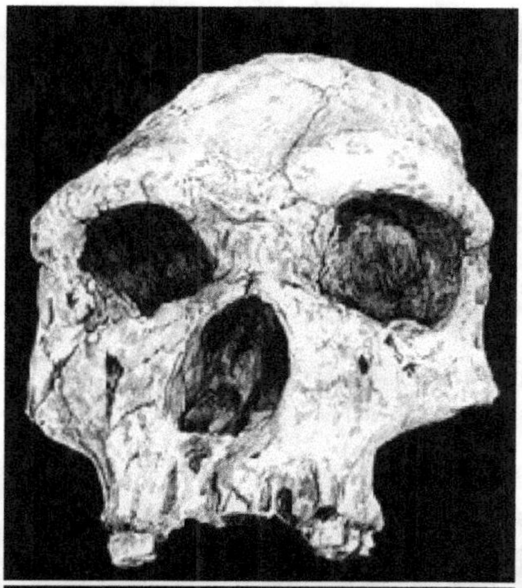

Figure 48:
Homo erectus

Africa, Asia, Europe: 1.8 million to 50,000 years ago. Similar to modern humans, but taller, stronger, and with a pronounced ridge over the brow. Thought to be the first to use rafts to travel the oceans. Tool maker.

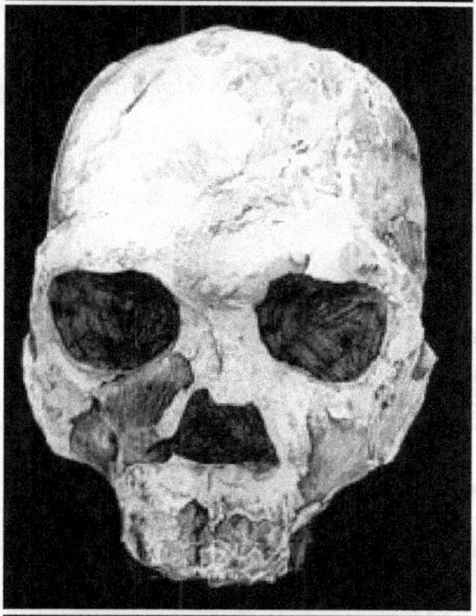

Figure 49:
Homo ergaster

Africa: 1.9 to 1.4 million years ago. Considered an early, exclusively African form of *Homo erectus*. Started making stone tools 1.6 million years ago.

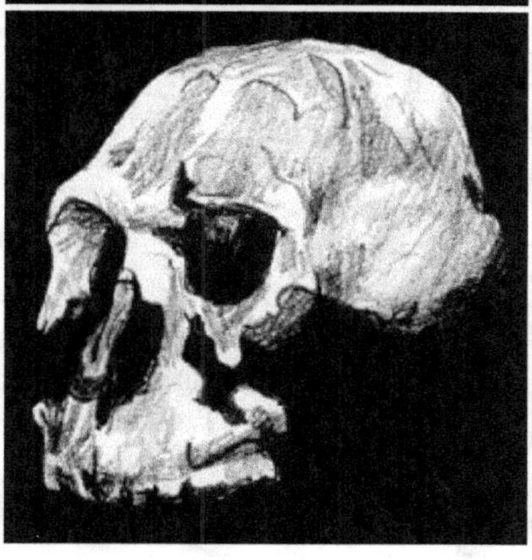

Figure 50:
Homo rudolfensis

East Africa: 2.5 to 1.9 million years ago. Known from a single specimen from Kenya. Bipedal, but apelike, with long arms and hips not adapted to long-distance walking.

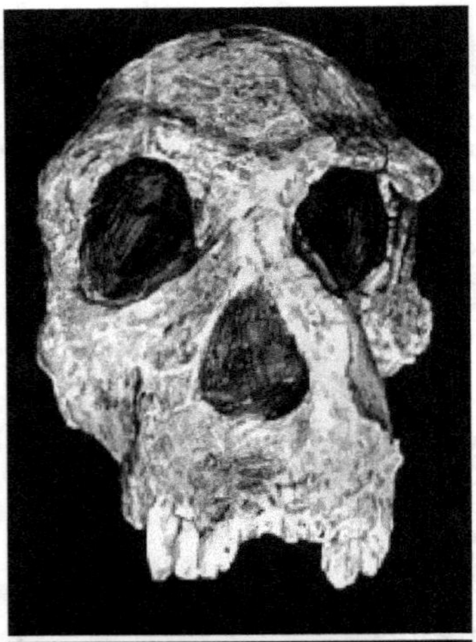

Figure 51:
Homo habilis

Africa: 2.3 to 2 million years ago. Known from only a few specimens; the first of the *Homo* genus known to use stone tools.

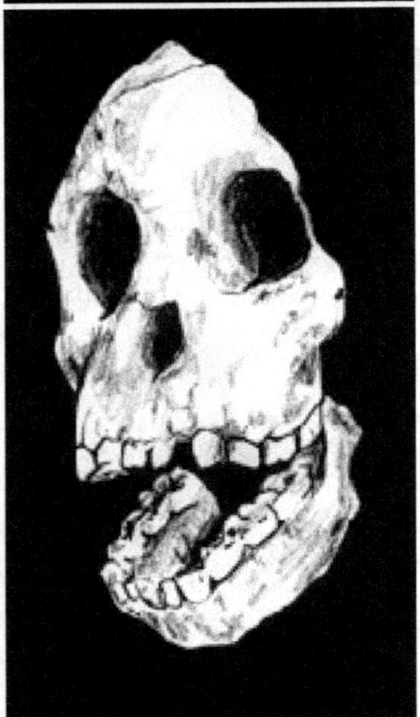

Figure 52:
Australopithecus africanus

South Africa: 2.9 to 2 million years ago. Compared to the older *A. afarensis*, this species has a rounder skull and slightly larger brain.

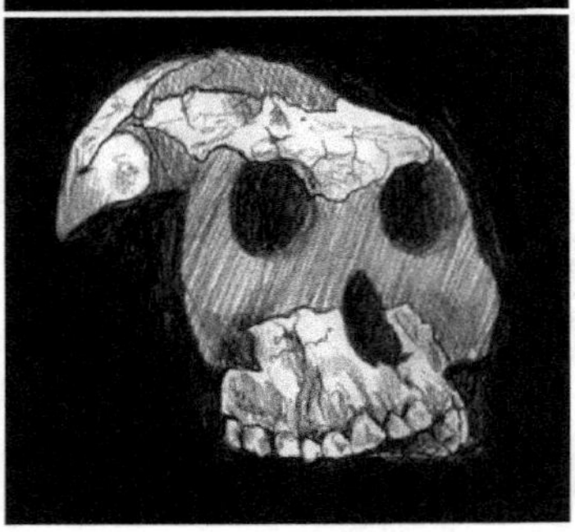

Figure 53:
Australopithecus garhi

Africa: 2.6 to 2.5 million years ago. The first tool-making hominids known. Probably not a direct ancestor in the human lineage.

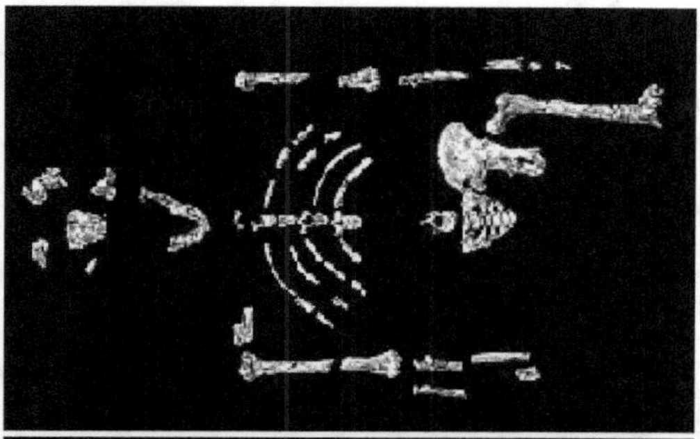

Figure 54:
Australopithecus afarensis

Africa: 4 to 2.8 million years ago. More than 300 specimens known. Bipedal with apelike features, including a prognathic (forward-protruding) face.

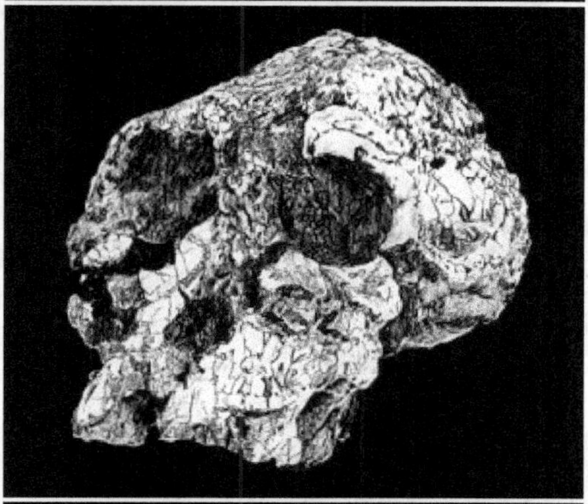

Figure 55:
Australopithecus anamensis

Africa: 4.1 to 3.9 million years ago. Similar to *Australopithecus afarensis*; probably the direct ancestor of this group.

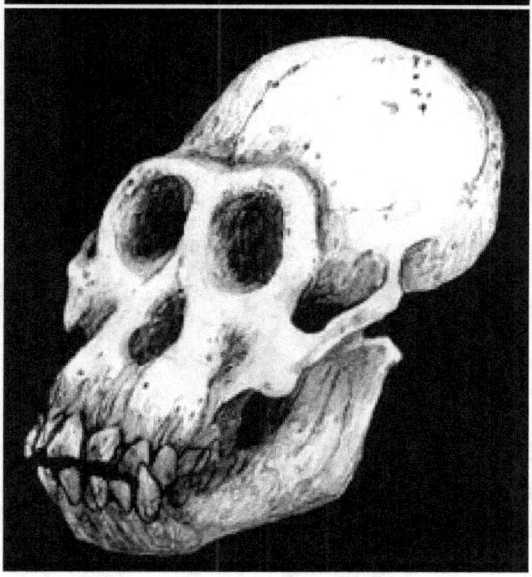

Figure 56:
Ardipithecus ramidus

Africa: 5.8 to 4.4 million years ago. One of the earliest proto-humans known.

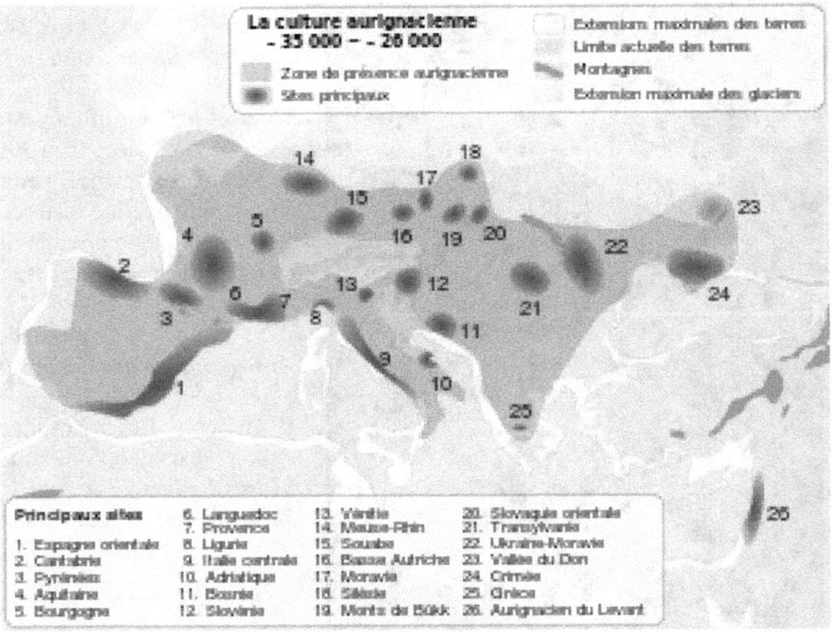

Figure 57: Earliest cultural site of Hss

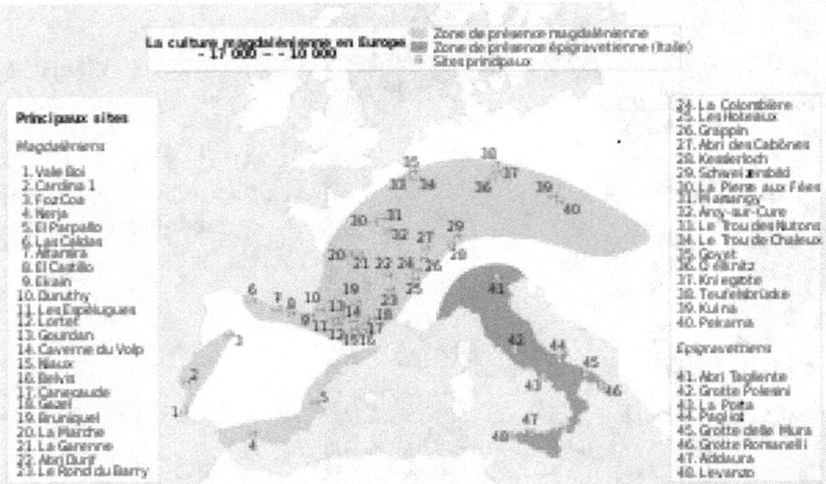

Figure 58: Map of the distribution of the main (Last Glacial Maximum) Aurignacian (c. 33,000-20,000 BP) sites

Above is a map of Magdalenian sites at the transition between the Upper Paleolithic and the Mesolithic (c. 12,000 BP). Magdelenian is the final cultural tradition before the present Holocene seasonal climatic displacement.

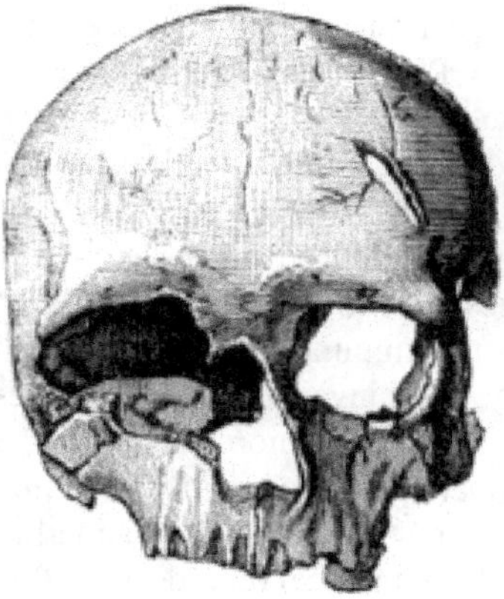

Figure 59: Cro-Magnon 2, the female skull (c.30,000 B.P) found at Cro-Magnon, France (1884 drawing)

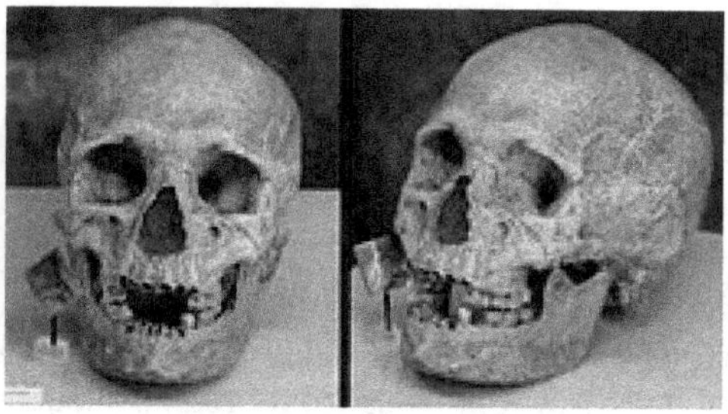

Figure 60: Skull of the Pataud woman

The Abri Pataud shelter shows human habitation throughout the Aurignacian to Solutrean, but was abandoned in the early Magdalenian, about 17 ka. The "Pataud woman" is the name given to the remains of a young woman, about 20 years old, deposited with the body of a newborn child, and is about 21,000 years old. The woman's skull was buried separately, about four meters from the body, lying protected between stones.

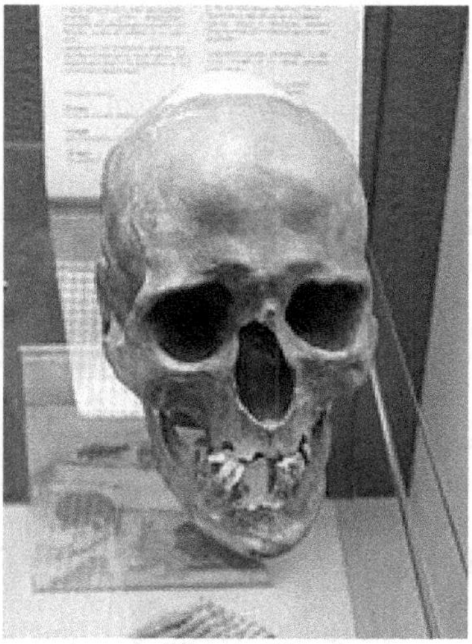

Figure 61: The Chancalade skull

Chancelade man, a short and stocky older man buried in Chancelade, France, was found with Magdalenian tools, c. 15,000 BP.

Denisova hybridizations

In 2008, archaeologists working at the site of Denisova Cave in the Altai Mountains of Siberia uncovered a small bone fragment from the fifth finger of a juvenile member of Denisovans. Artifacts, including a bracelet, excavated in the cave at the same level were carbon dated to around 40,000 BP. As DNA had survived in the fossil fragment due to the cool climate of the Denisova Cave, both mtDNA and nuclear DNA were sequenced.

While the divergence point of the mtDNA was unexpectedly deep in time, the full genomic sequence suggested the Denisovans belonged to the same lineage as Neanderthals, with the two diverging shortly after their line split from the lineage that gave rise to modern humans, possibly as far back as 200,000 BP. Modern humans are known to have overlapped with Neanderthals in Europe and the Near East for possibly more than 40,000 years, and the discovery raises the possibility that Neanderthals, Denisovans, and modern humans may have co-existed and interbred (see Chapter 1). The existence of this distant branch creates a much more complex picture of humankind during the Late Pleistocene than previously thought. Evidence has also been found that as much as 6% of the DNA of some modern Melanesians derive from Denisovans, indicating limited interbreeding in Southeast Asia.

Alleles thought to have originated in Neanderthals and Denisovans have been identified at several genetic loci in the genomes of Cro-Magnons outside of Africa. HLA haplotypes (the HLA complex helps the immune system distinguish the body's own proteins from proteins made by foreign invaders such as viruses and bacteria) from Denisovans and Neanderthals represent more than half the HLA alleles of these Cro-Magnon Eurasians, indicating strong positive selection for these introgressed (hybridized) alleles. Corinne Simoneti, at Vanderbilt University in Nashville, and her team have found from medical records of 28,000 people of European descent that the presence of Neanderthal DNA segments may be associated with a likelihood to suffer depression more frequently. Does this finding throw clarity on the possible shock into extinction suffered by the Neanderthals in their encounter with Cro-Magnon moderns?

> The flow of genes from Neanderthal populations to modern human was not all one way. Sergi Castellano of the Max Planck Institute for Evolutionary Anthropology in Leipzig, Germany, has in 2016 reported that while Denisovan and Neanderthal genomes are more related to each other than they are to us, Siberian Neanderthal genomes show similarity to the modern *Homo sapiens sapiens* gene pool, more so than to European Neanderthal populations. The evidence suggests that the Neanderthal populations interbred with these modern humans possibly 100,000 years ago, probably somewhere in the Near East (Qafzeh Skhul in Israel?).
>
> (Callaway, 2016)

H. floresiensis: living fossil

H. floresiensis, a group of related fossils of beings that lived in their Indonesian habitat, from approximately 190,000 to 13,000 years before present, has been nicknamed *hobbit* for its small size, possibly a result of insular dwarfism. *H. floresiensis* is intriguing both for its size and its age, being an example of a recently discovered species of the genus *Homo* that exhibits derived traits not shared with modern humans. In other words, *H. floresiensis* shares a common ancestor with modern humans, but early on, split from the modern human lineage and followed a distinct evolutionary path. The main find was a skeleton believed to be a woman of about 30 years of age. Found in 2003, it has been dated to approximately 18,000 years old. The living woman was estimated to be one meter in height, with a brain volume of just 380 cm^3 (considered small for a chimpanzee and less than a third of the *H. sapiens* average of 1400 cm^3).

However, there is an ongoing debate over whether *H. floresiensis* is indeed a separate species. Some scientists hold that *H. floresiensis* was a modern *H.*

sapiens with pathological dwarfism. This hypothesis is supported in part, because some modern humans who live on Flores, the Indonesian island where the skeleton was found, are pygmies. This, coupled with pathological dwarfism, could have resulted in a significantly diminutive human. The other major attack on *H. floresiensis* as a separate species is that it was found with tools only associated with *H. sapiens.*

The hypothesis of pathological dwarfism, however, fails to explain additional anatomical features that are unlike those of modern humans (diseased or not) but much like those of ancient members of our genus. Aside from cranial features, these features include the form of bones in the wrist, forearm, shoulder, knees, and feet. Additionally, this *sapiens* hypothesis fails to explain the find of multiple examples of individuals with these same characteristics, indicating they were common to a large population, and not limited to one individual. Van den Bergh et al. (2016) suggested derivation from a population of *H. erectus* that arrived on Flores about a million years ago (as indicated by the oldest artifacts excavated on the island) and rapidly became dwarfed.

A phylogenetic analysis published in 2017, suggests that *H. floresiensis* was descended from the same (presumably Australopithecine) ancestor as *Homo habilis*, making it a "sister species" to either *H. habilis* or to a minimally *habilis-erectus-ergaster-sapiens* clade, and whose line is much more ancient than *Homo erectus* itself. On the basis of this classification, Floresiensis is hypothesized to represent a hitherto unknown and very early migrant out of Africa.

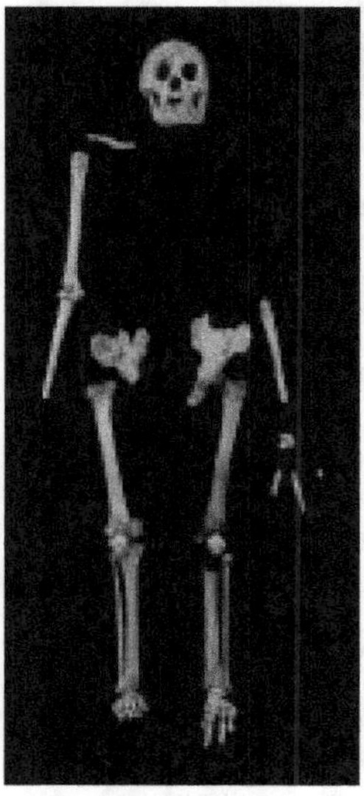

Figure 62: H. floesiensis

This female's head, like the rest of her body, was unusually small. Her brain was about a third the size of ours, but her **species** made and used stone tools, and hunted a variety of animals. The small body size of ***Homo floresiensis*** may have helped the species survive on an island with limited resources.

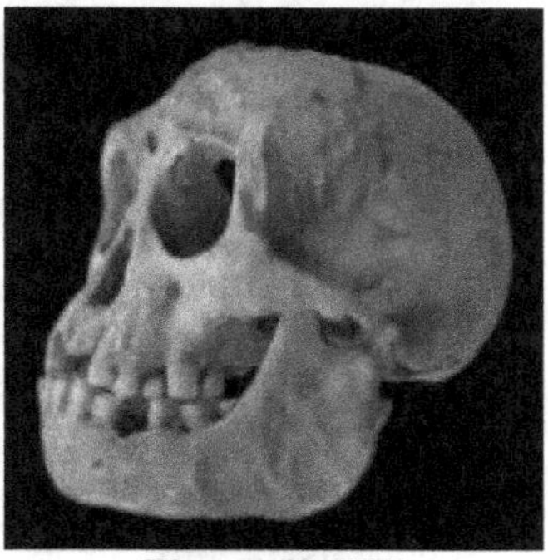

Figure 63: HH

Early european fossil

Probably the most influential site for paleontologists' most recent change in understanding evolutionary processes was the 430,000-year-old ***Homo heidelbergensis*** site of Sima de los Huesos in Spain. At this site, a large community of *hominins* was found to encompass a wider range of skeletal morphology than was previously considered within one species. This has led to a reassessment of "species" in general. What scholars should call the species identified within the site is still under review, concluded in 2016.

First analysis of nuclear DNA from Sima de los Huesos hominins, conducted by Dr. Matthias Meyer of the Max Planck Institute for Evolutionary Anthropology and co-authors, provides evidence of their early relationship to later Neanderthals.

Figure 64: Sima de los Huesos hominins lived in what is now Spain, about 400,000 years ago.

African *Eve* of modern humans—argument pro

The Out of Africa (OoA) or African Replacement Hypothesis is a well-supported theory that argues that every living human being is descended from a small group of *Homo sapiens* (abbreviated *Hss*) individuals in Africa, who then dispersed into the wider world, meeting and displacing earlier forms such as Neanderthals and Denisovans. Early major proponents of this theory were led by British paleontologist Chris Stringer and in direct opposition to scholars supporting the multiregional hypothesis, who argued that *Hss* evolved several times from ***Homo erectus*** in several regions.

The Out of Africa theory was bolstered in the early 1990s by research on mitochondrial DNA studies by Allan Wilson and Rebecca Cann which suggested that all humans ultimately descended from one female: Mitochondrial Eve. Today, the vast majority of scholars have accepted that human beings evolved in Africa and migrated outward, likely in multiple dispersals. However, recent evidence has shown that some

> sexual interaction between *Hss* and Denisovans and Neanderthals occurred, although at present their contribution to ***Homo sapiens*** DNA is considered fairly minor.
>
> (Hirst, 2018)

The above quote is by K. Kris Hirst, an archaeologist with 30 years of field experience. She is the author of ***The Archaeologist's Book of Quotations*** and her work has appeared in ***Science*** and ***Archaeology***.

Out of Africa fossils/archaeology

The oldest known supposed *Hs* site in the world to date is Jebel Irhoud, in Morocco, where the skeletal remains of five archaic ***Homo sapiens*** have been found alongside Middle Stone Age tools. At 350,000–280,000 years old, the five hominids represent the best-dated evidence of an early "pre-modern" phase in ***Homo sapiens*** evolution. The human fossils at Irhoud include a partial skull and lower jaw, which although they retain some archaic features such as an elongated and low braincase, are thought to be more similar to *Hs* skulls found at other presumed Hs African sites.

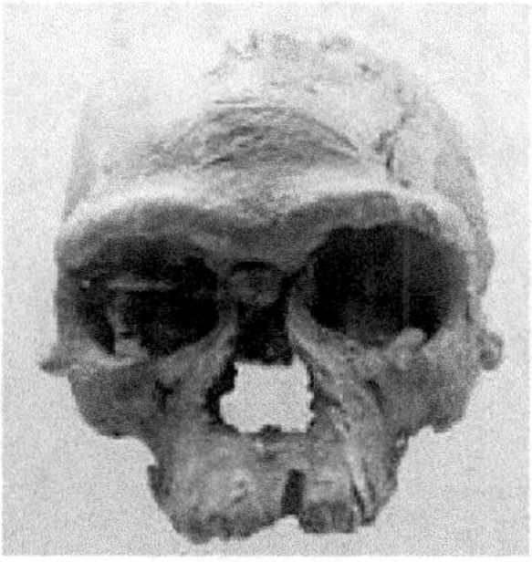

Figure 65: Jebel Irhoud-1, approx. 315,000 years old, Morocco, found with Mousterian tools, possible Capoid racial origins

Carleton Coon hypothesized on the basis of morphological affinities that Irhoud was Capoid (Boskopoid-Bushman) in racial heritage, but because of the associated Mousterian) Neanderthal) Neanderthal cultural affinities, but not on the road to Europe. Fred Smith follows Hublin and Coon in viewing the Irhoud fossils as being transitional sapients but not similar to the moderns in Europe many thousand of years down the line. Schwartz follows the view that Irhoud as well as a number of other North Afrcan candidates are advanced, parallel, but different than the northerly Neanderthals, but no sign of an evolved Irhoud type's latterly advance into Europe.

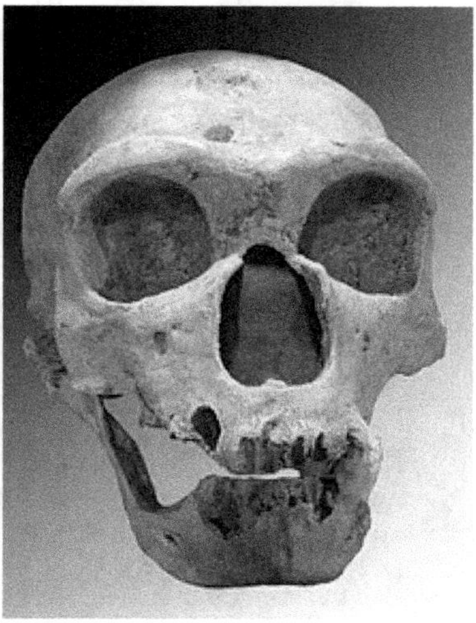

Figure 66: Late Neanderthal skull (La Chapelle-aux-Saints 1), c. 60,000 BP

As developed here in successor chapters, this writer sees the taxonomic affixture *Homo sapiens* to be the exclusive appellation of groups of humans able to speak in a declarative mode. The Neanderthals probably did not, and likely, Irhoud did not either.

Laetoli footprints in Tanzania

The Laetoli footprints were formed and preserved by a chance combination of events—a volcanic eruption, a rainstorm, and another ashfall. When they were found in 1976, these hominid tracks, at least 3.6 million years old, were some of the oldest evidence then known for upright bipedal walking, a major milestone in human evolution.

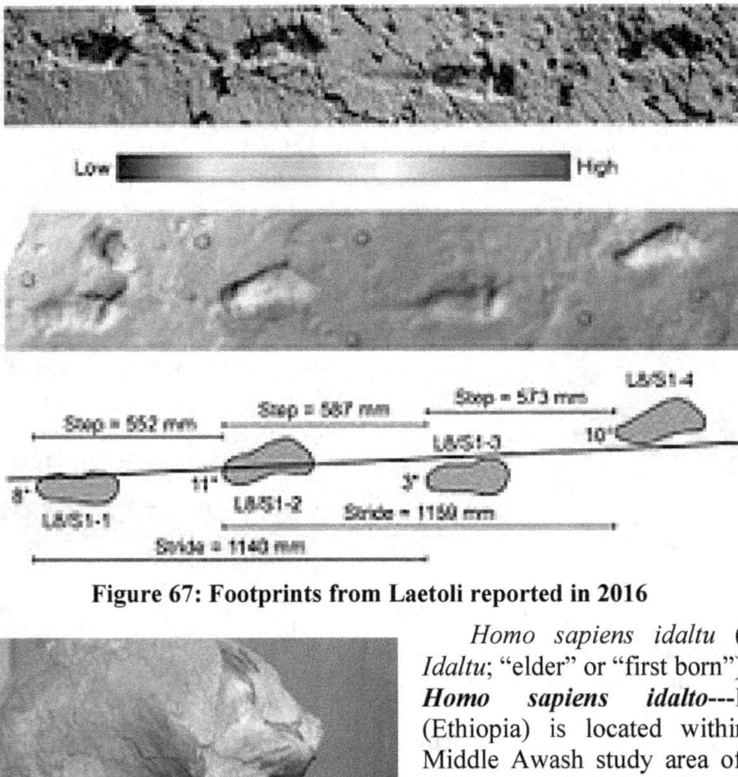

Figure 67: Footprints from Laetoli reported in 2016

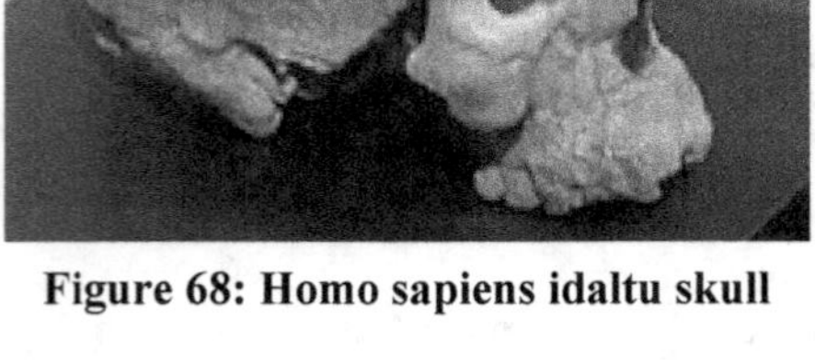

Figure 68: Homo sapiens idaltu skull

Homo sapiens idaltu (Afar: *Idaltu*; "elder" or "first born"), also ***Homo sapiens idalto*---Bouri** (Ethiopia) is located within the Middle Awash study area of East Africa and includes four archaeological and paleontological-bearing members dated between 2.5 million and 160,000 years ago. The Upper Herto Member (160,000 years BP) contained three hominin crania identified as *Hs*, associated with Middle Stone Age Acheulean transition tools, including hand axes, cleavers, scrapers, Levallois flake tools, cores, and blades. *Homo sapiens* *idalto*, 160,000 BP Although not considered *Hs* because of its age, Bouri's Herto Lower Member (260,000 years ago) contains earlier Acheulean artifacts, including finely made bifaces and Levallois flakes; no hominid remains were found within the Lower Member, but it will likely be reevaluated given the results at Jebel Irhoud.

These specimens supposedly represent the direct ancestors of modern *Homo sapiens sapiens* which, according to the "recent African origin (RAO)" or "out of Africa" model, developed shortly after this period.

Recall that Khoisan mitochondrial divergence dated from other proto-sapiens models takes place not later than 110,000 BCE in Eastern Africa. All these African relatives (Capoid/San) seem to be unrelated to Eur-Asian Cro-Magnons, *Hss.*

Take their taxonomic identification as *Hss* with a grain of salt.

Formerly called "Bodo," Klein saw this human as classically *Homo erectus*, while Schwartz saw both skulls as basically archaic *Homo sapiens*, again, **with** no relationship to modern humans in Euro-Asia.

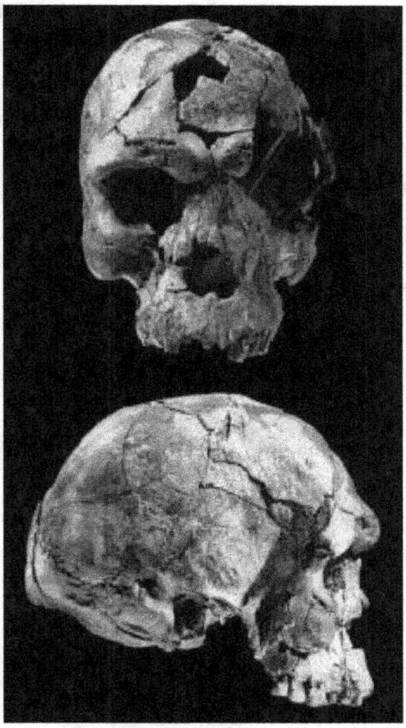

Figure 69: Herty Ethiopia, c. 160k BP

Some new fossils from Herto in Ethiopia, are the oldest-known supposedly modern human fossils, at 160,000-years-old. The discoverers have assigned them to the new subspecies, *Homo sapiens idaltu*, discussed above, and say that they are anatomically and chronologically intermediate between older archaic humans and more recent fully modern humans. Their age and anatomy is cited as strong evidence for the emergence of modern humans from Africa, and against the multiregional theory which argues that modern humans evolved in many places around the world.

This writer comments that the above description and analysis of the generic *idaltu* form reveals the early capoid advances toward *Hs*. But until we have proof of discursive language ability, we ought not describe them as *Homo sapiens*, in any sense.

Three of these Ethiopian skulls were found: most recent, BOU-VP-16/1 is an almost complete adult. The conclusion of the authors is that the Herto skulls "sample a population that is on the verge of anatomical modernity but not yet fully modern." They therefore assigned it to a new subspecies *idaltu* ("elder" in the local Afar language).

Omo Kibish (also Ethiopia) contained the partial skeleton, the partial skull with no face, of a supposed *Hs* who died approximately 195,000 years ago, alongside Levallois flakes, blades, core-trimming elements, and pseudo-Levallois points.

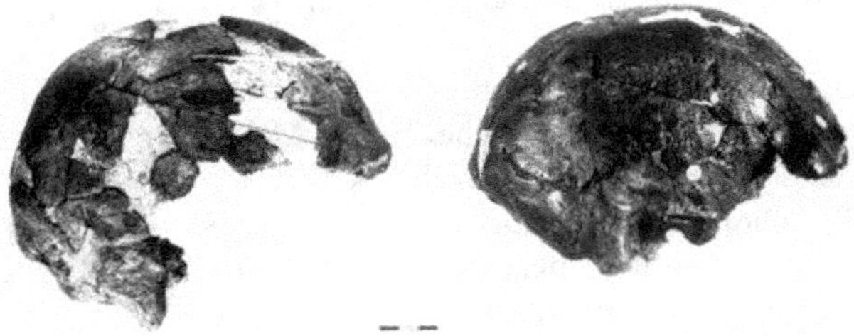

Omo 1 Omo 2

Figure70: Omo 1&2—Ethiopia, c. 195,000 BP

Discussion: Frank Brown and Ian McDougall, along with John Feagle, recently returned to the Omo Valley and re-dated mineral crystals from volcanic tuffs where Omo I and II came from. Why? One reason is that the 160,000-year-old *Homo sapiens idaltu* crania (BOU-VP-16/1, BOU-VP-16/2 and BOU-VP-16/5) from Herto, Ethiopia, dethroned their claim to the oldest known *Homo sapiens* specimens. The team dated the Ar/Ar feldspar crystals from pumice clasts above (Member III) and below layers (Member I) of river sediments that contained the early human bones. The argon-argon dates yielded a 195,000-year-old age. They published their results in *Nature*, **in 2005, under the title** "Stratigraphic placement and age of modern humans from Kibish, Ethiopia."

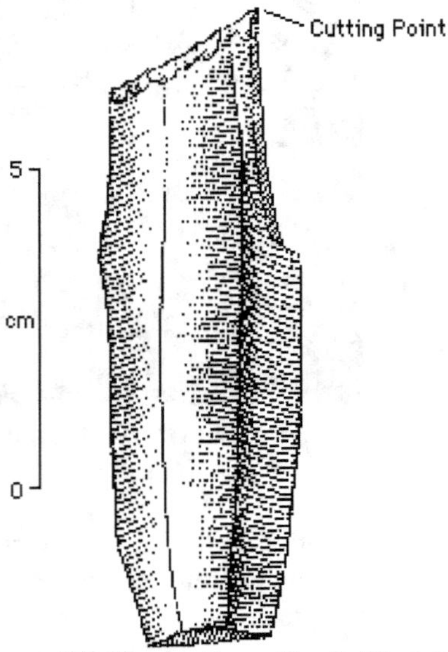

Figure 71: Illustration—Burin Made on a Blade

Brown says that pushing the emergence of *Homo sapiens* from about 160,000 years ago back to about 195,000 years ago "is significant because the cultural aspects of humanity in most cases appear much later in the record—only 50,000 years ago (Europe-*Hss*)—which would mean 150,000 years of *Homo sapiens* without cultural stuff, such as evidence of catching and eating fish, use of harpoons, anything to do with music (flutes), needles, even tools."

The tool (left) shows itself very late in the archaeological literature, regarding the first appearance of modern aspects of behavior such as bone carving for religious reasons, or tools (harpoons and things), ornamentation (bead jewelry and such), drawn images, arrowheads. They only appear as a coherent package about 50,000 years ago (Cro-Magnon), and the first modern humans that supposedly left Africa between 50,000 and 40,000 years ago seem to have had the full set. As modern human anatomy is documented at earlier and earlier sites, it becomes evident that there was a great time gap between the appearance of the "modern skeleton" and "modern behavior." (Sciences.**University** of Utah)

African Eve: "Weak Hypothesis"

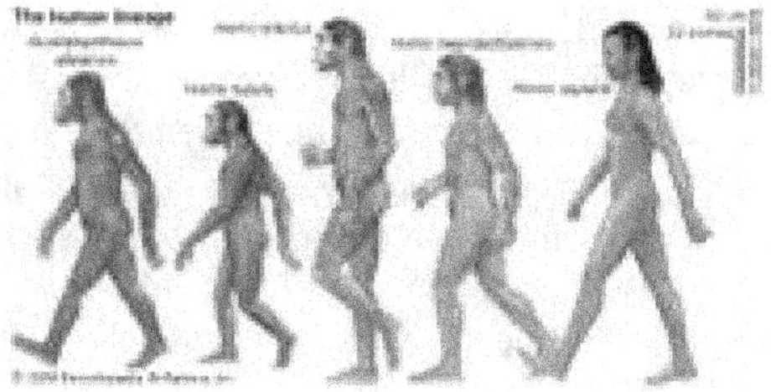

Figure 72

All laboratory or office theories eventually need concrete material confirmation: facts on the ground. Messrs. Brown and Fleagle, in their discussion above, make some good points. Here, we have concrete fossils at the 150,000 years and earlier that are hypothesized to have transitioned to modern human beings at the 50–40,000 level, according to the abstruse molecular calculations of the inventors of this "out of Africa" hypothesis.

Usually, evidence produced by organic molecules from human organic bits are then compared to other organic bits, such as from Neanderthals, ancient and modern humans, all in the general area of the finds which have generated much interesting info. Using easily available evidence about the bio-molecular nature of current human dwellers, we could reach a reasonable sense of the successors and hybridizations of a slice of humanity, from say 50,000 years ago and possibly before.

We have nothing like this with regard to the migratory advance of ancient, northerly African humans into Eur-Asia, as full-fledged moderns with the appurtenances of modern culture. Where are the intermediate fossils in African in situ, just before their colleagues made the jump? As the Japanese say about the Koreans, "they are the Japanese who stayed home," and the Koreans naturally, replied bout the Japanese, "they are the Koreans who *swam* across the seas." Or to quote Carleton Coon's quip, himself no mere novice in this discipline, "the Capoids/Boscopoids/Bushman of Africa are the Mongoloids who stayed home, c. 700,000 BP." Coon believed he had the physical evidence to so theorize.

We do know that at the end of the Pleistocene ice age, c. 40–15,000 BP, the Boskopoid ancestors of the Bushman were in control of north Africa, and their interesting art is everywhere. They even made an expeditionary invasion of southern Spain, repulsed by the Cro-Magnons, who then occupied North Africa. These *Hss* are seen today in the Altai Mountains of Morocco, as the Berbers. All the scientists working on these ancient fossil people described above agree that the resident culture in northern Africa c. 150,000 BP was similar to the Mousterian of the Neanderthals who were extant at that time, mostly in Eur-Asia.

Can we speculate that a hybrid Cro-Magnon/Neanderthal group such as Skhul and Qafzeh (Israel, c.100k BP), they with Mousterian artifacts, possibly embarked on a tourist trip, the reverse of Moses, down the eastern edges of the Mediterranean Sea, then west, all the time sharing some of their genes with the indigenous North African folk?

Take *Homo floresiensis,* a female hominin who may be an australopithecine or very primitive *Homo*, 3–2 mya. Can you imagine these creatures wandering the earth for such a long swath of evolutionary time? From Africa to a small Indonesian Island, having successfully evaded, outwitted the many animal and human predators that could have destroyed them, before her racial/ethnic band of survivors arrived safely 190,000 BP. Up to now, we have had no evidence of such distant relatives of ours leaving their African homeland. This migratory exploration of theirs so far has left no trace. So, there is a possibility that some of the folks discussed above and possibly their more modern descendants will someday show up as fossils or genes in the Danube basin or some such pathway that the Cro-Magnons might have traversed as they entered Europe some fifty thousand years ago, but from where?

As for African Eve and her relatives, some 50–40,000 BP and their supposed move into Eur-Asia, we don't have any concrete physical evidence that these folks, their fossil ancestors illustrated above, made a transition into *Homo sapiens sapiens.* In reality, *Homo floresiensis* didn't modernize, given its long hegira. Certainly, the Africans could have developed to the Neanderthal level and attained the latter's great endocranial capacity. We know that modern *Hss* types do carry in their genoplasim a small percentage of Neanderthal genes, reflecting the universal hypbridization trends in the genus *Homo*. Finally, remember that the North Africans could well have turned into our sapient Bushman, a tragic ending, as with *Homo floresiensis* and the innumerable species of mammals/primates, now extinct, that were our ancestors.

What we have here is a nice-sounding philanthropic equalization of the origins of humanity, such that true *Homo sapiens sapiens* originated not in Eur-Asia but, in Africa. Wonderful idea, *but where's the beef?*

Chapter 7

A Civilization in the Making—1

Evolutionary progressives and conservatives

We have noted earlier in this essay of the dialectic, the *yin* and *yang*, between the evolutionary adventurers and the defensive seekers of the secure. Also explored, is the constant reversals in which former predators become conservative stasis seekers, and defensive types become the pursuers.

Thus, stasis and dynamics take place under a basic umbrella, with two principles of life survival:

1. Counter Carnot's second law of thermodynamics, don't dissipate but keep metabolizing available energy.
2. Use that free energy to reproduce your form of life, so that the ever-new generations mature to their own reproductive fecundity.

Here, the two great uni cellular Domains, Bacteria and the Archaea, are conservatives, and have fulfilled their survival requirements for the eons. The Archaea gave birth to the Eukarya, a progressive dissident who was able to take "advantage" of environmental change, so as to metabolize energy in differing environmental circumstances. Then, it advanced from one unicellular form of life to the multi-cellular, with a nuclei providing the genetic variability to counter external changes in energy availability and still maintain its formal integrity.

Remember however, we multi-cellulars are in ever danger from that ancient uni-cellular predator, bacteria, and that possibly even more ancient and dangerous unicellular enemy—the virus

This new group of progressives departed from unicellular conservatism, to grow into metazoan complexity. They gave birth to new forms of pugilists, fighting against entropy, energy degradation, dissolution. Here, the three Eukaryote Kingdoms, fungi, plants, animals, were created. Here too came many secessions from the living world of the conservative stabilizers. The dialectic was between a living world of yin and yang alternations, always under the evolutionary supreme court decisions by the external environment.

The decisions as to who will survive into tomorrow, and yet be a relatively important element in the hierarchy of life, pursuing the ever restructuring of free energy, is really out of the hands of our own human pretentions. Remember a fact about the jellyfish, a lowly and ancient invertebrate—its sting can sometimes cause vertebrate death.

Conservative forms, though well-integrated into their environments, have means of defense against predatory progressives, i.e: horns, hoofs, sharp edges, to name a few. Importantly, in each case of development in the progressives, great size, untenable physical show boats of sexual attraction, etc., can also go under. Else the environment can throw up new obstacles that render progressive attributes selectively negative, even unto extinction.

More to the home front, the mammals waited in defensive conservatism for over one hundred million years, while the magnificent dinosaurs romped, a canopy of exuberant progressivism. The mammals, or at least one line of them, nurtured the complexity of placental reproduction, endothermy, day/night visual skills, and a brain oriented towards the behavior of defensive avoidance, even miniaturization. Then, changes in climate or geology, provided a chance for growth and aggression, and the placental mammals became the progressives of the animal world.

Who or what will be the next progressive life form? And, will the mammals go quietly into the dust?

From defensive conservatism to offensive progressivism

Defensive adaptive conservatism characterizes the position of the early to mid-evolutionary development of the primates. Their adaptive stance in the beginning stages of the placental outbreak was to be alert, an integration of the senses and the morphology (physical attributes) towards day/night and ground/trees.

From before the Paleocene period, c. 90–50 mya, they were honing this ability early on to feed themselves, and nurture generations of quiet, vulnerable neonates. The brain was the integrator of this sensory information, giving ever greater time/space for this information to be digested and then acted upon. There was less and less momentary instinctual "go and get them" or quick "out of the way" responses in the primate clade. Instead, there was lots of observation by all the senses involved and then instinct/cognitive guided action.

Clearly, it worked such that by the Miocene c. 20 mya, the primates themselves had split up into many forms and pathways of adaptive and selective success. The tarsiesrs, Old and New World monkeys, and the anthropoids, were all lured by tempting adaptational options to spurt in one direction or another, to occupy this world of relatively neutral progressive or conservative adaptive profiles. Probably, the ancestors of the hominins were the most recessive. Individuals "took their time" to make a physical decision. It worked. The brain grew in an orthoselective genetic direction, ever more reticenct in behavioral specialization, its physical size ever more a product of selective success. They then began to transit from defense to offence, pushing their closest primate competitors aside, the hominids, formerly the Miocene apes. These fellows either migrated out of sight into Europe or Asia, or as in the case of the ancestors of the chimpanzees and gorillas, evacuated themselves to the West African jungles.

There were then probably a mass of hominin varieties, given our successive digging up of fossil varieties on the hominin theme, traced to a period of many millions of years ago. Our authorities have finally abandoned the DNA myth of chimp/human separation of 6–5 mya. Now, it points to at least ten million years ago. And it would not surprise this writer if we begin to find odd and suspicious fossils, as he has long predicted, hinting at little preliminary hominin guys and gals at the 20 mya frontier.

The alternate *Hominin* pathway

The early hominins, *Ardipithecus*, *Australopithecus africanus, etc.*, and the new primitive form, *Flores*, exemplify the gracile quality of the early appearing forms, here not to forget the robust line of *Australopithecus* which survived to the dawn of *Homo erectus* and beyond, c. 1.4 mya. These latter erectine humans, from c. 2 mya, were well-established. They were solid, large humans with an awareness of the world to be worked on with self-manufactured tools to do the job, all for the sake of survival. There were already many racial exemplifications, but were always a sturdy, heavy browed and boned people. These oncoming erectines, from c. 2 mya, do not give one the impression that they were defensive animals delicately on the brink.

They were conservatives, apparently here on Earth to stay, hunt, gather, procreate. But while they progressed in the direction of *Homo sapiens*—ever larger skulls, finer tools—they were still not, in our estimation, manipulators of the environment. They hunted, and the animal population still seemed in balance.

Many of the exemplifications discussed in the previous chapter were candidates for promotion to modernity, whether from Africa or Eur-Asia. They remain as well-situated and integrated animal representatives of the rules for natural selective and adaptive survival, to long perdure in their respective terrains. They could still be wandering the lands without a perceived negative impact on nature to this day, whether or not they are incorrectly labeled *Homo sapiens idaltus*, or *Homo* sapiens n*eanderthalis*

The true revolution

Homo sapiens sapiens, Cro-Magnon, appears in Europe c. 45,000 BP. The rich, transformational cultural creativity they represent appears in full-blown maturity a few thousand years later, throughout Europe and Western Asia. These migrants, seemingly from the southeast, are unique physically (morphologically), in tandem with their sharp deviation from the cultural efforts of earlier people.

The only exception are the Khoisonoid/San peoples, sometime associated with the Boskpoid discoveries in South Africa. They can also be seen as a racial sub-species of humankind, the Capoid race. Their colorful drawings throughout Africa, north and south, seem to date back to the 70,000 BP period. We will discuss their status as a possible "out of Africa candidate," below.

Neanderthal and other late Pleistocene hominin skull thickness is 10–13 millimeters. In Cro-Magnons, skull thickness is 3–4 millimeters. Neanderthal males are about 5'4"–5'6" in height; Cro-Magnon males stand at 5'8" –5'11". These two sub species were now living together, at least for a short period, c. 5–20 million years (highly debatable at this point). The Cro-Magnons had an unusual, gracile body formation, were relatively hairless, and with a delicate, soaring skull over its face. However, as with the Neanderthals, they were Caucasoids in terms of skin color, sharing light colored hair, with probably a tendency towards blue eyes.

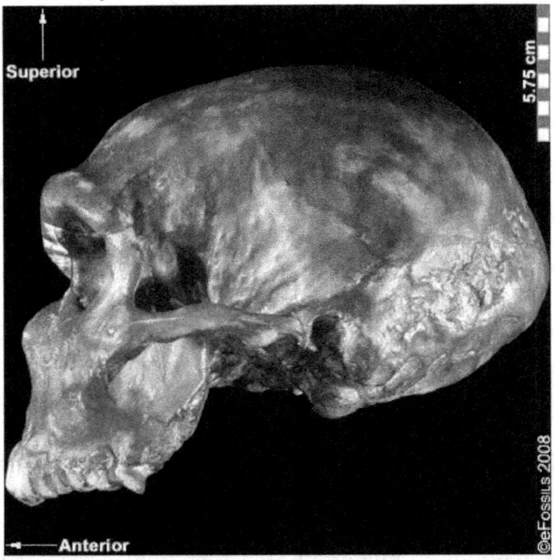

Figure 73: Neanderthal skull, France, c. 50,000 BP

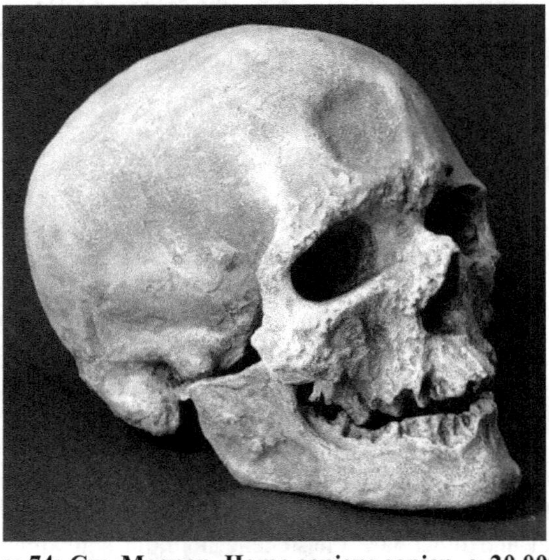

Figure 74: Cro-Magnon, Homo sapiens sapien, c. 20,000 BP

The contemporary perspective of scholars and discoverers of the Cro-Magnon art work is enough to describe the revolutionary impact of these peoples. This cultural efflorescence was spontaneously achieved in their fragile, climatic Ice Age weather setting. It has sent a puzzling, if transcendent message to us from these distant millennia.

"Aurignacian body ornamentation [beads, pendants] explodes onto the scene in southwest France during the early Aurignacian (i.e. between 35,000–33,000 B.P). It appears to have been complex conceptually, symbolically, technically and logistically right from the very beginning. Thus, it is totally unreasonable to see Aurignacian body ornamentation as somehow formative of or transitional to a Magdalenian florescence. This sudden, intrusive, and complex character of the earliest body ornamentation remains one of the greatest explanatory challenges in all of hominid evolution." (White, 1989, p. 385)

Marlise Simons describes the archaeological analysis and writing of Jean Clottes at Niaux in the Pyrenees caves of 13,000 B.P. All the caves and their art are described in books by Clottes: *The Dawn of Art* and *The Cave beneath the Sea.*

Figure 75: Cave art from Lascaux, France (left and right) and Altamira, Spain, (center)

Niaux cave: in the foothills of the Pyrenees

> About one mile into the Niaux cave, there is a crossroad, marked by ancient geometrical signs: an orange-red dart, some stripes, a cluster of red dots. They seem to point to a great gallery where more than 60 paintings are concentrated. 'I have wondered why this hall has become the main sanctuary,' Mr. Clottes says 'Is it because it is at the end of the passage, because it is round, because it has this great echo?' As his flashlight moves along the wall, bison, ibexes, horses appear. They are galloping, fleeing, charging, falling. There is no visible order, no unity of size. Tests have showed the images of this panel were made 13,000 years ago.
>
> (Simons, 1996)

"Sunghir, east of Moscow, is the earliest example (35–30,000 BP) we have of complex ritual funeral traditions in which the status of buried, decorated individuals is not necessarily dependent on their actual achievements, is two children buried in highly decorated fashion with a c. 35–45-year-old adult, also highly decorated. The children's skeletal remains show they may have been afflicted with a disease or genetic impairment. While the adult grave—which contains about 3,000 mammoth ivory beads, pierced fox canines, and ivory armbands—is stunning, that of the manufacture of mammoth ivory spears, ivory disks, and pierced cervid antlers were found with the skeletons."

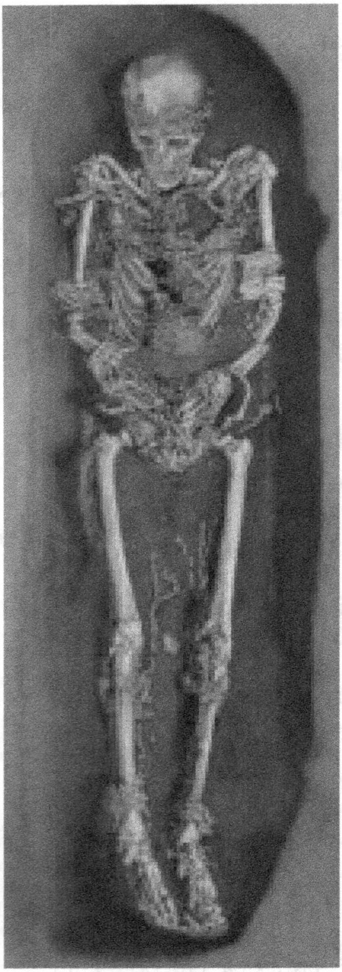

Figure 76: This photo montage shows the highly decorated remains of a 35- to 45-year-old adult male who was buried at Sunghir, east of Moscow. K. Gavrilov/Trinkhous and Buzhilova/Antiquity

"Why this paleolithic burial site is so strange (and so important)"

"Sunghir: More than 13,000 beads were found (which would have taken 10,000 hours to produce). Red ochre, an important ritual material associated with burials at this time, covered the burials." (Surugue, 2018)

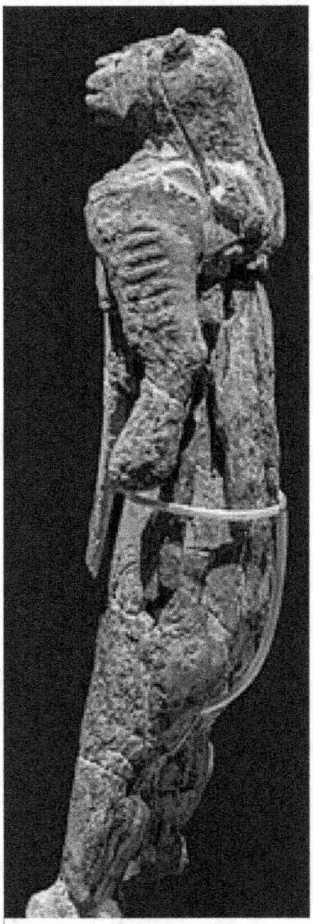

Figure 77: Löwenmensch, a lion-headed mammoth ivory figurine found in Germany, dating to the early Upper Paleolithic of about 35,000 to 40,000 years ago

Also in the Aurignacian period in Germany: the lion-headed figurine (below) is the oldest-known zoomorphic (animal-shaped) sculpture in the world, and the oldest-known uncontested example of figurative art. It has been determined by carbon dating of the layer in which it was found to be between 35,000 and 40,000-years-old, and therefore is associated with the archaeological Aurignacian culture of the Upper Paleolithic. It was carved out of mammoth ivory using a flintstone knife. Seven parallel, transverse, carved gouges are on the left arm.

After several reconstructions that have incorporated newly found fragments, the figurine stands 31.1 cm (12.2 in) tall, 5.6 cm (2.2 in) wide, and 5.9 cm (2.3 in) thick. It currently is displayed in the Ulm Museum, Germany.

The Löwenmensch figurine or Lion-man of the Hohlenstein-Stadel is a prehistoric ivory sculpture that was discovered in the Hohlenstein-Stadel, a German cave in 1939. The German name, Löwenmensch meaning "lion-human," is used most frequently.

Neanderthal and Cro-Magnon contacts—recent to earlier

Paleolithic Stage of Development	Beginning (years ago)	Tool Tradition
Upper Paleolithic (in Europe)	17,000	Magdalenian
	21,000	Solutrean
	27,000	Gravettian
	33,000+	Aurignacian/ Chatelperronian
Middle Paleolithic (in Europe)	75,000+	Mousterian
Lower Paleolithic (in Africa)	1,500,000	Acheulian
	2,500,000	Oldowan

"The Mousterian is defined by the appearance of a method of stone-knapping or reduction known as the Levallois Technique, named after the type site in the Levallois-Perret suburb of Paris, France." (Eren and Lycett, 2012) Traditionally, the Levallois Technique was dated to 300 kyr, helping to define the very beginning of the Middle Paleolithic. Closer analyses reveal that Levallois may have developed from Acheulean tools themselves. A longitudinal study of Kapthurin Formation of Kenya, for example, examined a sequence of Acheulean and Middle Stone Age hominin sites aged approximately 200–500 kyr and suggests that the Levallois developed directly from local Acheulean technologies in a mosaic fashion (Tyron et al., 2006). Levallois involves, very basically, the striking of flakes from a prepared core. A knapper would take this core and trim the edges by flaking off pieces around the outline of one's intended flake. After many uses, the core would acquire a distinctive tortoise-shell appearance [Whitaker 1994]. This technique allows greater control over the size and shape of the flake products, but it also indicates a great leap from the cognitive requirements of previous Acheulean technologies.

[Cassandra M. Turcotte - The Center for the Advanced Study of Hominid Paleobiology (CASHP)]

Figure 78 and 79: Mousterian Stone Tools

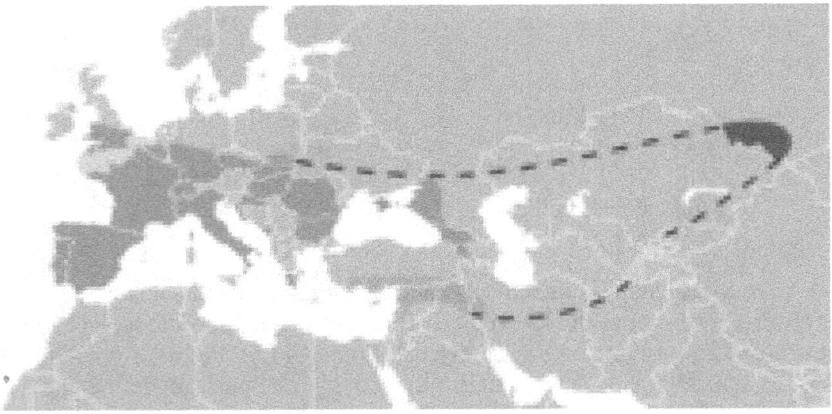

Figure 80: Known Neanderthal range in Europe (blue), Southwest Asia (orange), Uzbekistan (green), and the Altai mountains (purplele) skeletal remains, not tools

Figure 81: Neaderthals above in red—

There is controversy concerning the admixture of Neanderthl and Cro-Magnon cultural detritus. An early cultural tradition, the Chatelperronean has been associated with Neanderthal fossils hinting about a possible early relationship. This and the racial similarities in color (northerly light colored skin to allow for the improved absorption of Vitamin D) give some credence to the shared Chatelperronean hypothesis, (discussed below), also the fact that the Capoid/San in Africa do have a light chocolate skin color and some very different morphological attributes which constitute negative evidence against the "out of Africa," 60–40,000 thesis.

The Châtelperronian is a claimed industry of the Upper Palaeolithic, the existence of which is debated. It represents both the only Upper Palaeolithic industry (possibly) made by Neanderthals and the earliest Upper Palaeolithic industry in Central and Southwestern France, as well as in Northern Spain. It derives its name from the site of *la Grotte des Fées*, in Châtelperron, Allier, France.

It is preceded by the Mousterian industry, and lasted from c. 45,000 to c. 40,000 BP. The industry produced denticulate stone tools and also a distinctive flint knife with a single cutting edge and a blunt, curved back. The use of ivory at Châtelperronian sites appears to be more frequent than that of the later Aurignacian, while antler tools have not been found. It is followed historically by the Aurignacian industry.

Scholars who question its existence claim that it is an archaeological mix of Mousterian and Aurignacian layers. The Châtelperronian industry may relate to the origins of the very similar Gravettian culture. French archaeologists have traditionally classified both cultures together under the name Périgordian, *Early Perigordian* being equivalent to the Châtelperronian and all the other phases corresponding to the Gravettian, though this scheme is not often used by Anglophone authors.

Ana B. Marin Arroyo, et al. (2018), argues with colleagues that there is no cultural association between Neanderthals and anatomically modern humans. She argues that Mousterian, Chatelperronean and cultural Arignacian remains coincided for less than one thousand years. Also, no relevant fossils during this period appear in the Cantabrian region of Spain.

There is evidence that Neanderthals produced the Mousterian and Châtelperronian, and modern humans produced the Uluzzian. If this was the case, then it is possible that interbreeding took place in Europe.

It is also possible (although not proved) that Neanderthals could have borrowed the ideas for the Châtelperronian and symbolic activity from modern humans. Some suggest that this would mean Neanderthals were cognitively inferior to modern humans.

"But it must be borne in mind that we have very few hominid fossils from the period. It is possible that some of the relationships between hominid stone tool industry forms may change when new discoveries are made. This latest study provides a dataset to assess what the implications of these finds may be." [**"Early humans lived with Neanderthal neighbours"** [August 20, 2014 4.16pm EDT Tom Higham and Katerina Douka August 20, 2014 4.16pm EDT *The Conversation*]

1.

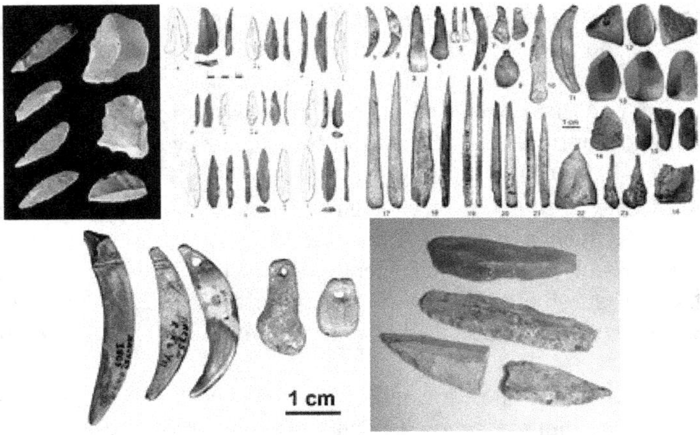

Figure 82: Images of chatelperronian tools

Peştera cu Oase Cave

Peştera cu Oase

Peştera cu Oase (Romanian pronunciation: [ˈ**peʃtera ku o̯ase**], meaning “The Cave with Bones”) is a system of twelve karstic galleries and chambers located near the city Anina, in the Caraş-Severin county, southwestern Romania, where some of the oldest European early modern human (EEMH) remains, between 37,000 and 42,000 years old, have been found.

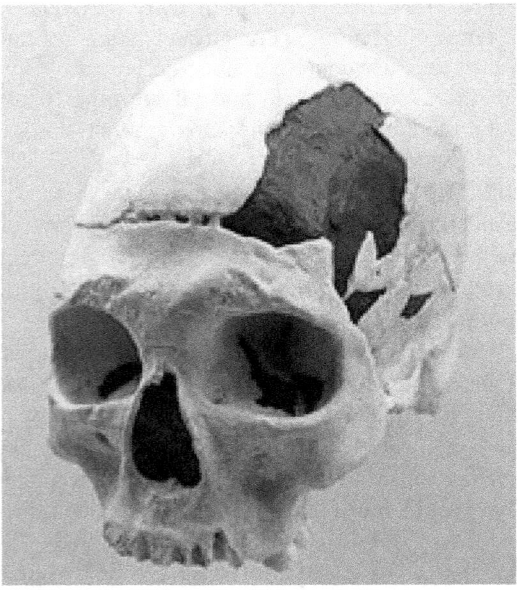

Figure 83: Oase 2 skull

Neither “Oase 2” nor “Oase 1” are particularly close, genetically, to any modern human populations. This evidence increases the speculation that these Romanian fossils may be the product of Neanderthal/Cro-Magnon interbreeding.

An older vision of the relationship between Cro-Magnon and the Neanderthal speaks to the ultimate impact of the coming of the Cro-Magnons and the consequent disappearance of all traces of Mousterian, Neanderthal culture from the Eur-Asian evolutionary scene, c. 40–25,000 BP.

> One can imagine Neanderthals watching from the end of a cliff how the Cro-Magnons hunted in the valley. To the brutish, fear-ridden Neanderthalers these almost hairless giants with jutting chin and towering forehead must have appeared to be superhuman spirits. He saw them using a miraculous technique which was utterly unfathomable and incomprehensible to him. He saw the animal herds from which he for thousands of years picked what he wanted being ingeniously rounded and driven into traps, over precipices or through a narrow gorge into the fire, the clubs and spears of the hunters who jabbered excitedly and made strange cries while they waited for the bag. As he gazed at the elaborate preparations they made for the chase, Neanderthal Man must have felt an eerie terror and sickening weariness at the futility of his own poor efforts. He must have realized that against these—what we would call God-like beings he had not the slightest chance to compete. More and more he must have drawn away from such terrifying creatures, not because they threatened his life, but simply because their miraculous powers were too much for him to contemplate and filled him with primeval panic.
>
> (Simeons, 1962, p. 276)

Out of Africa—San, Bushman, Khoisan, Boskop

There is tangible, physical evidence of an "out of Africa" migration north into Europe and within the molecular time scale of the supposed invasion and conquest. Only, this invasion was soon repulsed and the invaders moved back into Africa. These invaders were the Capoids/Khosinoids/San people, today on the verge of racial extinction in South Africa. At one time they were the dominant clade of humans in Africa and they can make a claim for being the first to be practitioners and lovers of the arts.

The ancestors of the hunter-gatherer Sān are considered to have been the first inhabitants of what is now Botswana and South Africa. The historical presence of the San in Botswana is particularly evident in northern Botswana's Tsodilo Hills region. In this area, stone tools and rock art paintings date back over 70,000 years and are by far the oldest known art.

But the art is not only to be found in southern Africa, it is widespread in the north when the Sahara desert area was a more hospitable ecological zone. Most of North Africa's rock art is today found in the Sahara Desert, in Egypt, Chad, Libya, Niger, Algeria, Mali, Morocco and Mauritania. The richest concentrations are found in mountain ranges such as the Tibesti (Chad), Ennedi (Chad), Akakus (Libya), Tassili-n-Ajjer (Algeria), Air (Niger) and of course Atlas Mountains (Algeria and Morocco). Richest of all is Algeria's Tassili-n-Ajjer where some of the world's most diverse and extraordinary rock art occurs. (Tara Trust for African Rock Art)

Figure 84: Game Pass Shelter in the Drakensberg Mountains in KwaZulu-Natal Province. Here the fammous "Rosetta Stone," a 25-meter-long panel with astonishing images was found and therefore it is seen as the most significant site in Southern Africa. The rock art site is first described in 1915. Other sites close by are Giant's Castle, Battle Cave and Junction rock shelter. Motives found here: shaman figures standing behind eland, therianthrope shaman images as they transformed via trance dance, into an animal in the spirit world. Shaman holding the tail of a dying animal.

Carleton Coon in one of his later writings, 1982, asserted that many of the North African fossils that he worked on in the *Ternefine* Algeria area date back to the 700,000 BP horizon. Others including more recent and sapient like Jebel Irhoud group reflect this Capoid-San- Khoisan, racial or sub-species affinity. They are a very different clade of humans than the then contemporary Neanderthals of Eur-Asia, c. 200–40,000 BP.

Henry Lohte was able to indicate the primitive military roads that crossed the region in primitive ages. The north-south Africa route had been a busy, popular road. Caucasoids or Europeans from the last Ice Age; light-skinned orientals and southern Europeans pushed forward as far as the Transvaal. The remains of human skulls and limb bones that were found at Boskop in the Transvaal, belonging to a primitive human race that probably immigrated from the north. They may have beaten a trail before the last Ice Age incursion, a repeatedly trodden, popular road stretching from North Africa to the Cape, the length of the African Continent. (Schenck,1961; ONAPSA, 1987; Grandvaux, 2004; Vérité, 2011)

The Capoids came north c. 75–70,000 BP, probably during a warm interstadial. They stayed until c.15–12,000 BP. In the east central part of Spain, the area around Valencia, we find both fossil and cultural examples of this habitation, sometimes interspersed with Cro-Magnon Solutrean tools and cultural examples.

In a French museum, there is a copy of a painting made on the side of a cliff, earliest dating, 15,000 BP. It is a picture of a cage on the side of a cliff held up by some kind of rope, else a promontory, probably painted by a Cro-Magnon. It is a derisive racial caricature. Inside the cage is a clearly delineated representation of a terrified Capoid/San male.

The Capoids went back to Africa about this time, eventually to be pushed further south by both oncoming Caucasoids and then the Bantu Negroid people of West Africa. They are today, a fragile remnant. Their potential talents are exemplified in the hybrid heritage of the moral intellectualism of Nelsen Mandela.

Capoid/San genetic presence is still observable in the present populations of North Africa, some say in Phaoronic Egyptian iconography as well. The Cro-Magnons must have crossed the Straits of Gibraltar at some subsequent period, as their ancient presence is still seen in the populace of the Atlas Mountains of Morocco.

One broad study of African genetic diversity completed in 2009, found that Sān people were among the five populations with the highest measured levels of genetic diversity among the 121 distinct African populations sampled. This points to them being the oldest, one of the most distant of languages from other linguistic profiles, both African and non-African heritage. It argues that they may have been one of the earliest true *Homo sapiens* on the planet (Cavalli-Sforza, L.L. et. al., 1993).

Language

No scientifically minded person could assign the taxonomic moniker *Homo sapiens* to a language deprived clade of *Homo* types. There is no argument about *Homo* with regard to fossils in the 2–.5 mya period. But the affixing sapiens (Latin—"wise") to humans who cannot express a declarative sentence? Sadly, there may be a touch of political persuasion to this taxonomic enthusiasm!

Spoken declarative language is the desideratum of what it means to be human. We have long made the distinction between the triggers that make animals, usually placentals or birds, give off vocalizations: fear, predation, defense, sexual communication, etc. and the voluntarism of human vocal expression. For placentals and birds it all comes out automatically, built into the genes when the senses communicate the relevant information. For humans it is built into our morphology, also precipitated by our genes.

Human discourse is, however, unusual. There is choice involved, also, many ethnic languages, same biological appurtenances. Language is a mode of communication underlined by choice of reference, a subject of discourse that is voluntarily sent forth as articulated sound. We call it symbolic communication, rather than instinctive reactions. It is a mark of our cultural capacities. The subjects of our contemporary communication might also controversially constitute the differentia between *sapiency* and *sapient sapiency*.

The evolutionary controversy over language has concentrated on the relationship of the Neanderthals and the Cro-Magnons with regard to the former's ability to both articulate consonant and vowel sound, and then to organize them into sentences of objective, external discourse, i.e. to "speak about something." In the old days of this inquiry into two adjoining sub-species, much of the interest involved the capacity of all the parts of the sound producing machinery to be in morphological place when each sub-species was in the sociality and tool making phase of their evolution.

Today, the argument has evolved in the direction of inquiries into brain structure. The issue is the same. Is the brain in question outfitted with the advanced neurology and connectivity so as to produce more than grunts, panicky or joyful yells, etc.? Several skulls have recently become central in this debate.

Pro-Neanderthal "speakers about the world" say that the Neanderthalers in "no way, are lacking the brain structure."

The others argued for the centrality of the progressive fully modern immigrants "from Africa," c. 50,000 BP. Thus, thumbs down on the indigenous Eur-Asian Neanderthals.

An excerpt from the statements of position of the two groups of authorities gives us interesting insights. First, the pro-unique progressives, Cro-Magnons even if unwilling to be ethnic specific.

The evolution of the faculty of language largely remains an enigma. In this essay, we ask why. Language's evolutionary analysis is complicated because it has no equivalent in any nonhuman species. There is also no consensus regarding the essential nature of the language "phenotype." According to the "Strong Minimalist Thesis," the key distinguishing feature of language (and what evolutionary theory must explain) is hierarchical syntactic structure.

The faculty of language is likely to have emerged quite recently in evolutionary terms, some 70,000–100,000 years ago, and does not seem to have undergone modification since then, though individual languages do of course change over time, operating within this basic framework. The recent emergence of language studies from the paleo-anthropological evidence suggests that the appearance of symbolic thought, our most accurate proxy for language, was a recent evolutionary event. For instance, the first evidence of putatively symbolic artifacts dates back to only around 100,000 years ago, significantly after the appearance on the planet of anatomically distinctive ***Homo sapiens*** around 200,000 years ago.

"A Neanderthal hyoid from Kebara in Israel eventually proved very similar to its ***H. sapiens*** homologue, prompting the declaration that speech capacity was fully developed in adult ***H. neanderthalensis*** This was soon contested on the grounds that the morphology of the hyoid is both subsidiary and unrelated to its still-controversial position in the neck, we are left with objects from the African Middle Stone Age (MSA) such as pierced shell beads from various approximately 100,000-year-old sites (e.g.,) and the approximately 80,000-year-old, geometrically engraved plaques from South Africa's Blombos Cave as the earliest undisputed symbolic objects. Such objects began to be made only substantially after the appearance, around 200,000 years ago, of anatomically recognizable ***H. sapiens***, also in Africa. To be sure, this inference from the symbolic record, like much else in paleontology, rests on evidence that is necessarily quite indirect." (Bolhuis, Tattersall, Chomsky, Berwick, 2014)

The oppositional statement follows: for the Neanderthals, "although new genetic data suggest the existence of some cognitive differences between Neanderthals and modern humans—fully expected after hundreds of thousands of years of partially separate evolution, overall our claims that Neanderthals were fully articulate beings and that language evolution was gradual are further substantiated by the wealth of new genetic, paleontological and archeological evidence briefly reviewed here... full linguistic teaching led to the most efficient technique and good tools, followed by gestural teaching. The reasons are that efficient knapping relies on not easily observable skills, like knowledge about platform angles just under 90 degrees. If one considers all of the cultural skills needed to survive in ecologies from the Arctic to game-poor Mediterranean littorals, it is difficult to argue that Neanderthals lacked complex linguistic codes, capable of communicating about spatial locations…" (Dediu and **Levinson**, 2018, pp. 49–55)

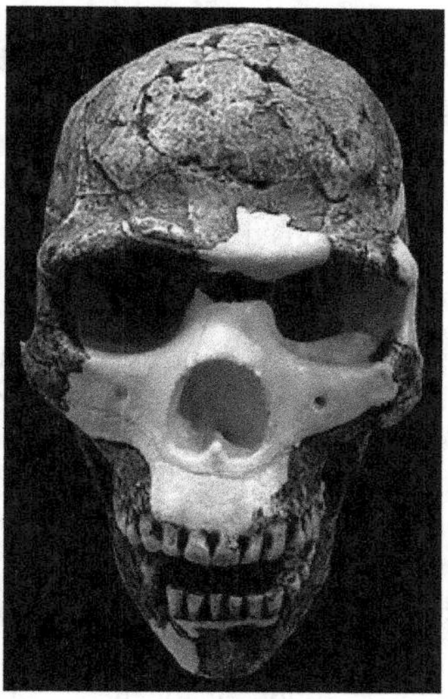

Figure 85: Skhul 5 replica

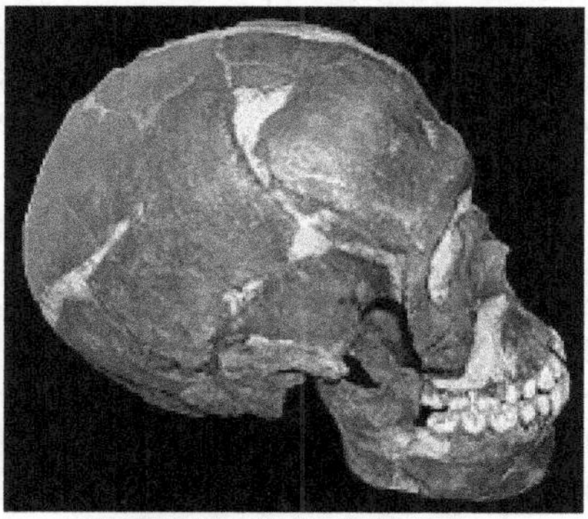

Figure 86: Qafzeh 9 replica

Even though Dediu and Levenson are in the school of those who believe humans becoming modern left Africa long before the 50,000 that the Out of Africa folks have argued for, here, they are in need of greater attention to the complex evolutionary interactions of Neanderthals and Cro-Magnons. How do the hybrids affect their theory, both those who are empirically probables and in those wholly theoretical combinations of the two species?

The Skulls: The *Skhul* and *Qafzah* fossile from Israel, Mt Carmel, have been dated to c. 90,000 BP. They seem to be a mixture of Neanderthal and Cro-Magnon characteristics, with largecranial capcities, but surrounded by Mousterian (Neanderthal) cultural detritus (see also illustration in Chapter 1).

Closer in time as discussed above is *Peştera cu Oase* ("The Cave with Bones"), a system of 12 karstic galleries and chambers located in southwestern Romania, where some of the oldest European early modern human (EEMH) remains, between 37,000 and 42,000 years old, have been found. These various fossils have been found to have significant Neanderthal genetic traces (see illustration above).

Then, there are the *Mladec* (c. 30,000 BP) and *Prdmost* (c. 26,000 BP) fossils whose skulls are from Moravia in the Czech Republic. These skulls are almost identical in character compared to the ancient Israel skulls, Skhul and Qafzeh, only 30,000 BP later, and now surrounded by Cro-Magnon cultural remains (see discussion, illustrations, below).

Now consider the totally Neanderthal fossil remains found in France, Grotte du Renne, Arcy-sur-Cure, surrounded by Chatelperronian technology seemingly of a mixed Mousterian/Aurugnacian tool kit, 50–35K BP. The tools however are sophisticated enough to be classed with later Aurignacian, Cro-Magnon tools. Was this Neanderthal find a part of a larger Cro-Magnon social group? Was he or she buried with special significance, as were the three Russian bejeweled individuals, one adult two youngsters at Sunghir in Russia, c. 35,000 BP(see above illustration). Were the tools of the "cave with bones" a Cro-Magnon gesture, or were these unique findings representative as of a yet hidden Neanderthal intellectual empowerment?

In all these cases, would we not see some sure signs that symbolic meaning could be found in the behavior of the Neanerthals, here indicating language, syntactical expression, the capacity to create propositional references?

The arguments of Tattersall, Chomsky, et al. while aligned to their "out of Africa" allegiances, and then a parallel modernistic replacement of the Neanderthals, are based on a comparative neurological analysis of Cro-Magnon/modern and Neanderthal fossils. It is more persuasive, mainly because language ability, a universal skill embedded in all current *Homo sapiens,* comes "on the heels" of that enormous morphological and developmental revolution in the Cro-Magnon body plan.

This has to be causative, at least partially, in the fact of the Cro-Magnon development of language. All the while, in spite of large Neanderthal endo cranial volume, no widespread and significant developments are seen to be present in their socio-cultural existence as evidenced in the cultural/fossil remains. The caucasoid Neanderthals just seem to be blown-up examples of the archetypical *Home erectus* morphology and and its long-existing cultural technologies.

Language and the Capoids

There is however, a really interesting question about the possible language development in another evolving human clade, the San/Bushman/Boskopoids/ Khoisinoids, a race of humans now at the edge of extinction, yet well-hybridized into other African ethnicities/races. As noted above, the San ancestors, at least at the 70,000 BP level (before the Cro-Magnons entered Europe en mass), had developed art decoration forms, and the use of paints. They kept this skill in their cultural armament until recent times. Did they have the use of language at that time level, 70,000 BP?

And if they did, why not the Neanderthals at 50,000 BP? A clue comes with a carved bone shaped into a harpoon head, and presumably bound to a wooden shaft. These *Semlike* harpoon head(s) date to 90,000 BP, and are possible evidence of Cro-Magnon wanderers into southern Africa. Found in the Central African Republic (Zaire), it is very similar to Magdelenian culture bone harpoon heads dated to the Cro-Magnons, at 17–12,000 BP (both illustrated below). This could be a representative exemplar of the wandering spead of the Cro-Magnon genes throughout the world over a long time range, at least from c. 200,000 BP (see Qafzeh and Skhul also c. 100–90,000 B.P), and why not down into Central and Southern Africa.

(Left) Figure 87: Semlike harpoon head(s) carved from bone, dates to 90kBP Possible evidence of Cro-Magnon wanderers into southern Africa.. Able to harpoon large 150lb catfish with these harpoons

(Below) Figure 88: Similar Magdaeanian (Cro-Magnon) harpoon heads, c.17-12K BP

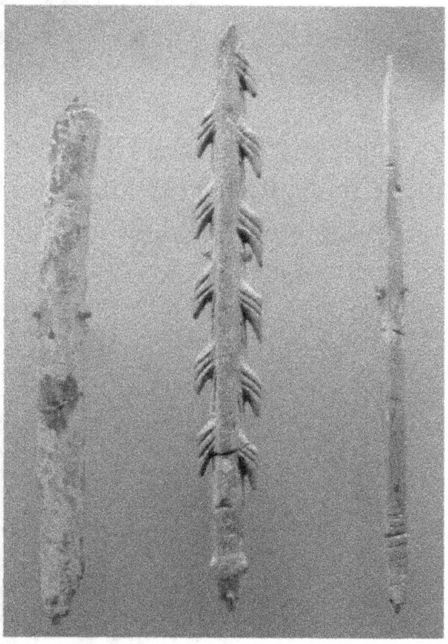

As Schenck and Lohte have argued (see above), the route from Northern Africa to the South was in constant use far back in human evolutionary history, and also probably by the ancestors of the San. All that is needed are a few Cro-Magnons making a typically human wandering journey, around the mid-East, then south, also into East Asia, sharing their technological knowledge and their genes. Those 90,000 BP harpoons snaring a delectable and nourishing 150-pound catfish would easily provide a welcome to the newcomer by the indigenous folks.

The real issue in the study of the origin of language may lie with the origin of these ancient click languages, and how early the Capoid race was sharing language skills, and symbolic behavior in general, with that of the apparent wanderers from the north. Mladeč is an Aurignacian archaeological site with remains of early modern humans directly dating to about 31,000 radiocarbon years. The site contains remains of at least half a dozen individuals, including children. They are the oldest bones that indicate a human settlement, or community in Europe.

Back in Europe

The Czech/Moravian Mladec/Predmosti collection also includes Aurignacian tools and art associated with early modern humans. The remains appear to show some Neanderthal-like features.

Předmostí 3

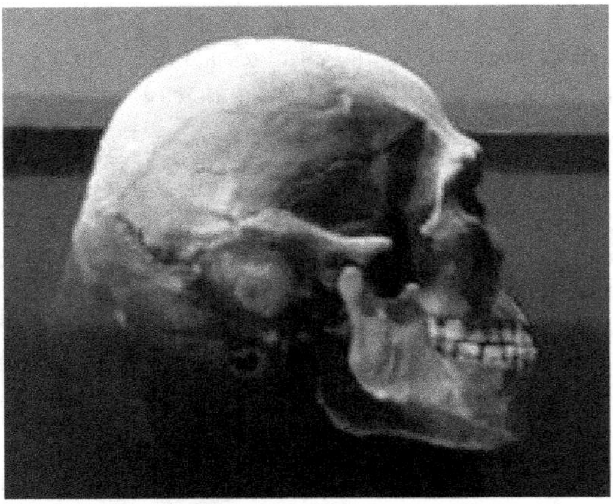

Figure 89: Cast of Předmostí 3

Předmostí 3 was an Upper Paleolithic, *Homo Sapiens* skull find from central Europe in the Czech republic, and is geologically dated as Late Pleistocene.

It was found in Předmostí, Moravia, and dated to 26,000 BP.

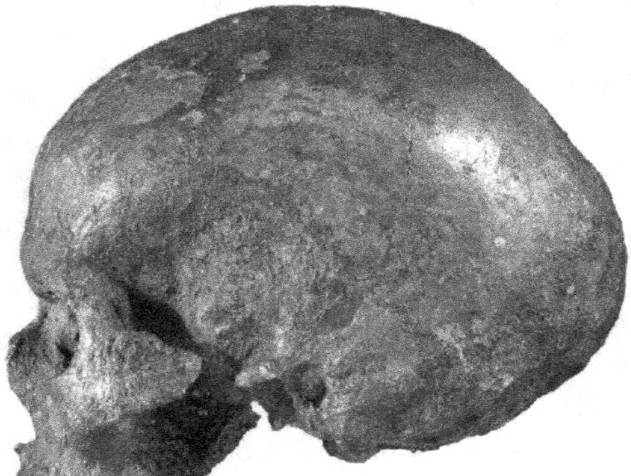

Figure 90

Mladec 1, Moravia, Czech Republic: They are an important archaeological site and probably even the oldest, largest and most northern settlements of the Cro-Magnon people in Europe (Cro-Magnon lived here as long as 31,000 years ago).

Cro-Magnon Habitation

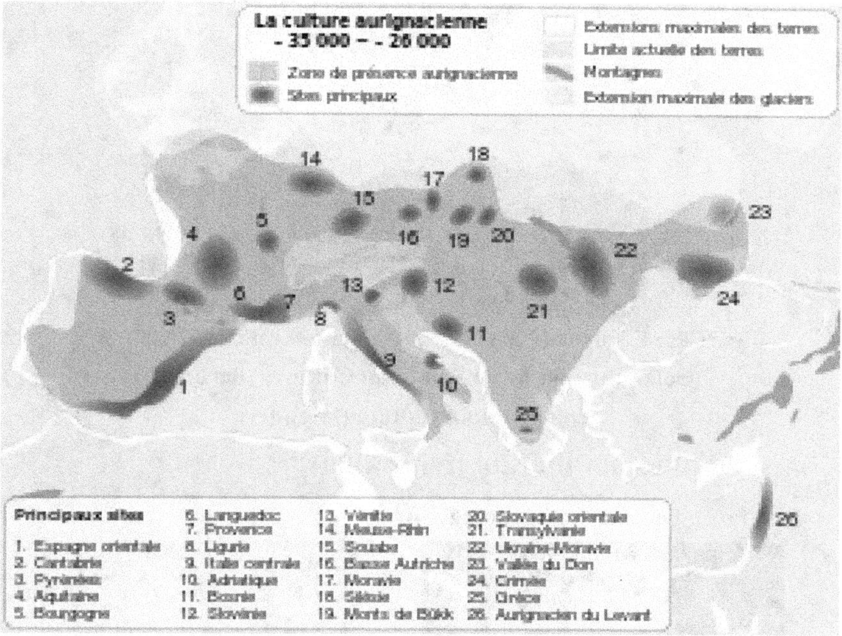

(Above) Figure 91: Map of the distribution of the main pre-LGM Last Glacial Maximum, Aurignacian sites.

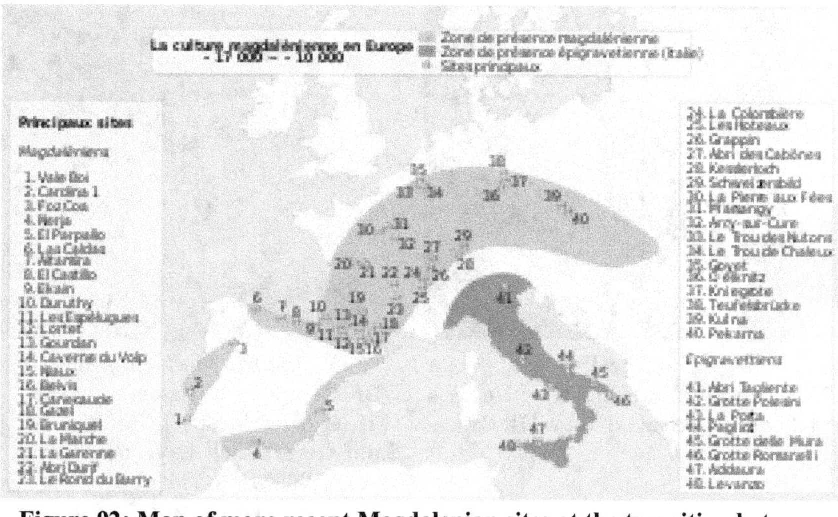

Figure 92: Map of more recent Magdalenian sites at the transition between the Upper Paleolithic and the Mesolithic, c. 16k BP-12k BP.

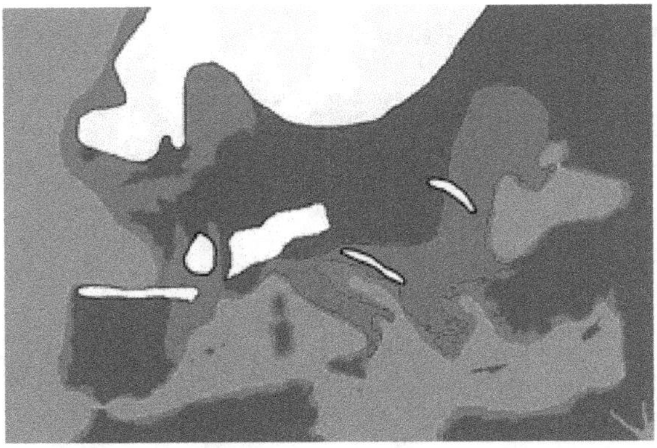

Figure 93: European Last Glacial Maximum refuges, 20,000 BP.

Solutrean and Proto Solutrean Cultures (dark red)

Epigravettian Culture (lavender)

A symbolic, non-literate civilization

Sex, Love, Fecundity, Technology, Art

Figure 94:
Venus of Laussel, France, an Upper Paleolithic (Gravettian, c.17k BP)

Figure 95: The Venus of Brassempouy is preserved in the Musée d'Archéologie Nationale at Saint-Germain-en-Laye, near Paris, c. 22,000 BP. Its height is 3.65 centimeters.

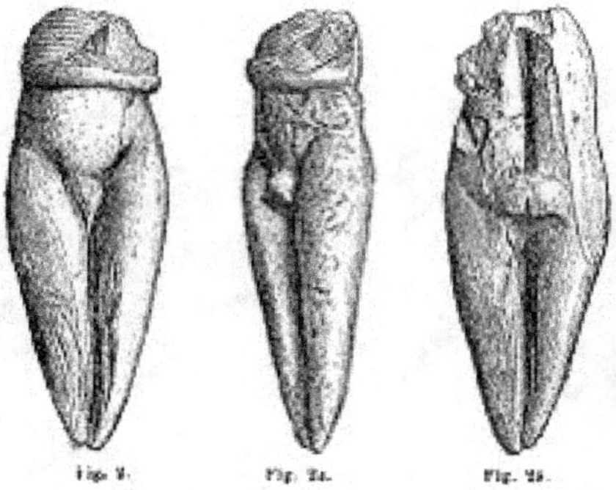

Figure 96: "La figurine à la Ceinture" (The Figurine with a Belt), one of several Venus figurines discovered alongside the Venus of Brassempouy

Figure 97: Cave wall painting, Lescaux, France

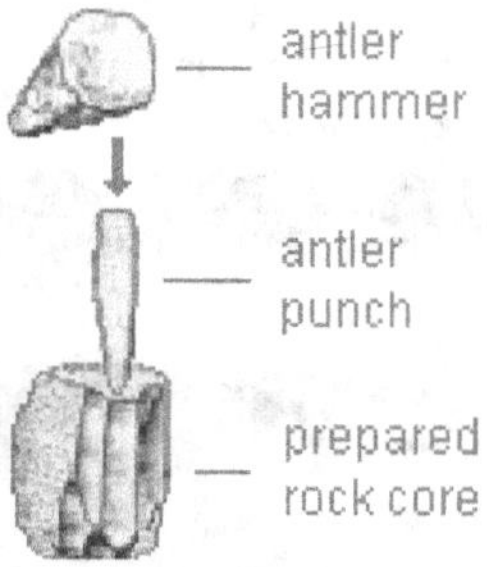

Figure 98: Punch flaking technique used to make blade flakes

Figure 99: Blade flakes and the "spent" core from which they came

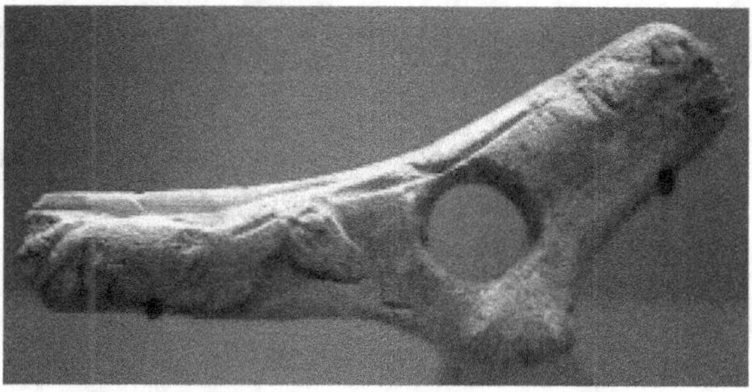

Figure 100: Ornamentally carved spear shaft straightenermade of bone from La Madeleine Rock Shelter, France (late Magdalenian Tool Tradition—12,500 BP).

Figure 101: Bone flute (Upper Paleolithic) (Geissenklösterle Cave, Germany

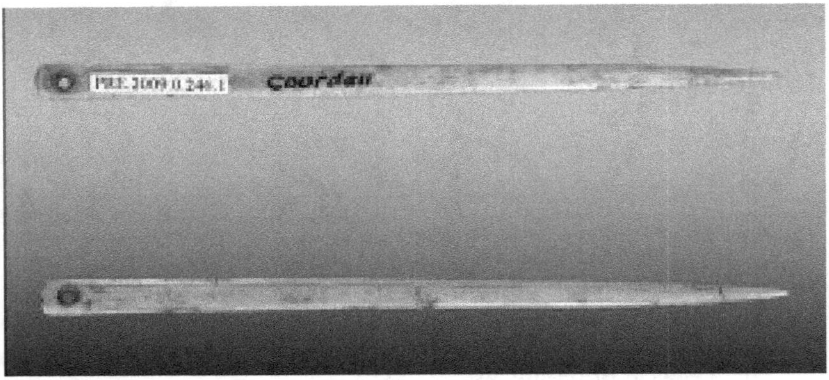

Figure 102: c. 13,000 BP—Magdalenian bone sewing needles from Gourdan Cave, France

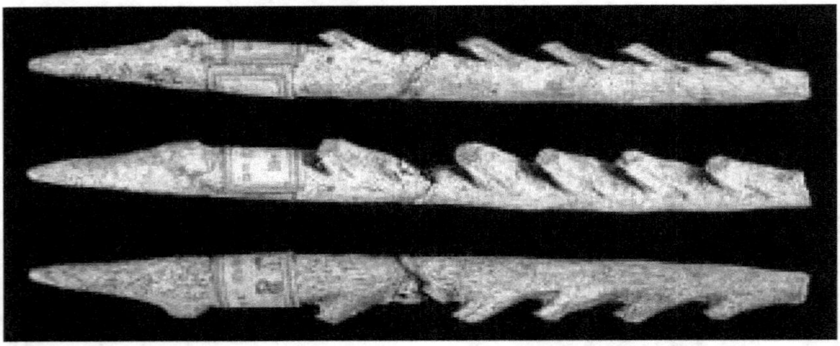

Figure 103: Magdalenian bone harpoon points with barbs (their broken tips are on the right)

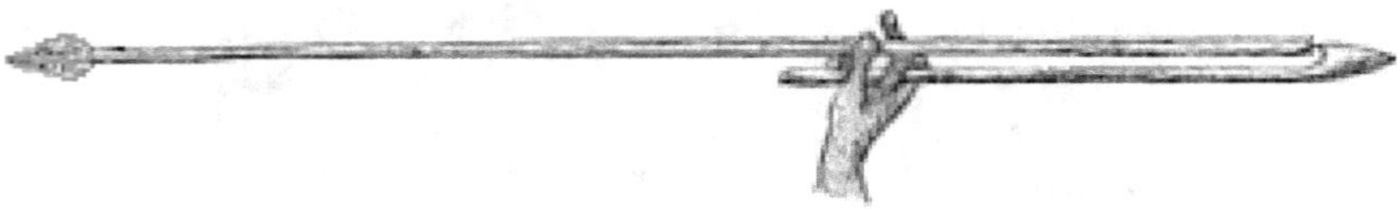

Figure 104: Spear thrower Major at-a-distance hunting device, invented by Cro-Magnons

Chapter 8

A Civilization in the Making—2

The progressive road?

In the primeval taxonomic domains, the Archaea, Bacteria, the early dissident Eukaria, the life rhythms were: birth, metabolism, reproduction, death. Not much choice from beginning to end. Response to the external world was: you may die, down the road, another will live. There was much biochemical variability for the moment, and not much of an individual future.

This changed as the Eukarya distanced themselves from the Archaea. A more complex nuclear cell developed with individual flexibility to metabolize diverse food availabilities, to sexually reproduce and produce new variants, again more able to find their way around an ever variable, changing environment.

And so we have the four derivative kingdoms of life making their play for immortality, protists—paramecia; plants; fungi; and animals, the latter with more optionality for life under change, movement, and predation of members of their own kingdom and the other three. Within the kingdoms, life and death, metabolization and reproduction continued this basic equation. Even so, within the kingdoms the dialogue between progressivism and conservatism always reshaped the dynamics.

Genetic change and consequent phenotypic (surface appearance) change, in structure (morphology), and in adaptive behavior, presented themselves as opportunities awaiting the ultimate adaptive move. Then and inexorably, the ultimate selective decision was made by mother nature. With the passing of time, constant external challenges to the survivability of all kingdom members, most lines of life, many, once progressive in their thrust forward into adaptive opportunism, became conservative. They may have found a plush metabolic position and grabbed onto it, becoming ever more fixed into this opportunity, hopefully for immortality.

It has worked out for many. They have been around for a billion or so years. Most others became extinct, unable to resource the genetics of change. Others, often outliers of the main line of the phylum, class, order, genus, found in the challenge of the external world an opportunity for progressive change. In the animal kingdom, we find this dialectic with the progressive ones finding ever more environmental room to adapt to change. The chordate/vertebrate body and brain plan become one of the leading life forms, as the billions of years played into the maturity of this kingdom's experiment with Darwin's progressive trinity, change (now genetic), adaptation, natural selection.

Conservative *Hominins*

With the appearance of the placental mammals we see a new vertebrate response to the ever changing environments of our planet. There is an almost inevitable development of a line of moving about animals responding to the adaptive and thus survival requirements of change within the generations. Thus, a brain was selected by nature that allows for much environmental discrimination, freedom of decision making without the close discipline of given instinctually determined behavior.

If the genetic profile of the line of animals worked, the mutational trends would become orthoselective. That means there would exist a dialectic between genoplasm and phenoplasm, such that the adaptive profile of the line, here the placentals, would produce mutations, variations accommodative to the required adaptive challenges of the particular line of animals. Here, the brain would become ever more powerful as an aid to procure food, metabolize energy, allow for the successful maturation to puberty of the young, and usually facilitate demographic expansion.

The placental guys and gals who held back from being too successful and specialized into an adaptive niche were the primates. And after some 40 to 70 million years the hominids, anthropoids in general, and the *hominins*, our own anthropoid line began to make their adaptive play. Here, as always, former progressives were turned by opportunity into conservative specialists. But, as always for many success stories, they were really awaiting their extinction moment in the setting sun.

This is basically the story of the evolution of the *hominins*. Like a fast expanding bush, they were hustling into a variety of adaptive niche to fulfill the Darwinian drama. We had gracile australopithecine, robust paranthropus cousins. Eventually, we encounter a wide range of robust erectines all throughout Africa, Europe and Asia. Mostly they were migrants out of Africa, as with the gibbon and orangutan many millions of years before.

Then there are the so-called transitional *Homo sapiens*. The taxonomy of these candidates is controversial. Were they early sapiens or just still large-brained erectines, all with cultures Oldowan, Achuellian, Lavellois, and maturing with the Mousterian. The latter is mostly identified with the Neanderthals. All these transitionals share the large, thick-skulled, heavy-boned characteristic of those living forms eventually headed for the dust bin of evolutionary history.

Robust Winners

This expected mature and final stage of human evolution, the transition from gracile to successful robust stage seemingly lasted for several million years For example genetic analysis of the Spanish Sima de los Huesos fossils [Meyer et al. 2016] seems to suggest that *H. heidelbergensis* in its entirety should be included in the Neanderthal lineage, as "pre-Neanderthal" or "archaic Neanderthal" or "early Neanderthal," while the divergence time between the Neanderthal and modern lineages has hypothetically been pushed back to before the emergence of *H. heidelbergensis*, to about 600,000 to 800,000 years ago.

Some of these successful erectines

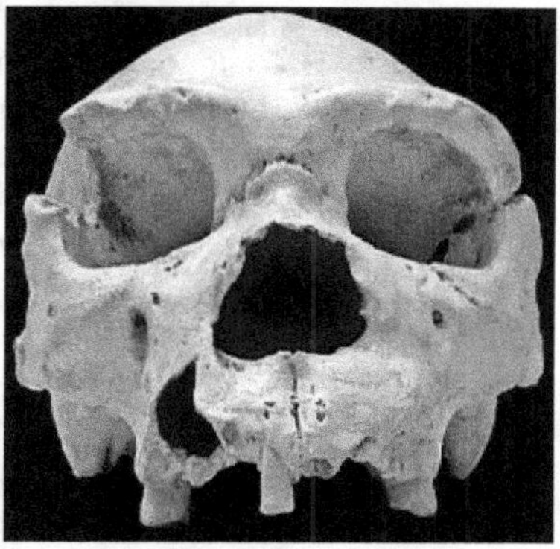

(Left) Figure 105: The Homo heidelbergensis *Cranium 5*, one of the most important discoveries; its nearly complete mandible was only found years later. Northern Spain, c.400K BP. This kind lived as long ago as 700,000 BP.

Heidelbergensis, c. 700-400,000 BP, now possible predecessor of Neanderthals, split off by 400,000 BP. See also, the *Homo dalii*, a Chinese fossil from 300,000 BP. This East Asian erectine had a morphological distinctiveness from its Eur-Asian and African erectines confreres. Some of its unique character is retained in modern Chinese skulls.

Some anthropologists, notably Chinese representatives, cite the characteristics of the Dali cranium and other similar Chinese fossils of that era as evidence for genetic continuity into modern *H. sapiens* today, as Dali's traits are commonly found in modern Chinese *H. sapiens* populations. In turn, it is often argued that modern Chinese humans did not evolve in Africa, but instead evolved in China from a separate lineage of H. erectus., c330-. 260k BP

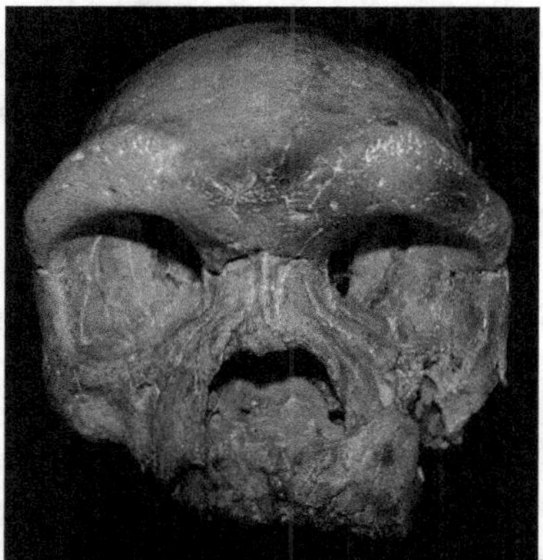

(Left) Figure 106: What can the Dali skull tell us?] c.300K BP

(Bottom) Figure 107: Comparison of Dali and Jinniushan archaic Homo sapiens?

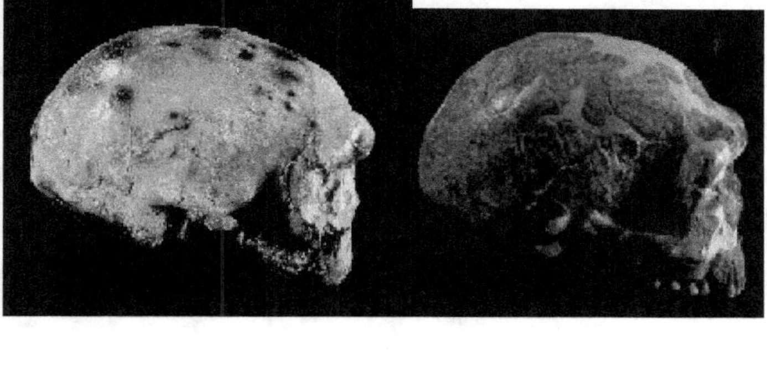

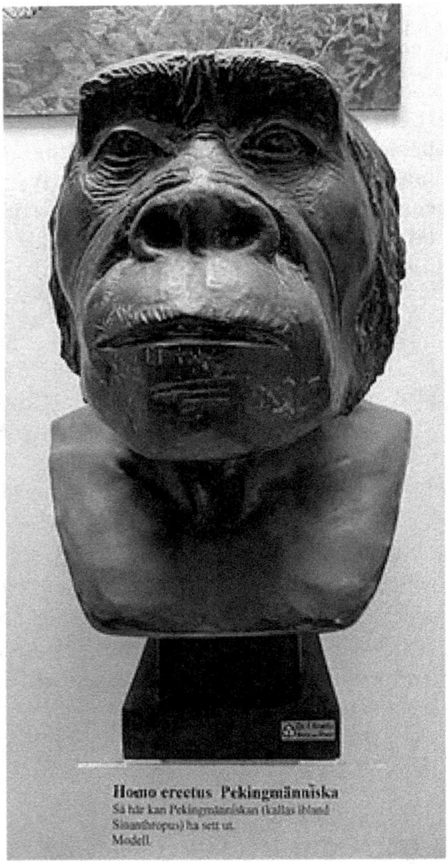

Figure 108: Peking *H. erectus*, c.700k BP

The Jinniushan female is part of a controversy over whether she is erectus or early sapiens, and whether they are true pre-Chinese or part of the Eur-Asian complex. The female skull is very large. Both fossils show body trait possibilities of extreme robusticity, large humans possibly adaptive to climate, but also characteristic of attaining successful food/and reproductive achievements. Appearing in the Beijing Man, c. 700,000 BP, from East China, are straight line brows over the eyes. The above Chinese skulls have brows over each individual eye. Puzzle?

No out of African 50,000 BP world wide wipeout here, as claimed by the out of Africa cum modern humans..

Both *Homo erectines* and Neanderthals used the same crude stone tools. Archaeologist Richard G. Klein, who has worked extensively on ancient stone tools, describes the stone tool kit of archaic hominins as impossible to categorize. It was as if the Neanderthals made stone tools, and were not much concerned about their final forms. He argues that almost everywhere, whether in Asia, Africa or Europe, before 50,000 years ago all the stone tools are much alike and unsophisticated, e.g. a million year old line of coming to be conservative evolutionary lines. [Klein, R. 1990 The Human Career: Biological and Cultural Chicago: University of Chicago Press]

However, as a possible originating heritage for the later Cro-Magnon "spear throwers," c. 15,000 BP, the following discovery is interesting and enigmatic. An archeological site in Schöningen, Germany, contained eight exceptionally well-preserved approximately 400,000-year-old spears for hunting, and various other wooden tools. Also, 500,000-year-old hafted stone points used for hunting are reported from Kathu Pan 1 in South Africa, tested by way of use-wear replication. This find could mean that modern humans and Neanderthals inherited the stone-tipped spear, rather than developing the technology independently.

The Schöningen Spears are eight wooden throwing spears, dated to before 300,000 years ago, discovered between 1994 and 1998 in the open-cast lignite mine, Schöningen, county Helmstedt, Germany, together with thousands of animal bones. They are regarded as the direct evidence for active hunting by *H. heidelbergensis* (pre-Neanderthals).

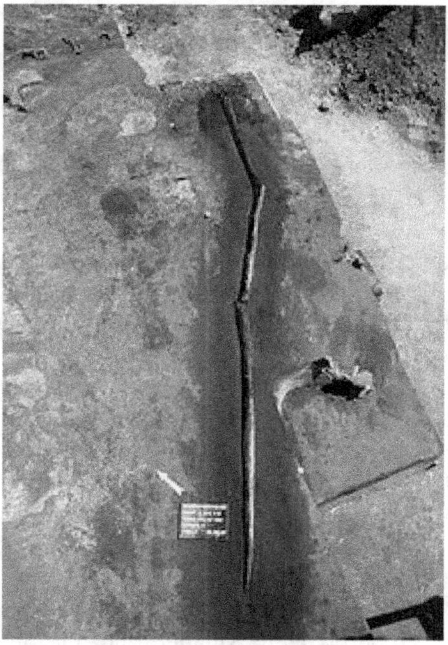

Figure 109: A Schoningen spear *in situ*

The social semi-instinctual patterns, of these transitional erectines seems not to have radically evolved. Evidence is, as Klein argues with regard to Acheulian, Lavallois, Mousterian tool kits. However, these hominins show some freedom from instinct. Heidelberg spears etc. seem to be an advance on even later Neanderthal culture. In general, these *Homo erectines* reveal their generationally evolved and embedded conservatism.

There is another evolutionary tale of possibly pre-Neanderthal Heidelbergensis cultural achievements. Could there have been an environmental or other crisis, c.700-500,000 BP that separated groups of Heidelbegensis northerners or southerners? Could one group have chosen the defensive adaptational route over a broad geographical panorama, leading to the Neanderthals soon becoming widely spread?

Another group, possible deviants were thrust into more challenging, even deadly environments creating a sharp challenge to the survival of the group, creating a few outlier bands surviving, through pure intensive brain and will power. Could this have been the funnel through which a few pre-Cro-Magnons followed. Invisible for so many thousands of years eventually, in true evolutionary outlier fashion they were yet to show their vitamin potential. The quiet before the storm.

Baby Faced Predators: Heterochrony

In the preceding chapter 7, we discussed the cultural relationship between the oncoming Cro-Magnon, *Hss*, and the other ethnic and ancestral hominins that they encountered. The cultural achievements, art, technology, their economy and the sense of individuality, beauty, reflect the revolutionary character of these mammals, primates, even anthropoids on this world stage. This, at that time, the late Pleistocene ice ages, they were traveling through a very tough and challenging environment, then, to make their presence felt.

But the cultural products and the competent adaptive behaviors which were reflected in their absorption into the moderns tells us about the dangerous surface events, perhaps even a clue into the modern extinction of the Tasmanians. The ongoing demographic shrinking of the Capoid/Khoisonoids, they, having been driven south into deepest Africa, augurs a new and tragic extinction. Their genes, however, will live on.

This new, Cro-Magnon revolutionary culture did have its origins in a new arrangement of genes which manifested itself in the physiology/morphology, phenotypic presence of these people. The puzzle of how and why the Cro-Magnons came to be and with such ultimate devastation to the status quo of the two life kingdoms of the world, plants and animals, needs examination.

The concept of *heterochrony*, changes in the time rate of individual *ontogeny*, development of the individual from birth, and the accompanying concept of *neoteny*, the retention of early age characteristics of individuals of the species into maturity, came to the forefront.

Concepts such as *paedomorphosis*, maintenance of infantile characteristics of the individual into adulthood, and *peramorphosis*, the extension of the developmental stages of the individual during his/her ontogenetic development, were prominent theoretical concepts in the attempt to explain what had happened in this portion of our evolutionary history.

Neotony

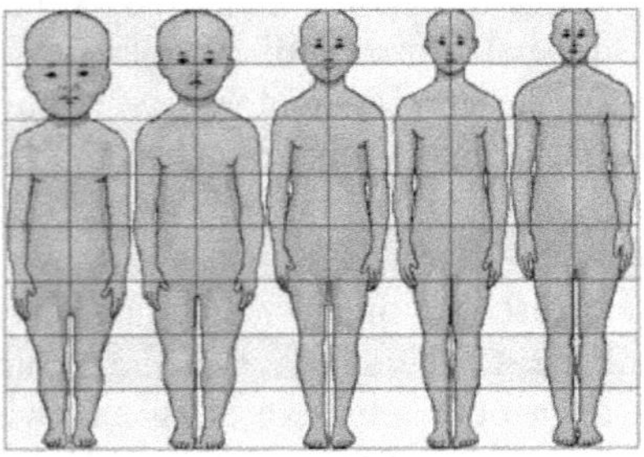

Figure 110

The human head becomes proportionately smaller and the legs become proportionately longer as humans mature. This implies that proportionately large heads and proportionately short legs would be neotenous features for adults.

Several heterochronies have been described in humans, relative to the chimpanzee. In chimpanzee fetuses, brain and head growth starts at about the same developmental stage and grow at a rate similar to that of humans, but brain/head growth stops soon after birth, whereas humans continue brain and head growth several years after birth. This particular type of heterochrony, hypermorphosis, involves a delay in the offset of a developmental process. Humans have some 30 different neotenies in comparison to the chimpanzee, retaining larger heads, smaller jaws and noses, and shorter limbs—all features found in juvenile chimpanzees.

Neoteny retards the development of the organism into an adult, and has been described as “eternal childhood.” In this form of heterochrony, the developmental stage of childhood is itself extended, and certain developmental processes that normally take place only during childhood (such as accelerated brain growth in humans), is also extended throughout this period. Neoteny has been implicated as a developmental cause for a number of behavior changes as a result of increased brain plasticity and extended childhood.

Walter Garstang, 1928, Leeds University

Garstang argued: there are many important differences between chordates and echinoderms. Most importantly, adult echinoderms show little likeness to chordates: echinoderms, c. 800–600 mya, are radially symmetric, possess calcium carbonate plates in their skin and have tube feet. Garstang made the radical suggestion that perhaps it was echinoderm larvae, not adults, which early on had given rise to chordates.

Echinoderm larvae, like chordates, are bilaterally symmetric. Especially notable are their similarities to larvae of hemichordates, which are a step closer to full chordates as they share two of the five most commonly noted chordate characteristics, namely a hollow neural tube and pharyngeal slits.

Garstang's idea has been expanded and is supported by many lines of evidence. Perhaps most interesting and compelling is the fact that some amphibians can stay in larval form and still reach sexual maturity—this shows that echinoderm larvae could, theoretically, have become sexually mature and simply stopped morphing into adults, instead, evolving into chordate ancestors. Species that show this refusal to leave the larval stage include mud puppies and other salamanders, which either partially or completely show neoteny (also called paedomorphism): retention of juvenile traits or phenotypes after sexual maturity.

Garstang' s Hypothesis of Chordate Larval Evolution

At the time of its discovery in 1869, the tadpole larva was considered a descendant of an ancient free-swimming chordate ancestor:

- **Adults were then considered degenerate sessile descendants of the free-swimming form.**

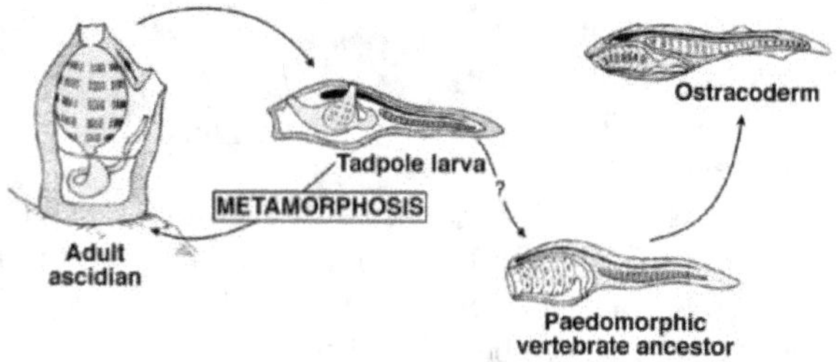

In 1928, Walter Garstang turned this sequence around, suggesting that the sessile adults were ancestral:

- **He theorized that the larva failed to metamorphose into an adult and developed gonads and reproduced in the larval stage.**
- **He termed this *paedomorphosis*, the evolutionary retention of larval traits in an adult body**
- **His suggestion that evolution may occur at larval stages matches embryological evidence.**
- **Paedomorphosis occurs in some amphibians (also called *neoteny*).**
- **Humans have also sometimes been described as paedomorphic, because our features are more similar to baby apes than they are to adults.**

Figure111

W.K. Gregory (1876–1970) believed the mammal monotremes were the result of a paedomorphic survival attempt out from their marsupial early heritage by retaining infantile characteristics into maturity. This presumably allowed for the platypus, echidna to survive in a marginal, noncompetitive ecological area.

There is also a possibility that the gracile *Australopithecines, afarensis,* and *africanus* may have undergone paedomorphic processes in their contemporaneous competition with *A. robustus, A. boisei* , and *A. Paranthropus*. Thus, this genetic revolutionary process may have become recessive in our genetic *hominin* heritage.

Chimps and Humans [[G. Milburn, 3/30 2009]

Figure 112

One of the biggest "human" questions is "where did we come from?" While the mechanisms of evolution are well established, the route humanity took to get to its present state is not as well determined. It's the difference between knowing the rules of chess and being able to figure out the personality and play style of a grandmaster from a few snapshots of a very long game in progress.

One proposed mechanism for the evolution of humans from primates is neoteny, where juvenile traits are retained and adult adaptations lost. This has been observed in foxes subject to behavioral selection. For instance, look at this young chimpanzee.

This picture is from a 1926 study by the German anthropologist Adolf Naef. He describes it as "the the most human-like picture of an animal, of any that is known to me." The little guy does seem to have a rather regal and refined air about him, but we can't just wave our hands and call it case closed at this point. Can we look at the development of a chimpanzee and see if there are any quantifiable parallels?

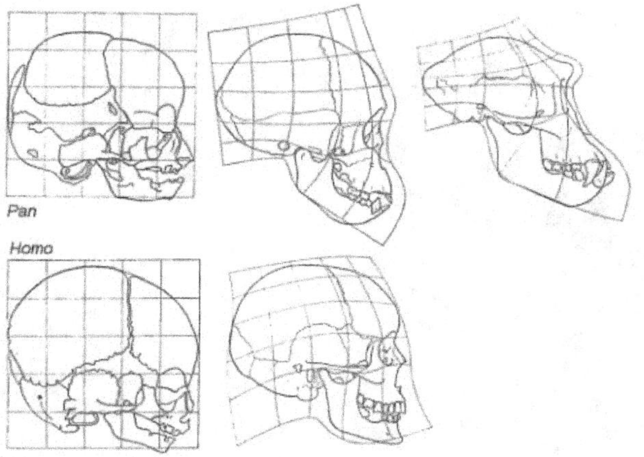

(Left) Figure 113: The two skulls on the far left are those of an infant chimpanzee (top) and an infant human (bottom). Bone structure and shape are very similar, with the classic huge head and tiny cute face we seem programmed to love. The two skulls in the middle are of an adolescent chimpanzee (top) and an adult human (bottom). We can see the jaw start to lengthen in both, and their overall similarity. The final picture on the top right is of an adult chimpanzee, who has a significantly larger and more powerful bite than any adult human.

Figure 114

Baby chimps will eagerly mimic a human caretaker – sticking out their tongues, opening their mouth wide, or making their best effort at a kissy face. Not only is the basic mechanism of learning there (imitation), it appears to be very focused on social relationships. And this ability decreases with age!

Causes of neoteny

The underlying mechanisms for this includes heterochrony (change in features during development). Paedomorphosis is common in many animal species domesticated by humans, including dogs, chickens, pigs and cattle. It is believed to be a side-effect of the selective pressure of human-directed breeding for juvenile behavioral characteristics such as docility and cuteness

Research indicates that neoteny occurs, because the hypothalamus of the brain fails to produce the hormone that causes the pituitary to stimulate the thyroid gland, to produce growth hormones that trigger metamorphosis. Some scientists think that neoteny may have evolved as a response to the hazards of life on land. (IASZoology.com)

What is Neoteny? Dmitri Belyoev, d.1985[2/24/2010]

Figure 115: Dmitry Belyaev, the brains behind the breeding. Photo by Institute of Cytology.

Figure 116

Foxes have been observed to evidence dramatic changes in look and behavior in as few as 20 years of targeted breeding. Individuals were selected to breed that exhibited "tame" behavior, or the neotenous features of cooperation, playfulness and displays of affection. Coats changed, ears flopped down, barking emerged and even estrus varied, prolonging reproduction periods. This happened in less than a dozen generations. When humans selected cooperative individuals to mate, cooperation became a feature of descendants. Physical changes accompanied selection for behavior.

Bands of proto humans drifted over time in neotenous directions, transforming quickly in just the same way that we breed dogs. Still, there is an overall human trend in a neotenous direction. Playful, curious, creative, social and affectionate adults become more and more highly valued over time.

Lightning-fast human evolution is a direct result of sexual selection and environmental effects influencing testosterone and estrogen to adjust maturation rate and timing to speed up physical and behavioral transformation. Wolves, foxes and salamanders evolve in just this way, and so do humans. (www.neoteny.org)

Humans prolong, rather than reduce, rates of growth and development of body parts; the brain, for example, ceases growing later in humans than in apes, but during this prolonged period of early ontogeny, it grows at a rapid pace.

Perspective

It is a great puzzle. The first tangible sign of the coming of this neotenous creature, Cro-Magnon, *Hss*, was at the Danubian gateway to Europe some 45,000 BP. Such an enormous deviation from hominin/hominid/anthropoid, mammal orthodoxy could not have happened in a moment. We can look back to Africa, some two million years ago. We see hominins *Homo ergaster*, *Homo rudolfensis*, and possibly even *Homo habilis*, making their way north as probably did the Georgian erectines, at 1.8 mya, they too being massive in bone. Carleton Coon laid forth some hypothesis that one of these African hominins might have undergone heterochronic development as they, along with all hominids, anthropoids, mammals do. They wander in search of better horizons 'on the other side of the mountain'

Remember *Flores*, in Indonesia, they on an evolutionary level 2.5-2 mya, probable wandered out of Africa during this time period and many hundreds of thousands of years later wound up in Southeast Asia, very un-evolved from its originating profile.

Researchers investigating the genome of modern Europeans say they have found evidence of a genetic blockade, a moment of small number creativity, when a group came together in genetic homogeneity, some 200k BP or earlier (400k BP), and then spread out like a funnel. But what sprung the neotenies loose in the existing populations? First, a tendency, of hominins migrating from the south, caught in a crisis—the Riss—climate six-descent, 190–130,000 BP? A few juvenile-looking members of the clade with lots of energy, and being a bit more adventurous, they separate out. Away from the group, early Heidelbergensis/Neanderthal types and their mates secure an economy that got them through a time of frigid stress, eventually breaking up into small wandering groups, while sharing their genes with more primitive erectine types. They may have thus precipitated so-called boundary sapiens and other ethnic/racial groups to pivot over the edge into sapiency. They, being the early Cro-Magnons, later *Homo sapiens sapiens*.

[FOOTNOTE: On Eur-Asian origins of Cro-Magnon, genetic blockade, climate challenges, see *TIDBM* Footnotes #15-22, PP 326-327:. P321 origin in oxygen isotope period #6= 190-130k BP-Footnote 38,39 P.327]

[This ice age 190-130 was not only long but intensely fierce, as with the final ice age before the Holocene,c.14K BP. Riss/Wolston;,the Wurm, 29-14k BP, was also fierce. The present Holocene, similar in mildness , seasonal as with other interstadials since onset of the recent series of ice age descent, from 2.6mya, the entire Pleistocene. See also Ch1 listing of footnote.

TIDBM---P.314-316 The Inevitable Domination by Man, (2000)---Locating the Genetic Bottleneck

Footnotes 14-22, quote, #21, P.315-316; Footnotes P.326-327---#14-22}

Race and the Cro-Magnons

But you will, as we all do, ask the question, why do we Europeans look so much alike, as is the case with the Neanderthals? Cro-Magnons were not especially hairy, nor did they have big noses, and were short and thick-boned. Both had white skin, and hair of blond or ruddy coloration. If they came from Africa even 500 thousand years ago, would the Cro-Magnons have bleached? Consider, Carleton Coon has made a good case, even though few have followed his research. He argued that the Capoids today of South Africa have many morphological characteristics similar to the far-away East Asian Mongoloids (Chinese, Koreans, Vietnamese, Japanese). The skin coloration of these peoples are today only somewhat lighter than the current Capoids/Khoisinoids.

These Mongoloid wanderers from Africa have been in East Asia for at least a million years. Again, take the example of our little Flores primitives in Indonesia. They are a model of the hominins at the 2.5 million years ago morphological and cultural levels. Beijing, in the vicinity of where the Mongoloids seem to have been centered is at the latitude level of Philadelphia, USA. There is here time enough for them to have completely whitened after that ancient trip from Africa.

With the same light hair, skin, blue eyes as their cold-adapted Neanderthal cousins and Heidelbergensis ancestors, who had been in Eur-Asia for about one million years or more, it can be argued that Cro-Magnons have been hanging around the north latitudes long before the Riss glaciations. This does not mean that their paedomorphic/neotenous transformation did not take place during Riss glaciations (a really tough one), but that they probably started separating from their Neanderthal cousins long before this late time frame, c. 200,000 BP, and of course well before the advent of their spectacular cultural, as well as demographic entrance into Europe, c. 45,000 BP.

The guess here is a separation from the erectines before c.500k BP.

It should be noted that this final and late entrance into Europe, c.45K BP, took place during a warm period of the Wurm descent, i.e c. 57–29,000 BP. The probable appearance of an easy climate, much huntable wild life, and even wild fruit and berries, all could have signaled "come hither."

Paleoanthropologists O. Bar Josef, AI Ammerman, and L.L. Cavalli-Sforza argue that the origins of the Cro-Magnon probably occurred in the Anatolian Plateau, in the Zagreb Mountains of Iraq and Iran, the Caucasus. The most easterly findings are three Cro-Magnon fossils in the Hotu cave at the shores of Iran's Caspian Seas. [PP 158-261, TIDBM] Also, there are the Russian Cro-Magnon fossilsn at Kostinki, c. 45–35,000 BP and Ust-Ishim, c. 45,000 BP. Note that there was intact DNA found at Afontova-Gora, c. 17,000 BP (near Kranoyarsk—females with blond hair).

In sum, it is all hypotheses as to when, where, how this event took place to create a creature whose brain was largely freed of the traditional reigns of instinct that mother nature has applied to all forms of life. Does mother nature's disciplinary law still apply?

Don't throw off that conservative leash of instinct. Watch out for radical progressivisms, uncontrolled, unrestrained success==evolutionary disasters!

Venus Symbol: Sex, Beauty, Love

They are found throughout Western Europe and into the East. The eastern Venuses are less impressionistic in creativity and less pronounced in their sexual expression. Think of the vast geographies over which these creations are to be found, the limited transportation means between this very thinly scattered population of Cro-Magnons, the challenges of climate and the stone technology available to their hands and minds. Such a distant symbolic communication of a cultural vision has to confirm the reality of a long ago genetic blockade of a few brothers and sisters still carrying the memory of affiliation, a thousand miles apart, by foot. This transcontinental theme, this 'song' of life has to confirm consanguinity.

What the similarity of purpose and the differences in artistic solution of these Venuses tells us is another and current disputation. One thing is clear, these creations have no predecessor art objects for a model. They came out of the naïve, uneducated brainstorming of the minds of the Cro-Magnons. Clearly, there is a rumbling sexual passion in the hunters who likely carried the individual carvings in hand or in pocket as memory icons of almost pornographic lust. It immediately tells us something about the human brain's lack of sexual periodicity. The lower brain here gives a surge of animal energies into the thinking cortex.

Oh yes, the Bonobo are sexually active all the time. But they cannot say "NO" or cannot change their minds about sexual activity, momentarily, else as cultural fashion as in our own era. Individual humans can choose celibacy, by virtue of our *Hss* sapient cerebral cortex. This same cortical neurology can be reversed by a change of mind. That celibate could become interpersonally sexual.

In 1981, Patricia Rice studied a multitude of female clay figurines found at Dolni Vestonice, believed to represent fertility in this society. She challenged this assumption by analyzing all the figurines and found that "it is womanhood, rather than motherhood that is symbolically recognized or honored." (Rice, 1981)

But there is more than the graphic exhibition of breasts, buttocks and vulva. There is also love as expressed in the tiny carving of the Venus of Brasesempouy. Her hair is carefully designed as with other Venuses'. There is also abstraction, a conceptual artistic realization of what could be blatant sexual stimulation, "porno." The passion we see here, is that powerful emanation from the new human brain and neurological system. Its widespread geographical creators hit on something essential to their new bio-cultural essence.

(Left) Figure117: Venus of Brassempouy (23,000 BCE). First venus to display facial features: first Stone Age Portrait. . 1.5 inches, France

At Dolni Vestonice, c. 31–27,000 BP, they developed a mixture of clay and ground bone that often produced a self-glazing finish in the kiln, causing many of the figures found at this site to have an unusual, shiny, black appearance. The Black Venus of Dolni Vestonice is the most well-known example of a figurine made with this material. While fairly unique in terms of her color and material, the Black Venus shares the curvaceous appearance of her counterparts found throughout Eurasia. Height, 4.4 inches.

Figure118 & 119: Below Left and Below Right

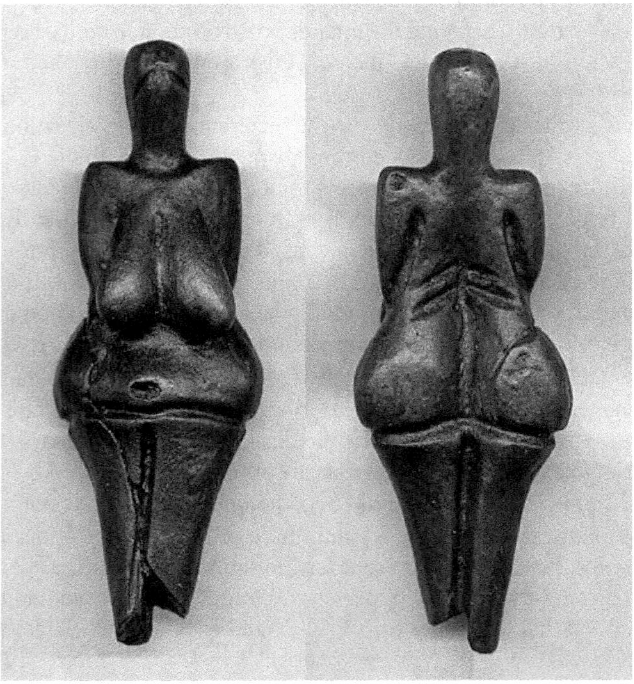

Figure 120 &121: Below Left and Below Right

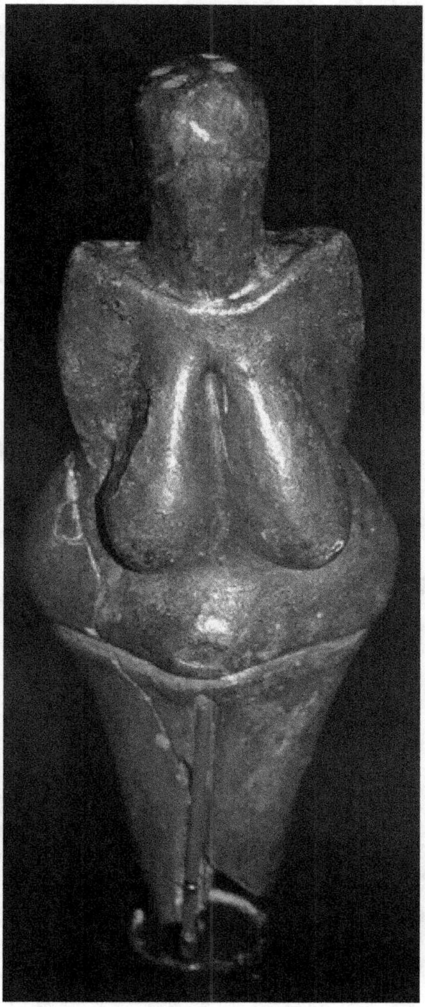

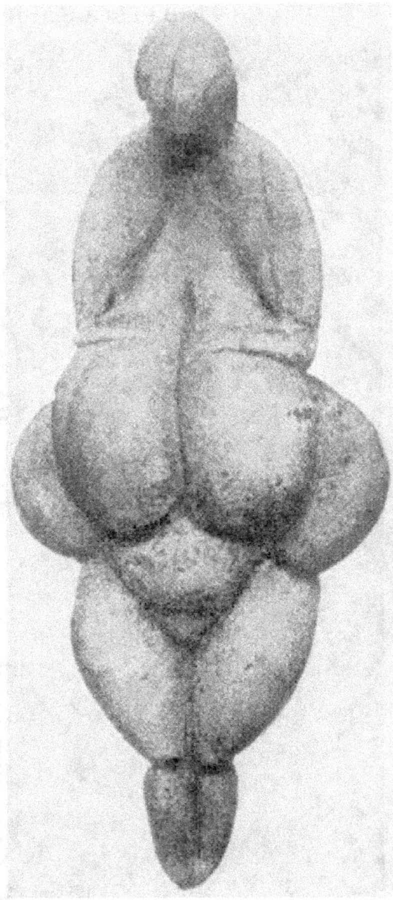

(Above and Right) Figure122 & 123:
Imagined Restoration of Venus of Lespugue

Venus Lespugue

It was discovered in 1922, in the Rideaux cave of Lespugue (Haute-Garonne) in the foothills of the Pyrenees by René de Saint-Périer (1877–1950). It dates from c. 24,000 BP. Approximately six inches (150 mm) tall, it is carved from tusk ivory, and was damaged during excavation.

Of all the steatopygous Venus figurines discovered from the upper Paleolithic, the Venus of Lespugue, if the reconstruction is sound, appears to display the most exaggerated female secondary sexual characteristics, especially the extremely large, pendulous breasts.

According to textile expert Elizabeth Wayland Barber, the statue displays the earliest representation found of spun thread, as the carving shows a skirt hanging from below the hips, made of twisted fibers, frayed at the end.

Figure 124: Venus of Hohlefels, the earliest Venus figurine, c. 35,000 BP

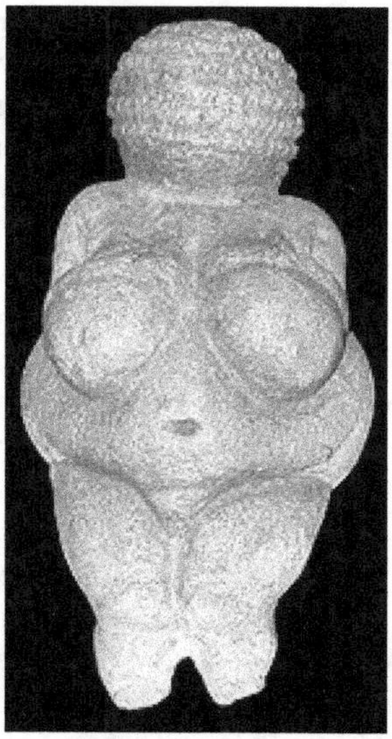

Figure 125: Venus of Willendorf

Venus of Hohle Fels	35–40,000 BP	Swabian Alb Germany	Mammoth Ivory
Venus of Dolni Věstonice	27–31,000 BP	Moravia, Czech Republic	Ceramic
Venus of Lespugue	24–26,000 BP	French Pyrenees	Ivory
Venus of Willendorf	24–26,000 BP	Lower Austria	Limestone, Red Ochre
Venus of Brassempouy	c. 23,000 BP	Brassempouy, France	Ivory, Red Ochre

The drama of life and play

Johan Huizinga, a noted Dutch historian wrote in 1938, "*Homo Ludens*. The play element of culture."

Our point of departure must be the conception of an almost childlike play-sense expressing itself in various play-forms, some serious, some playful, but all rooted in ritual and productive of culture by allowing the innate human need of rhythm, harmony, change, alternation, contrast and climax, etc., to unfold in full richness.

"In play there is something "at play" which transcends the immediate needs of life and imparts meaning to the action. All play means something. If we call the active principle that makes up the essence of play, "instinct," we explain nothing; if we call it "mind" or "will" we say too much" [Huizinga]

The geneticist and evolutionist, Theodosius Dobzhansky, once wondered how we could ascribe a biological, Darwinian selectionist view of the world and life of Johan Sebastian Bach. Indeed, Bach had many, many children for which his musical efforts brought to reproductive maturity. But what selective meaning does all this music composing, playing, and singing have in the animal world of competition and survival.

And that is the question we must ask of the wall cave paintings, they created with vegetable and found clay dyes mixed together, and then with their stone and animal fat lamps explored the innermost areas of the cave world around them; exploration to create this vision in color of the livingness of things around them. But also the carvings, the sculptured friezes

No doubt down below these hillside redoubts, this a male club, were the living quarters. In the warmer seasons, they traveled about hunting and gathering, sleeping in round, tent-like protective quarters for four or five folks. Usually there were about thirty to forty in the group. Else at home in the cave dwellings, garbage dumps well away from living quarters, the paedomorphic, domesticated wolves hanging around for a free lunch, the Cro-Magnons seemed to live to create, The hunting was a sport if we can understand the carvings on their spear throwers, all done with stone tools: no iron, copper or steel here.

So too the Venus'. Why the need when on the hunt to carry along the often impressionistic reminders of their passion? The jewelry was munificent, for both female and male, the hair dressings and probably the sewn clothing kept the females occupied. Consider the reality of non-stop pregnancies and neonates that must be kept alive. Also consider the shaping of the female gender and the child birth surviving females over these many thousands of years. Motherly love, heterosexual passion, keeping the home front in shape as the men went away for the hunt, except on 'weekends' when they did their 'culture', lots of art.

Entertainment beyond the painting and carving, usually was a male preoccupation. Still the larder had to be filled, and they were not vegetarians or on a diet. And so examples of their dramas, men dressing up as animals along with other entertainments, music, singing? The Australian aborigines created rituals in which they acted out their dependence on animal life, and noisily too. Our example below seems more fun–like. This too with the lion man which goes back to the beginnings of their sojourn in Germany. Lion man, the sphinx, Richard of the lion hearted, also MGM and it grand roars.

The defecating or birthing Ibex, Deer, or Chamois is one of many examples of the careful stone tool transformation of this natural world into both the esthetic as well as the practicalities of making a living, here the spearthrower. These Cro-Magnon people could not resist the transformation of serious survival activities into something beautiful that stimulated their minds. The example of the "fawn" here illustrated below, is one of a number of such careful recreations of reality, now found in mostly Magdalenian era debris, c.16–11,000 BP.

Figure 126: Sketch of Breuil's drawing

The Sorcerer is one name for an enigmatic cave painting, found in the cavern known as "The Sanctuary," at the Cave of the Trois-Frères, Ariège, France, made around 13,000 BC. The figure's significance is unknown, but it is usually interpreted as some kind of great spirit or master of animals.

"The Sorcerer" is composed of both charcoal drawings and etching within the stone itself. Details, such as etching, are often difficult to view from photographs due to their size and the quality of the light source. Particularly celebrated prehistorian, Jean Clottes, asserts that Breuil's sketch is accurate ("I have seen it myself perhaps 20 times over the years").

The Löwenmensch figurine or Lion-man of the Hohlenstein-Stadel is a prehistoric ivory sculpture that was discovered in the Hohlenstein-Stadel, a German cave in 1939. The German name, Löwenmensch meaning "lion-human," is used most frequently because it was discovered and is exhibited in Germany.

The lion-headed figurine is the oldest-known zoomorphic (animal-shaped) sculpture in the world, and the oldest-known uncontested example of figurative art. It has been determined by carbon dating of the layer in which it was found to be between 35,000 and 40,000 years old, and therefore is associated with the archaeological Aurignacian culture of the Upper Paleolithic. It was carved out of mammoth ivory using a flint stone knife. Seven parallel, transverse, carved gouges are on the left arm.

After several reconstructions that have incorporated newly found fragments, the figurine stands 31.1 cm (12.2 in) tall, 5.6 cm (2.2 in) wide, and 5.9 cm (2.3 in) thick. It currently is displayed in the Ulm Museum, Germany

Figure 127: *Löwenmensch*, a lion-headed figurine found in Germany, dating to the Upper Paleolithic of about 35,000 to 40,000 years ago

Figure 128: Magdalenian, c.12k BP Spear thrower made of reindeer antler shows a fawn defecating (ibex chamois or deer) or giving birth), colored with manganese and hematite. Right eye is inserted amber, left eye, lignite, note bird theme on back of animal. Also note that amber is not a local product. It was probably traded for. Here, an example of the cultural life of the Cro-Magnon that needs further archaeological exploration. France.

Below are other exemplars of the passion for creating art objects out of practical hunting: spear throwers. Where did this mentality come from? What neurological structures could have preempted the survivalistic drives of a mammal, *Hss*, certainly always on the edge of survival, for this new mental obsession?

The swimming reindeer

Tip of a mammoth tusk carved as two reindeer depicted one behind the other. The leading, more detailed figure is a female with smaller body and antlers. Finely incised notching and shading outlines the variation of colors in her coat. The following male figure has larger antlers and equally detailed carving of the ears and eyes but its body is unshaded. Both animals are depicted with their chins up, the antlers laid back, the front legs extended forward and bent at the knee and the back legs extending back. The male genitalia are clearly represented.

Figure 129: The male reindeer is on the left; the female is to the right. Mammoth ivory, Magdalenian, France, 15-12k BP

Curator's comments

"The representation of the animals may have been determined by the tapering shape of the mammoth tusk although the interpretation that the piece depicts the animals swimming has parallels in painted art of the period. The presence of the antlers, the shading of the female coat and the male-female juxtaposition suggest this might be an autumn-winter depiction when animals migrate, cross rivers and mate."

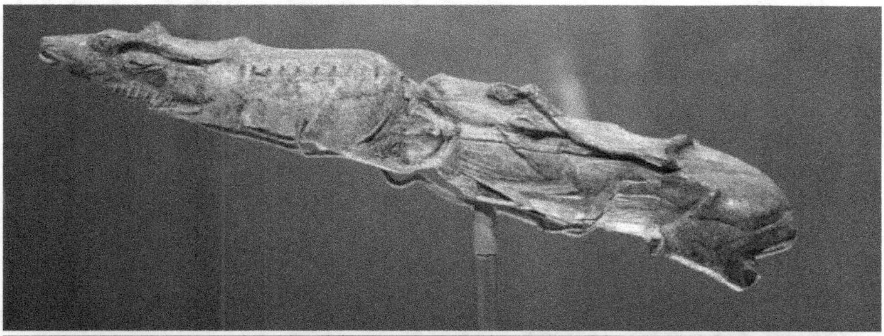

Figure 130 & 131: Mammoth ivory, carved, polished, incised. It dates to the late Magdalenian period, approximately 13,000 years ago. Found in France

Former Director of the British Museum Neil MacGregor says of the manufacturing process:

> If you look closely, you can see that this little sculpture is the result, in fact, of four separate stone technologies. First, the tip of the tusk was severed with a chopping tool; then the contours of the animals were whittled with a stone knife and scraper. Then the whole thing was polished using a powdered iron oxide mixed with water, probably buffed up with a chamois leather. And finally the markings on the bodies and the details of the eyes were carefully incised with a stone engraving tool. In execution as well as in conception, this is a very complex work of art.

And it seems to me that it has all the qualities of precise observation and interpretation that you'd look for in any great artist.

Figure 132

Cap Blanc

Figure 133: Cap Blanc Frieze (15,000 BCE)

The most important example of Magdalenian stonework. Originally, the stone frieze occupied roughly thirteen meters or roughly three-quarters of the entire back wall of the shelter. It includes images of reindeer, bison and horses, and the central horse measures about 2 meters (7 feet) in length, and dates to about 15,000 BCE. In total, the frieze contains 14 horses (or parts thereof), 13 bison and 2 reindeer, along with more than a dozen unidentified animals. As with all Stone Age friezes, traces of red ochre pigment were found, indicating that the colorful carvings must have been visible from some distance.

Travel forward 12,000 years. Same humans

Figure 134: ***Acropolis/Parthenon*****, Athens Greece, c.400 BCE**

Figure 135: The Acropolis of Athens, seen from the Hill of the Muses

Figure 136: The Erechtheum

Travel forward, 2500 years, same humans—Lincoln Center, NYC, USA

Figure 137: Lincoln Center of the Performing Arts

Chapter 9

A Civilization in the Making—3

Apprise

Before humans entered the world of choice, changing one's mind, saying no to one's seemingly biological desires or needs, animals were given genes by nature to behave within the rigid parameters for survival, always a test of the ongoing continuity of nature, their dependency on stasis, the external status quo.

Behavioral choice had been introduced into the evolutionary scene in times of great external ferment, either external physical or animal/plant competition. Choice was adaptive for quick decision making. "Do not wait for the adaptive sieve of natural selection to deliver salvation by a fortuitous outlier."

With the coming of *Hss*, the slow but sure directives of instinct are no more. Progressive choice has won out in a genetic explosion. Heterochrony, paedomorphism, and neoteny have upset the old balances between instinct and volition. Now the brain, the "high flying" cortex of *Hss*, directs humanity to search for the available sureties in the environment. Now, by creating inner mental regularities, perceived and integrated relationships, this thinking brain can lead individuals and their confreres to act and hopefully survive.

The new hominin neurological apparatus stands as a paean to a perennially changing external physical and biological world and the human ability to respond behaviorally in an adaptive fashion. The test is always given by the supreme arbiter, "Mother Nature."

Material symbolism

Take the cave paintings, the sculptures, Cape Blanc, and the spear throwers as a group. What significance should we find in this seemingly passionate Cro-Magnon esthetic interest? This writer senses a triangular relationship: animal, human, symbolic meaning. Animals feel their biological needs, act on them in terms of perceptual signals, a food or reproductive object. Earlier humans perceived external entities in relation to self, and acted in a more flexible animal behavioral response.

The inchoate awareness of the world, ritual, tradition, and stability of environment all helped Neanderthals to hunt and survive, and create their traditional Mousterian technology. Outside of the art of the ancestors of the Bushman, the Cro-Magnon brain created a new reality in the natural world—symbolic meaning.

The Cro-Magnons as with ourselves could not help viewing animal life, their sustenance, with wonder, the inner dynamics of their search for metabolization, fecundity, the basic functions of all animal life, all mysteries of their physical attributes, somehow translated into what we would call, on the basis of history, a sense of *beauty*.

Their own social relationships were determined by their interactions with animal life, mostly, the hunt, the kill. Yet it was not merely the physical challenge of the hunt, the expected feast of blood and meat protein. They must have also competed socially in their artistic recreations. The raw and humanly dynamics of the prey was also mentally metabolized in the flickering darkness of the cave, an ancient Sistine Chapel, high above the riverside home sites

Consider the hundreds of hours devoted to the carving of the stone tools, the jewelry that they created to present to a loved one, a respected and departed leader of the band. Then there is the example of the spear throwers. From a distance this instrument of the hunt could wound an animal and then be followed in its death throes. Yet it was not merely a weapon of the kill. As with the aristocratic panoply of our medieval knighthood riding in individual combat towards glory or death, there was the symbolic significance of the event, the medieval duel, or the Cro-Magnon hunting expedition.

The weapon had to be carved gloriously, most often embellished with an animal subject, a binding appreciation even love, given the always deadly outcome. It is almost a ritual sanctification of the act of the hunt. Here find a male modernity, human, perennial. Given the many hours spent in fabricating this tangible and personable object, once completed, it became not merely a tool of death, but also a creation of beauty.

Not so hidden, of course, is the sense of power against a nature that is unrelenting. We have here the objects of victory over nature. From the standpoint of evolution, it is positive selection. But it is not only natural selection by dint of the victorious hunter, but also the talented artist. There is, in addition the fact and the external power in domestication. Certainly, throughout the long ice-age tenure of Cro-Magnons, a splinter of deviants from the wolf packs began to hang around the Cro-Magnon garbage dumps, an easy, non-combatant morsel. Probably the most paedomorphic of the species gradually became pet dogs.

If so the wolf, who else, thought the Cro-Magnon sage. The high respect if not love of the horse which we see in their artistry may have stimulated the domestication and riding of these noble animals; so did Richard Leaky think. Then, in the most frigid period of the final Wurm glaciations (19–12,000 BP), in the Magdalenian, there are indications of northerly bands of Cro-Magnons herding reindeer to protect and secure a resource supply for the band or tribe, a la Joseph of Egypt to the Pharaoh.

Venus: sex, beauty, and love

For at least thirty thousand years, men, most probably, have shown an esthetic, as well as a biological interest in the nude female. It is interest centered in the eyes as well as the groin. Widely distributed throughout the domains of the Cro-Magnon, modern researchers have interpreted the Venus carvings as fertility symbols, the ongoing concern for the perpetuation of the social group, possibly showcasing power and influence of the female (similar to that in parta, Pharaonic Egypt and in Judaism too).

The Lithuanian archaeologist, Marija Gimbutas, wrote a number of books about her excavations in southeast Europe, mostly south of the Danube river and basin, through which, it is hypothesized, most of the Cro-Magnon migrants entered Europe. Gimbutas' excavations mostly involve much later Neolithic and Bronze age artifacts. Her theory was that the figurines which she described and illustrated in her various books were essentially an homage to the god-like fecundity of the female, even using some upper Paleolithic nudes as exemplifications of this theory. But since these much later figurines and the markings associated with them were too crude and ambiguous to so interpret, first as females, and then as having important denotational markings as von Petzinger, see below, is arguing, her view of the figurines, as pacific goddesses of a warless time in human social evolution, has been mainly disregarded. (Gimbutas, 1989)

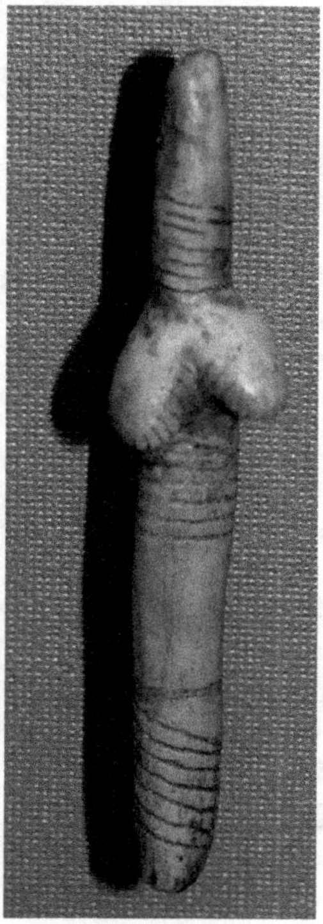

Figure 138: Phallic Goddess, Hungary, c. 5500 BCE

The upper Paleolithic Venuses (see Chapter 8), mostly four to six inches in size, even in the case of the Venus of Dolni Vestonice, c. 31–27,000 BP, found in Moravia, were fire-hardened and finished as a work of art, and were clearly for hand holding, else nestled in the hunter's pockets away from camp, perhaps as a stimulus to finish the kill and get back home to his beloved. The sexuality is pronounced, breasts, buttocks, vulva all clearly delineated and appreciative of his protein-rich beloved. Most important from the esthetic point of view is their modernistic abstraction. There is a universal tonality to these nude figures, not a specific image of she-who-is-home, more like the import of a magazine cover of Marilyn Monroe or a Peter Paul Rubens Flemish nude, but without the concrete facial features.

The exception, thus far found, is the Venus of Brassimpouy (see Chapter 8), her under-two-inch facial characteristics, clearly carved, hair, as with all the Venuses', clearly coiffed. The rest of her sculpture, if found, we would wager is not in the nude. Here in the Venus nudes, are the universal and timeless hints about maleness and its great attachment and never ending sexual attraction (love) for the female.

Remember, these women were mostly pregnant year after year. In the ecological context of those ice ages, the survival of the band or tribe depended on healthy females giving birth to healthy infants, then nursing and caring for them even while pregnant with an oncoming infant. This is underlined by the numerous grave sites with young, “teenage” women, sometimes buried side by side with the fossil remains of a tiny neonate. These babies were born with great, but soft-boned heads, and the female birth canals were still relatively unexpanded, an unchanged inheritance from a more ancient evolutionary stage.

A healthy, voluptuous companion in life had to have given females a special social and economic status. Think of the recovered sewing (Magdalenian period—17–12,000 BP, needles). Perhaps woman also participated in jewelry fabrication, all with stone tools. Women had to have (a) say in things as they were the fertile foundation for the continuity of the life of the community. The Venus figurine becomes not only a symbol of memory and desire, but a unique esthetic transformation of the physical elements of life into the conceptual. Perhaps, most importantly, these effortful and loving fabrications represent the creative, esthetic potentialities of this new evolutionary mentation.

“Ancient phallus unearthed in cave”

[By Jonathan Amos

BBC News science reporter 7/25/05]

A sculpted and polished phallus found in a German cave is among the earliest representations of male sexuality ever uncovered, researchers say.

The 20cm-long, 3cm-wide stone object, which is dated to be about 28,000 years old, was buried in the famous Hohle Fels Cave near Ulm in the Swabian Jura.

The prehistoric “tool” was reassembled from 14 fragments of siltstone.

Its life size suggests it may well have been used as a sex aid by its Ice Age makers, scientists report.

Abstraction

Here, we leave the pictorial world of symbolic representation. We encounter a new mental dimension of *Hss*, the representation of ideas by agreed-upon, symbolic markings. Every social group could in theory and fact say, “this mark means this and only this. Our social community will know what this mark means.” And so, as in wars, we create a series of cryptic symbols representing secret ideas, only known to the social group. Spoken, discrete languages are the closed door to the ethnic sense of uniqueness in *Hs* beings.

But if the social group was as close to each other in cultural identification as the Cro-Mgnons seem to have been throughout Europe, then it is possible as in our world of flags, road signs, computer and typewriter symbols (= /equals, ./period, etc.) that whole world community would

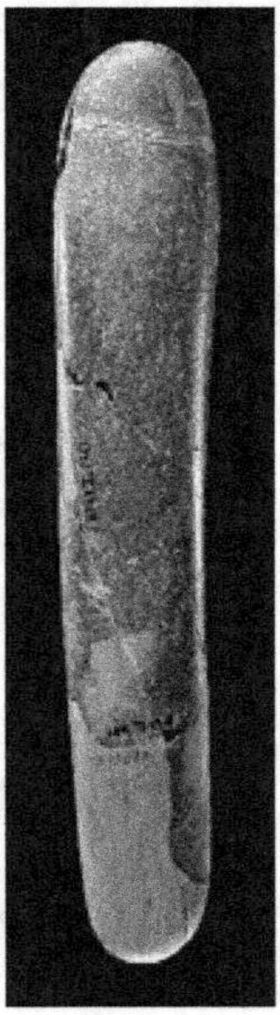

Figure 139: Phallus may also have been used to knap, or split,

have known what these symbols meant. So it goes with language, as once private conversations or orthography become part of one's national, religious, or cultural identification, translators are produced, allowing for wider communication.

Of course, that is what newly recognized Cro-Magnon abstract markings were, the communication of spoken ideas, directions, warnings—unlimited in nature—now simply referenced by a unique mark, possibly the thirty two symbols that G. von Pretzinger has identified; she has here supported the vision of Alexander Marshack's view that some of these markings were carved into various materials to allow the Cro-Magnons to see visually and abstractly what was going on in the evening sky so clearly seen in those days. They were not mere random punctures or slashes by bored hunters. Like our own hunting riflemen, the notches in the wooden stock of the rifle abstractly symbolized something, here, the killing of a prey.

The next step, of course, is the standardized markings of spoken into written language, alphabetic or logographic. But here, you need a different demographic and socio-economic environment. In the long term, the transition to a visual rendering of spoken language itself was probably stimulated by a great change in the climate.

Alexander Marshack rose to public prominence after the publication of *The Roots of Civilization* in 1972, where he proposed the controversial theory that notches and lines carved on certain Upper Paleolithic bone plaques were in fact notation systems, specifically lunar calendars notating the passage of time.

A.Marshack, 1972----

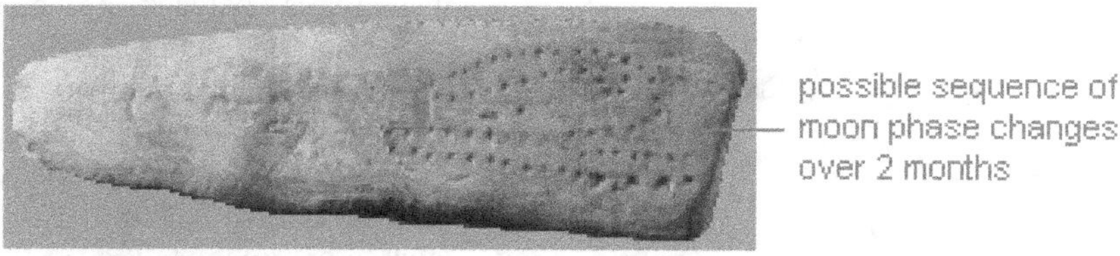

Figure 140: Southwest France, 38–35,000 BP [Alexander Marshack]

The incisions were structured in subsets of 29 notches, thus suggesting that they were used to mark the duration between two lunations. Using microscopic analysis, Marshack suggested that seemingly random or meaningless notches on

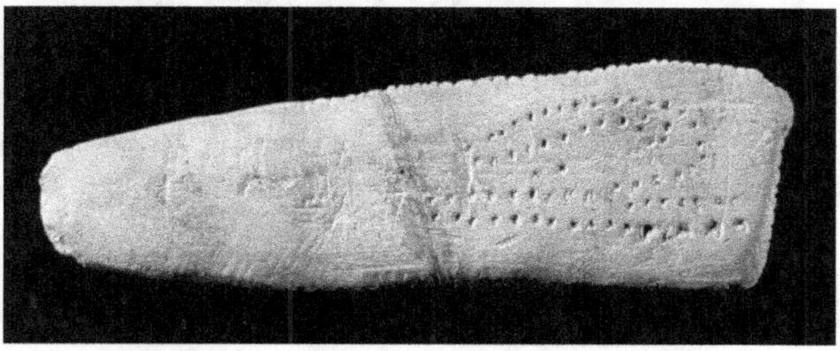

Figure 141: (Left) Abri Blanchard Cave, Eagle wing bone, 35–30,000 BP, lunar/solar calendar?

bone were sometimes interpretable as structured series of numbers.

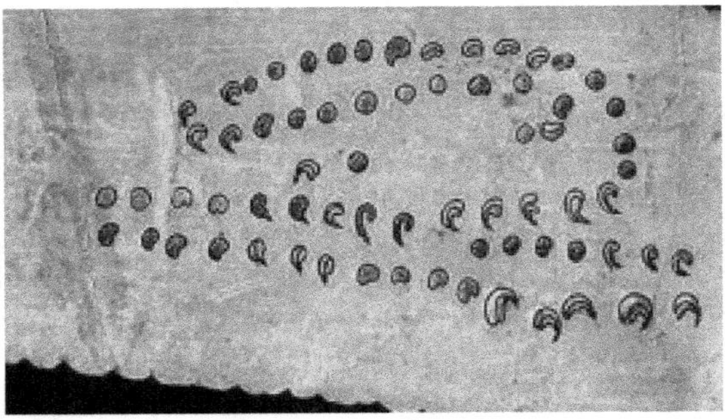

Figure 142: (Above) Recreation of Abri Blanchard Calendar: Five Planets

Zhuānxū calendar begins: "at the start of spring when the sun, moon and five planets met in Yíngshì." Liú Xiàng, c. 25 BCE (Fawcett, 2011).

Figure 143: Whole year lunar calendar, antler bone plaque incised with possible lunar calendar, 12,000 BP, France.

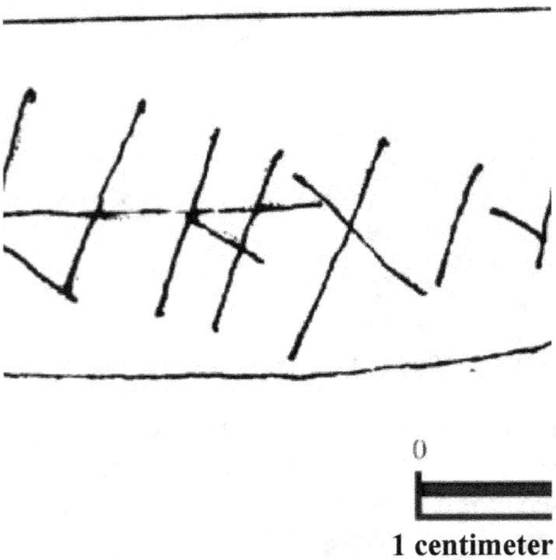

Figure 144: Bone markings from Dolni Vestonici, c. 31–27,000 BP

Using microscopic analysis, Marshack suggested that seemingly random or meaningless notches on bone were sometimes interpretable as structured series of numbers. For instance, Marshack hypothesized that notches on the bone plaque from southern France (which dates to approximately c. 35,000 BP) were structured in subsets of 29 notches, thus suggesting that they were used to mark the duration between two lunations. (Marshack, 1972, p. 445–477)

Prior to Marshack's work, many Paleolithic archaeologists focused their work on art such as the cave drawings at Lascaux, but paid little attention to the abstract notches and marks on plaques and other artifacts found at these sites. Marshack's work has been criticized as having over-interpreted many artifacts, finding numerical and calendrical patterns where none exist (White, 1989, pp. 366-390).

Lascaux Cave

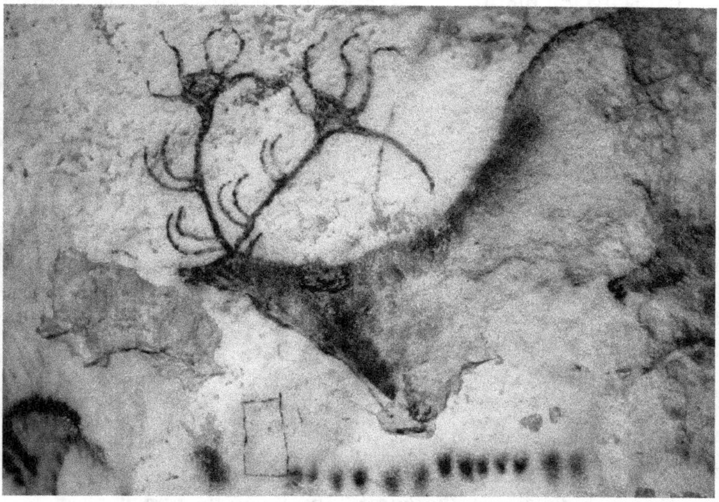

Figure 145: Photograph by Sisse Brimberg, Nat Geo Image Collection

In the famous French cave of Lascaux, ancient artists did more than decorate the walls with images of animals. They also left mysterious geometric signs such as the black quadrangle and dots below this Ice Age deer, c. 17,000 BP.

"Mysterious Markings May Hold Clues to Origin of Writing"

Geometric signs on cave walls and ancient artifacts may be some of humanity's earliest graphic communications.

Research by Genevieve von Pretzinger

Article By HEATHER PRINGLE

May 29, 2016

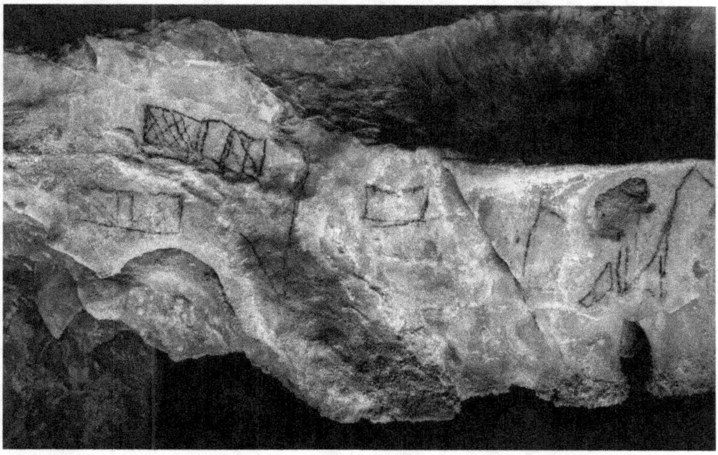

Figure 146: Photograph by Sisse Brimberg, Nat Geo Image Collection

Paleoanthropologist Genevieve von Petzinger studied rock art sites across Europe and discovered that Ice Age Europeans employed just thirty-two distinct geometric sign types from 40,000 to 10,000 years ago.

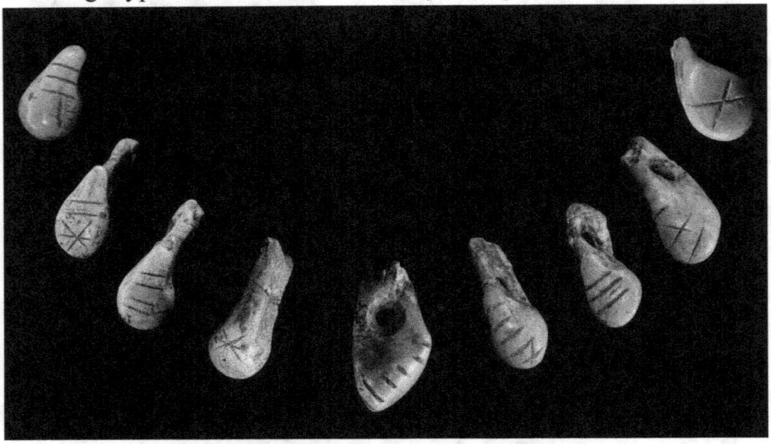

Figure 147: Photograph by D. von Petzinger

Once strung on a necklace, these deer teeth were discovered in the grave of an Ice Age woman who lived nearly 16,000 years ago. The markings on the teeth may have been memory aids to help the owner recite an important tribal story.

The First Signs

Unlocking the Mysteries of the World's Oldest Symbols by Genevieve von Petzinger NEW YORK Simon and Schuster, 2016

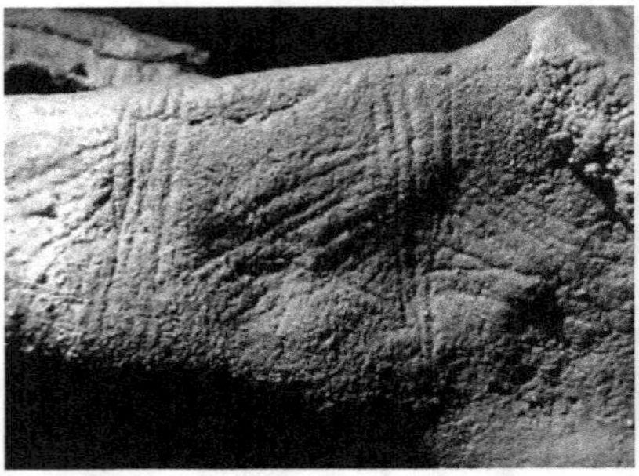

Figure 148: Tectiform signs Bernifal Cavel—Photo by G. Delluc

Figure 149

Figure 150: Indeterminate signs—Abri de la Ferrassie—Exposé au Musée des Eyzies-de-Tayac—Photo by C. Regnier

Figure 151: Indeterminate signs—Abri de Fongal—Musée de Préhistoire des Eyzies-de-Tayac—Photo by C. Regmier

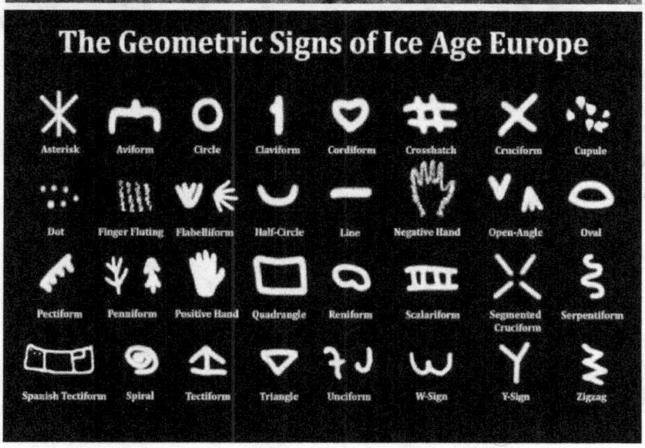

Figure 152:The thirty-two signs that von Petzinger has catalogued in Ice Age cave art across Europe. They account for the vast majority of non-figurative imagery found across the continent during this 30,000-year time span, suggesting that they were used with purpose and were meaningful to their creators. Each of the thirty-two signs has their own distinct pattern of use.

Figure 153: The Las Chimeneas cave in northern Spain. Pictured here is a Spanish tectiform sign (left), a local version of the tectiform (meaning "roof-shaped") sign. The Spanish tectiform is indigenous to northern Spain, and it's been hypothesized to be a representation of a Paleolithic dwelling or a boat or to be a more abstract interpretation of a local clan or tribal sign.

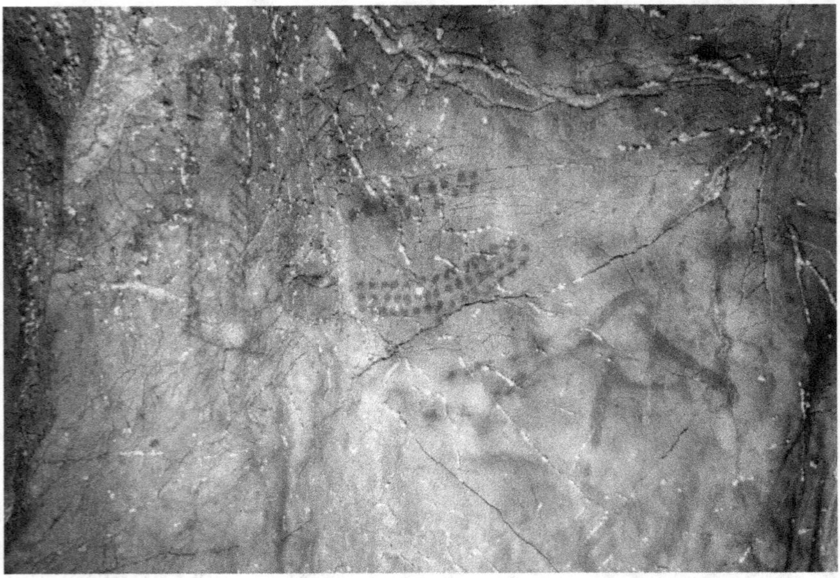

Figure 154: A Spanish tectiform sign—dots and a horse—found in La Paseiga, Spain, a richly decorated cave site featuring engravings, animal drawings and over 100 geometric images. Given La Pasiega's stylistically diverse selection of cave art, it is likely that various groups visited the cave over hundreds of years.

Figure 155: Up a side passage off the main pathway in the El Castillo cave in Spain, you will find a single panel consisting of a penniform (named after the Latin word for "feather") sign in black, surrounded by five bell-shaped signs in red. The ledge framing the bottom of this tableaux gives it an almost altar-like appearance.

"How to find the common signs that bridged dozens of caves and 30,000 years of human habitation." Over seven months, in 2013 and 2014, von Petzinger and her husband visited and documented fifty-two cave sites in Spain, Portugal, France and Italy. She selected sites where researchers had noted geometric signs but hadn't recorded specific details. After her fieldwork was complete, she analyzed all the signs she'd catalogued. What she saw made the hair on the back of her neck stand up: across 30,000 years and the entire continent of Europe, a mere thirty two signs repeated themselves over and over. How could she explain this continuity? "That really small number tells you that they must have been meaningful to the people who were using them, because they were replicating them," she says (D'Arcy, 2017).

Symbolics

The Venus of Laussel represents a combination of visual and abstract symbols, denoting not merely the pregnancy of a loved one, but also its possible monthly lunar durations. We see the coming together in the visual symbolic world, of an important event in the lifetime of an individual, and also a desire to record symbolically and abstractly the timeline of this important event.

This carving represents a milestone in the evolution of *Homo*. Humans now have the mental capacity to create an abstract system of symbolic relationships, meaning in parallel with the physical representation of an important series of human events: the fertilization of life through the sexual act; the vulva; the consequent formation of a fetus, represented in the expansion of the female breasts and tummy; the carved lines on an apparently significant carved object, the horn, the concept of periodicity, the thirteen slashes.

This carving, along with the symbols found in the cave carvings, tells us that the human mind is on the way towards a written language—sets of mathematical symbols. In fact, we have entered the halls of civilization—a unique biological creation by *Hss*—whatever it may mean for Darwinian "natural selection."

Figure 156: The Venus of Laussel
Venus of Laussel in Bordeaux museum.

Figure 157: Detail of the head.

Figure 158: Detail of the right arm and the horn.

Figure 159: Detail of the left arm and hand.

The Venus of Laussel is an 18.11-inch-high (46.0-centimetre) limestone bas-relief of a nude woman. It is painted with red ochre and was carved into the limestone of a rock shelter (*Abri de Laussel*) in the commune of Marquay, in the Dordogne department of south-western France. The carving is associated with the Gravettian Upper Paleolithic culture (approximately 25,000 years old). It is currently displayed in the Musée d'Aquitaine in Bordeaux, France.

The figure holds a bison horn, or possibly a cornucopia, in one hand, which has thirteen notches. Alexander Marshack said about the Venus of Laussel that "One cannot conjecture on the basis of one engraved sequence any meaning to the marks, but that the unusually clean horn was notated with storied marks is clear." She has her hand on her abdomen (or womb), with large breasts and vulva. There is a "Y" on her thigh and her faceless head is turned toward the horn.

Climate Change—Exodus

The withdrawal north of the final and sharpest millennia of the Wurm glaciations vastly accelerated the dynamics of human hybridization increasing the hidden movements of people as well as solidifying the character of the human species. Two great land bridges filled up with melted ice, new inland passageways opened up., As before, anthropoid hominins looked to the skies, to the green plants, consecutively warmed by the sun, frosted by the moon. A new cycle of climate began, a new series of cultural foci were experimented with, agriculture.

In Europe, the upper Paleolithic cultures' flexibility in the economics of hunting and gathering, and the resultant vacation time for art disappeared. While much of those thirty or so thousands of years were not terribly cold below the southern extent of the ice flows, the summers likewise were not warm, autumnal like, good for the foraging of the great animals for hunting and not overly demanding in terms of the need for shelter. The new Holocene climate of cold and snowy winters, the summers, hot and dry, changed much.

The new Mesozoic period saw the landscape changing with new dense forests hiding their prey and drying up their leisure time for the old arts. Many groups picked themselves up in much greater numbers than the ever migrating humans had experimented with during the Ice Age, c.45-12k BP. As much as there was love for the old way of life, the fun of the hunt, the leisure in the caves with the paints and one's woman, the word came back year after year to the Mesozoic stay at homes that the weather, food sources, opportunity, lay south, to coasts of the Mediterranean and the rivers of Africa and Asia.

Some had wandered to the east before the melt down (the Amero-Indians), and others travelled north over Sicilyan or Gebralter pathways. The Khoisanoids/Capoids moved into Spain. After the melt, we have the Eskimos, Aleuts, Laplanders, Scandinavians (Vikings) heading to the natural resources of the Nordland. Going in the other direction, south, Cro-Magnons moved into the Atlas Mountains of Morocco, eventually to be called Berbers. Some time into the future, Europeans would sail west into the Atlantic Ocean to meet the Amero-

Indians, then from the "new continent," would return east to enslave the people of West Africa.

A great transformation

Let us go back in time to the era, some forty five thousand years ago, when the Cro-Magnons were entering Europe with a fully articulated and sophisticated culture, replete with abstract ideas, concepts, curiosities, arts, and advanced Aurignacian technology. They had already developed the full panoply of a non-literate civilization, in miniature. They were also on the way to literacy, if we follow the new interpretations of the non literal markings. The economics, climate, and demography of their circumstances precluded this final advance into symbolic written language.

The Neanderthals were driven into extinction, their genes absorbed into Europe and West Asia. The evidence from their culture and the analysis of scholars, whose arguments we have noted on this issue in earlier chapters, is rich. The current sense is that the Neanderthals did not have discursive speech—the ability to talk about their world. Other peoples at this point in human evolution might well have had speech—certainly the Cro-Magnons, and probably the Khoisanoids of Africa.

These Khoisan ancestors of the Bushman practiced art, painting jewelry with decorations. These cultural involvements possibly go back some eighty to one hundred thousand years, throughout their African homeland. Also in Africa, the tools of Cro-Magnon have been found deep in the jungle areas of the Central African Republic. It is quite possible that the genes of these Cro-Magnon wanderers were passed on. Carleton Coon, who has written the most definitive works on the evolution of the races of humanity, argued that morphologically (physically), the Caucasoids (Cro-Magnons) and the Negroids have the closest affinities.

Beyond the above hypothetical comments, we know little of the Negroid cultures at this remote stage of human history, 400–100,000 BP slice of time. However, at some later point, they emerged from the West African jungles moved south and north, this before the Caucasoid invasions of Africa in historical times. They had already begun to push the Khoisanoids aside, and soon dominated the contested landscapes.

We have earlier argued that there is good evidence that the so-called "genetic blockade" or the "funnel metaphor," marking the origins in Europe or West Asia of *Homo sapiens sapiens*, could have taken place some 400–200,000 BP. This estimate means *Hss* had much time to find their own basic nature, to split up and wander the earth in searching bands, always distributing their genes, as would any migrating group, wherever and whenever they planted their feet. Updated: First big efforts to sequence ancient African DNA reveal how early humans swept across the continent [*Science* Elizabeth Pennisi, Sept. 21, 2017

During this period, there were many groups of peripherally evolved humans, if we time log their fossil remains carefully. Some of these skulls are rather impressively on the way towards sapiency. But their tools are still transitionally fabricated, with little evidence of being on a par with the Neanderthal Mousterian, and were hardly superior.

The important question for a general understanding of human biological and cultural evolution is: were these ancient humans, 300–200,000 BP, on a par with those human folks living today, all referred to taxonomically by modern academic tradition as truly *Homo sapiens*? And if so, we must ask, what are the hard and fast scientific taxonomic requirements for gaining full standing as *Homo sapiens?*

Considering that we today do call all humans on this planet *Homo sapiens*, can we reach a clear understanding of human biological and cultural evolution by throwing all these variously progressive/conservative forms of *Homo* into one grab bag of a species? And then, there flitters in front of our eyes an even iffier issue. By what taxonomic criteria do we assign as we have throughout this writing, the *Homo sapiens sapiens* sobriquet!?

Sapiency and language

We have alluded to the possibility that a partial hybridization of the African sub-species of humankind by the Cro-Magnons took place early in their hegira out of the genetic blockade. They then underwent their heterochronic transformation into a paedomorphic transformation of *Homo*, c. 400–200,000 BP. This ancient hybridization may have given these African peoples the capacity for spoken language. Elsewhere in the world, and here represented by the non-spoken abilities of the European Neanderthals, argues for similar human transitions/hybridizations and extinctions around the world around the time of the invasion of the *Hss* Cro-Magnons into Europe, c. 45,000 BP.

Go forward some forty-five thousand years, and we have a completely different and now universal spoken language situation. All the humans around the world can speak in a discursive information/communicative manner. Can there be any doubt that language is the unique adaptive skill in modern humans that not only separates them from all other forms of animal life, perhaps even from their cousins whom we call hominins, and sometimes erroneously add the moniker, sapients?

Luigi Cavalli-Sforza has done extensive research not only into the genetics of human evolution, but especially into the study of the relationship of our contemporary languages to each other. The following quotes from his work supports the hypotheses about origins and relationships put forth above.

> The genetic information for this work came from a very large collection of gene frequencies for "classical" (non-DNA) polymorphisms of the world aborigines. The data were grouped in 42 populations studied for 120 alleles. The reconstruction of human evolutionary history thus generated was checked with statistical techniques such as "boot-strapping." It changes some earlier conclusions and is in agreement with more recent ones, including published and unpublished DNA-marker results. The first split in the phylogenetic tree separates Africans from

> non-Africans, and the second separates two major clusters, one corresponding to Caucasoids, East Asians, Arctic populations, and American natives, and the other to Southeast Asians (mainland and insular), Pacific islanders, and New Guineans and Australians.
>
> Average genetic distances between the most important clusters are proportional to archaeological separation times. Linguistic families correspond to groups of populations with very few, easily understood overlaps, and their origin can be given a time frame. Linguistic super families show remarkable correspondence with the two major clusters, indicating considerable parallelism between genetic and linguistic evolution. The latest step in language development may have been an important factor determining the rapid expansion that followed the appearance of modern humans and the demise of Neanderthals.
>
> (Cavalli-Sforza et al. 1988, p. 6002–6006)

A diagrammatic exposition of these linguistic and presumably genetic relationships, here put forth by Cavalli-Sforza, reinforces his views quoted above.

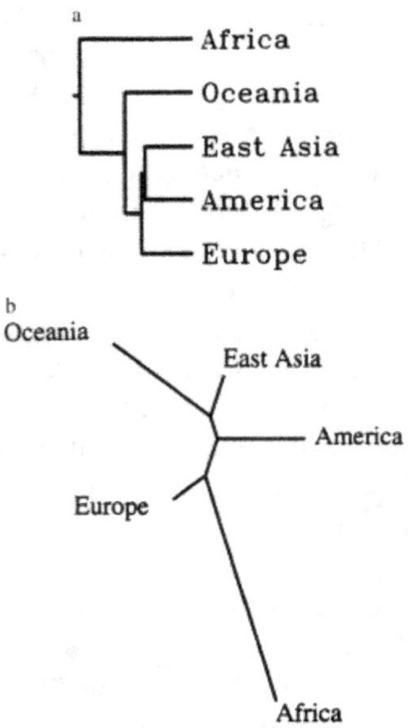

Figure 160: [Cavalli-Sforza—Genetic/linguistic trees—the longer the line the greater distance from the others, else, the oldest separattion' *PNAS*. July 22, 1997]

Clearly, the transition from guttural emotional feelings to intentional, factual alliteration reinforces the newer concentration of research not so much on the vocal structures, pharynx, phonological production but now to the brain structures which allow for syntax and grammar to form our cultural vision of reality. This research trend puts the onus of this communication and thus taxonomic revolution onto the genetics of brain structure and function.

There is an excellent discussion by J.J. Changeux of this non-linear evolution of the genetic relationships to our brain structure and function. One cannot see genes as indivisible entities. There is enormous connectivity with other neurological functions and relationships. Thus the expansion of the human sapient brain and its various interconnecting structures are almost impossible to deal with as surgical or singular explanatory or experimental entities. (Changeux 2005)

The heterochronic reconstruction of the sapiens "body and soul," in creating this neotenous/paedomorphic creature, Cro-Magnon, we now dare to call *Hss*, *Homo sapiens sapiens*. In nature's creation of human speech and then the written abstract symbol, we have seen a revolutionary reshaping of both our physical world and our biological nesting. In a sense we have through a new form of vocal communication and then after a great social (agricultural/urban) transformation of human life, experienced the creation of written and numerical symbols which followed in time.

These hetrochronic events have sharply separated us from the trend lines of our biological heritage and as a consequence, its general evolutionary disciplines.

Perspective

In Darwin's thesis evolutionary success is achieved when living forms are able to metabolize enough energy to reproduce and bring the young to reproductive maturity. The evolutionary dynamics that created *Hss* has as always been definitive as well as cruel. So many once functional forms of life have "bitten the dust." So too, this applies to the ancestors of *Hss*. For the past five to ten million years, a mammal, hominid/hominin form of anthropoid life, has been gaining positive selective attention by "Mother Nature." In the selective process, many once-viable lines of hominids, followed by all hominins, have departed this planet.

Sadly, the evidence is that the physical elements of life, climate, geology, have not been at the causal center of the demise of so many ancestors (cousins) of *Hss*. Indeed, it has been hominids, hominins in succession, with ever larger brains and more dynamic intelligence, all contending on this same ecological road for survival, that have pushed these many in to the ditch of selective obsolescence. The question is, as we work our way forward in evolutionary time, when will the next push into extinction take place within our clade? Our requirement for survival is that we understand, abstractly and then concretely, what were and are the dynamics that pushed less brainy forms of humans into selective extinction. It might be wise to examine carefully this suggestive taxonomic distinction, here proposed. That is, the explicit intellectual separation between *Homo sapiens* and *Homo sapiens sapiens?*

With *Hss*, the basic physical set of life-maintaining requirements, once undergirded by the surety and steadfastness of instinct and then social tradition, has been dissolved. We increasingly seem to exist and survive by dint of our symbolic, cognitive choices. Here, the adaptive survival equation is precipitated and determined by the abstract symbolic choices of our ever new social life style. Survival can now be negated by suicidal symbolic decisions which override the rational mind and the long distant cultural, social, and economic patterns and decisions within the matrix of our community life. Darwin's dynamic principles would still be vindicated by policies—decisions which could push the larger human species over the cliff. Here, the cause—a permanent, irrational loss of the basic awareness and survival needs of this new sub-species, *Hss*.

Knowing: the religious and philosophical question

This new mentality in nature, represented by the loss of our instinct, decisions made on presumably successful causal insights is revolutionary. What started out in the brain structure of mammals, to give them a wider field of behavioral choices within the generation, rather than the close adaptive honing of the genes, as in most other living forms, here, all four kingdoms, has broken with this billions of years old natural manifesto,

Can we venture to say, in terms of the import of this new *Hss* brain, is that amongst all previous life forms, excepting only the revolutionary eukaryote cell which allowed for a genetic nucleus, sexual reproduction and the creation of metazoan multi-cellular forms of life, only *Hss* has had an equally revolutionary impact on the nature of the struggle and destiny of living forms in the fight against the implications of Carnot's Second Law of Thermodynamics, *Entropy.*

We humans now live to change nature's world through the causal knowledge *Hss's* civilizations have attained. Could one have predicted that the evolution of that ancient notochord, perhaps 600 mya, could have created this modernity? Was this enormous powerhouse of a brain latent, back then in the implicit powers of the mind, to know. We know in order to predict beyond the ever altered conditions of our lives, both cultural and bio-chemical.

The dots and inscribed lines of the Cro-Magnons were the first attempts of this new mentality to question causes and regularities. One can call this search for prediction as the attempt to discover and record the universal laws undergirding our existence on this planet.

We are still searching to find the laws that will solve the various social and economic challenges to our decision making. When the first urban civilizations began to be formed and the complexities of life raised heads full of questions we created gods to sort things out, at the least mentally. This pantheon of powers ruling beyond our practical horizons quickly demanded human purveyors of these powers, priests, mullahs, rabbis and their tangible institutions, always needing the tangible refreshments of the blood and sweat of the small people. This need to know of humans in humility created brick and mortar enclaves, always a militant power structure. These blood and bones representatives of the gods always failed in their bloody predictions.

These theological questions about our humanness in the material world had only engendered institutional responses after humans had settled into a more stable urban and agricultural world. The Greeks had the genius to attempt to turn away from human or animal gods towards a more secular study of mankind in the universe. The scientific and philosophical questions there raised were epochal and demanded much from the working intelligence of those impacted populations.

Another mere fifteen hundred years (to the 17th century) and the method of scientific inquiry was introduced. Here these now perennial human questions about the nature of their own kind and the workings of this universe could be proposed, hypothesized and then put to an empirical, factual test. In this way an enormous dynamic of change was now precipitated, here, again amongst the *Hss* descendants of the Cro-Magnons. But still many philosophical questions about we humans and our relationship to ourselves and nature about us remained up in the air, blurred by ideology and theology. The scientific mentality could do only so much.

Darwinism and Existentialism

An example of the manner of thinking about this mysterious creation, *Hss*, in terms of the relative failures of both religion n and philosophy to clarify and becalm the turbulent waters of doubt and fear is the 20th century movement called Existentialism.

They propose: "We are alone, we are determined to think in terms of the human being as making decisions based on no outside authority." Jean-Paul Sartre, Simone de Beauvoir, Maurice Merleau-Ponty, and Albert Camus were the intellectual leaders. Existence precedes essence; human beings—through their consciousness—create their own values and determine a meaning for their life because the human is not possessed by any inherent identity or value. That identity or value must be created by the individual. By posing and engaging in acts that constitute identity and value, each person makes their existence more significant.

In the Darwinian and contemporary scientific analyses of living behaviors, genetics determine the responses to the outside world of the individual living cell. In humans up to *Hss*, the genetic structure is still specific and clear in its relationship with precipitated behavioral and social responses. There is a given set of limits, determined anticedently by the genes. Thus, in each functioning species, there exists a fixed instinctual quorum of responses that have worked adaptively and then selectively over the eons.

With *Hss*, this billions of years old calculus has largely been dissolved. An enormous and unfathomable complex repertoire of behaviors have, in *Hss* humans, been released into possibility, most seemingly having little to do with the day by day survivalistic (adaptive?) needs of the individual and society. Blame this on the enormous growth of the cortex, pulling an emotional atomic bomb, the limbic system, along with it.

What else about ourselves and the universe have we yet to discover? The answer one that we don't know, and can't predict. Human nature has produced vast changes in our historic cultures. This is our biological reality, the ever new symbolic meaning into the unknown.

Part II – The Age of Great Civilizations

Chapter 10

Sumer: *Hss* in the Holocene

Interstadial ecology

The years between the climax of the Magdalenian Cro-Magnon cultural epoch and the onset of the Holocene interstadial, a time of warming marked a great shift in the story of human evolution. The onset of distinct seasonal temperatures and climate set off a new dynamic. The period c. 12,000–10,000 BP must have seen a sharp increase in *Homo sapiens sapiens* migrations to the less snowy south and the magnet of warm water—the seas and the rivers. All were beckoning to these travelers, offering an easier life style. They did need to change their patterns of economic sustenance. Agriculture was the defining invention. It made possible stable urban life and a new direction for the *Homo sapiens sapiens*' mind, leading to the creation of a new form of civilization.

Origins

Perhaps it was being there at the right moment in the evolving human technological, economic, and social pattern of life. The Sumerians were game changers. The name comes from Akkadian, the language from the north of Mesopotamia, and means "land of the civilized kings." The Sumerians called themselves "the black-headed people" and their land, in cuneiform script, was simply "the land" or "the land of the black-headed people" and, in the biblical Book of Genesis, Sumer is known as Shinar. Excavations from al-Ubaid and other sites throughout southern Iraq have uncovered stone tools from the Ubaid people (probably non-Sumerians from c. 6000 BCE), such as hoes, knives, adzes and clay artifacts which included sickles, bricks, painted pottery, and figurines. Since the Mesopotamians believed that the gods had set everything in motion, and that human beings were created as co-laborers with the gods to maintain order and hold back chaos, the early writers of history in the region concentrated more on the links between the rulers and their gods. Whenever the Sumerian civilization was first established in the region, by 3600 BCE, they had invented the wheel, writing, the sail boat, agricultural processes such as irrigation, and the concept of the city (though China and India also lay claim to the title of "the first cities" in the world). It is generally accepted that the first cities in the world rose in Sumer, and, among the most important, were Uruk, Ur, Larsa, Isin, Adab, Kullah, Lagash, Nippur, and Kish. Eridu was acknowledged by the Sumerians as their first city (at their peak, c. 2100 BCE, populations of cities reached 50–80,000 people).

Trade was firmly established with foreign lands at this time (3600 BCE) and writing evolved from pictograms to cuneiform script. It is thought that trade was the main motivator in the development of writing as there now had to be some means for accurate, long-distance communication between the merchants of Sumer and their agents abroad.

In Samuel Noah Kramer's book, ***History Begins at Sumer***, he lists thirty-nine "firsts" in history from the region, among which are the first schools, the first proverbs and sayings, the first messiahs, the first Noah and the Flood stories, the first love songs, the first aquariums, the first legal precedents in court cases, the first tale of a dying and resurrected god, the first funeral chants, first biblical parallels, and the first moral ideas. The Sumerians essentially invented time, in that their sexigesimal system of counting (a system based on the number 60) created the 60-second minute and the 60-minute hour.

They also divided the night and day into periods of 12 hours, set a limit on a "work day," with a time for beginning and ending, and established the concept of "days off" for holidays. The historian Bertman writes, "The hand of Mesopotamia still determines the hourly length of the traditional workday and even the length of our electronic entertainment (half-hour or hour TV shows) when our workday has stopped." (Bertman 2005, p. 334) Bertman further notes that the modern day practice of checking one's horoscope comes from ancient Sumer and that the astrological signs one is born under were first noted and named by the ancient Mesopotamians. Ur-Nammu wrote the first legal code in Sumer (city of Ur, c. 2100 BCE), which became the precedent for the much later, and better known Code of Hammurabi of Babylon. The historian Paul Kriwaczek writes, "Ur-Nammu's universal legal pronouncements present a good example of the unifying drive of [city of] Ur's kings: the compulsion to regulate every aspect of life." (Kriwaczek 2010, p. 149) Sumer, under the unifying force of the Third Dynasty of Ur, (c. 2200 BCE) became a Patrimonial State ("meaning one constructed on the pattern of the patriarchal family ruled by a father figure," as Kriwaczek notes), in which the monarch served as the father figure who guided his children along a proper path toward prosperity.

Civilization-writing

Figure 161: The Kish tablet, a limestone tablet from Kish with pictographic, early cuneiform, writing, 3500 BC. Possibly the earliest known example of writing. Ashmolean Museum.

Smelting copper, board games, sailboats were created by c. 3500 BCE. Writing started with pictorial cuneiform, known as proto-cuneiform, which later became quite complex with an increased level of intangibility. The signs which indicated the name of cities, gods, nature, etc. were known as determinatives. Representations got a little simplified by 3000 BC. The strokes through stylus now conveyed concepts of words and not the particular signs. There was a reduction in the number of characters, from over a thousand to six hundred. With this deduction, the form of writing became more phonological. In the Archaic cuneiform, the direction of the early writing was from left to right. A wedge-topped stylus became dominant and gave the writing greater clarity. This cuneiform practice became orthodox in all the successor political transformations of Mesopotamia. People used it for more than three millennia, until the alphabetical form replaced it in the Roman Era.

(Left) Figure 162: The Mask of Warka (named after the modern village of Warka, located close to the ancient city of Uruk), also known as the Lady of Uruk, dating from 3100 BC, is one of the earliest representations of the human face. The carved marble female face is probably a depiction of the goddess Inanna. It is approximately 20 cm (8 inches) tall, and was probably incorporated in a larger wood cult image, though it is only hypothetical that the face is that of one of Sumer's goddesses. It is without parallels in the period. It is owned by the National Museum of Iraq, recovered undamaged during the US-Iraq conflict in 2003.

Sumer renaissance

The oldest surviving law in the world was written in the Sumerian language. The copy of this code was found in Nippur. It is the earliest existing legal text. It came into existence three centuries before the Code of Hammurabi. The laws are listed in a way that the crime is written after the punishment. This way of writing became the form of listing laws in the generations that followed. Moreover, the code gives a peep into the societal structure that existed during the times of Sumerian Renaissance, c. 2100 BCE. Listed below are some laws that the Code of Ur-Nammu had:

1. If a man commits a murder, that man must be killed.
2. If a man commits a robbery, he will be killed.
3. If a man commits a kidnapping, he is to be imprisoned and pay 15 shekels of silver.
4. If a slave marries a slave, and that slave is set free, he does not leave the household.
5. If a slave marries a native (i.e. free) person, he/she is to hand the firstborn son over to his owner.

6. If a man violates the right of another and deflowers the virgin wife of a young man, they shall kill that male.
7. If the wife of a man followed after another man and he slept with her, they shall slay that woman, but that male shall be set free.
8. If a man proceeded by force, and deflowered the virgin slave woman of another man, that man must pay five shekels of silver.
9. If a man divorces his first-time wife, he shall pay her one mina of silver.
10. If it is a (former) widow whom he divorces, he shall pay her half a mina of silver. (Adhikari 2018)

Figure 163: Below, during 2112–2004 BCE Third Dynast of Ur

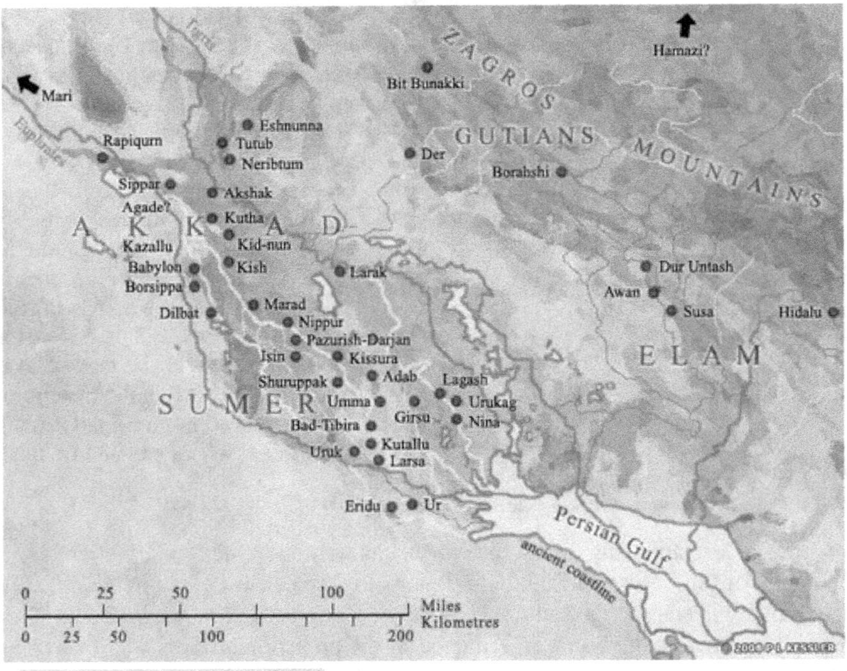

Figure 164: Gudea of Lagash, Lugal (Head Man), c. 2090 BCE

(Left) Figure 165: Portrait of Ur-Ningirsu, son of Gudea, c.2000 BC. Louvre Museum

(Top) Figure 166: Early writing tablet recording the allocation of beer, 3100–3000 BC

Figure 167: (Top) Reconstruction of a woman's headdress

(Top) Figure 168: A reconstruction in the British Museum, of headgear and necklaces worn by women, as found in some Sumerian graves

Figure 169: The Guennol Lioness (c. 3000 BCE) Limestone. Private Collection. Rare Sumerian sculpture of a lioness woman

Figure 170: Ram in a Thicket (c. 2600 BCE), British Museum. One of a pair excavated from the Great Death Pit, at Ur. A rare and exquisite example of Sumerian metalwork of the Third Millennium BCE. (Keiser Institute)

War and Peace: Civilization's Gift

Figure 171

Battle formation

The first war recorded in any detail was between Lagash and Umma (25 miles apart) in c. 2525 BCE on a stele called the Stele of the Vultures. It shows the king of Lagash leading a Sumerian army consisting mostly of infantry. The infantry carried spears, wore copper helmets with leather inserts, and carried rectangular shields. The spearmen are shown arranged in

what resembles the phalanx formation, which requires training and discipline; this implies that the Sumerians may have made use of professional soldiers. Bottom shows Lagash infantry following victorious Lagash King Eannatum riding on chariot.

(Left) Figure 172: Pear shaped mace head early military weapon, 3000 BCE.

Figure 173: The Standard of Ur War

The standard of war, made of lapis lazuli, Ur, c.2650 BCE

The armies of Sumer and Akkad represented the pinnacle of military development in the Bronze Age. No army of the same period could match the Sumerian's in military effectiveness and weaponry. The Sumerian civilization produced no fewer than six major new weapons and defensive systems, all of which set the standard for other armies of the Bronze Age and Iron Ages. Few armies in history have been so innovative. (Gabriel and Metz 1992)

Strategic Studies Institute
U.S. Army War College
(Online version by USAF Air War College)]

The Standard of Peace, Ur, c. 2650BCE

Figure 174: Hunting scene, 3100-2900 BCE

The Standard's Peace side has a completely different theme from the War side. Its two lower registers illustrate the bounty of the land. The bottom one depicts men carrying produce in bags on their shoulders and in backpacks supported by headbands, as well as men leading onagers by ropes. The second register shows men leading bulls and caprids (sheep and goats) and carrying fish—presumably the produce of the pastures, rivers, and swamps. The upper register depicts a royal banquet. The ruler, wearing a kilt composed of tufts of wool, is shown larger in scale than the others—the center of attention. The other banqueters, who wear plain-fringed kilts, face him and raise their cups together, while attendants provide food and drink. Banqueting in early Mesopotamia usually involved music. A lyre player and a singer, distinguished by their long black hair, stand to the right of the scene.

Figure 175

Evolution of Cuneiform

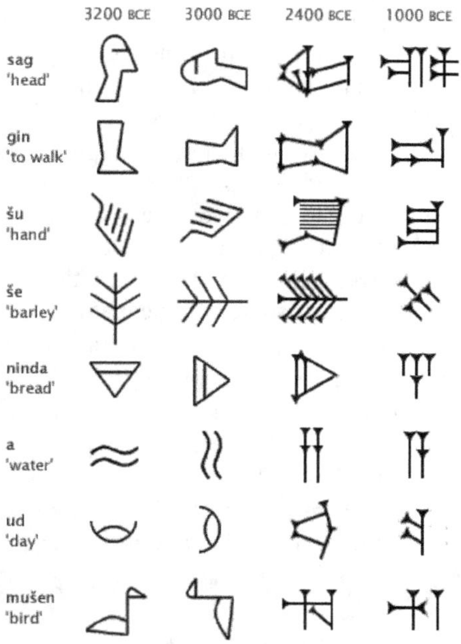

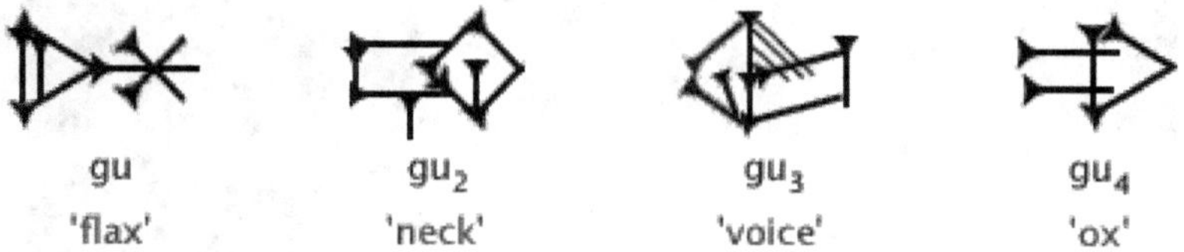

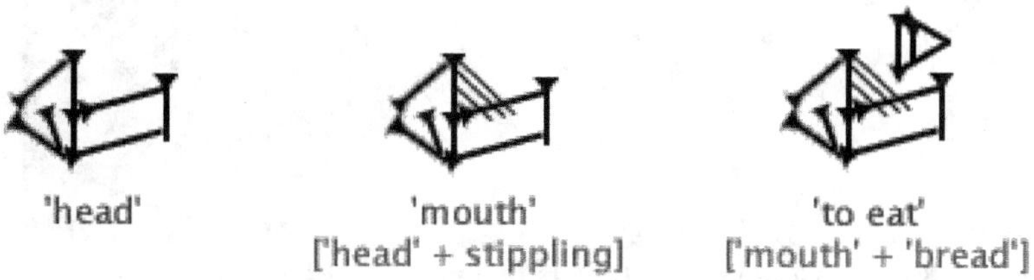

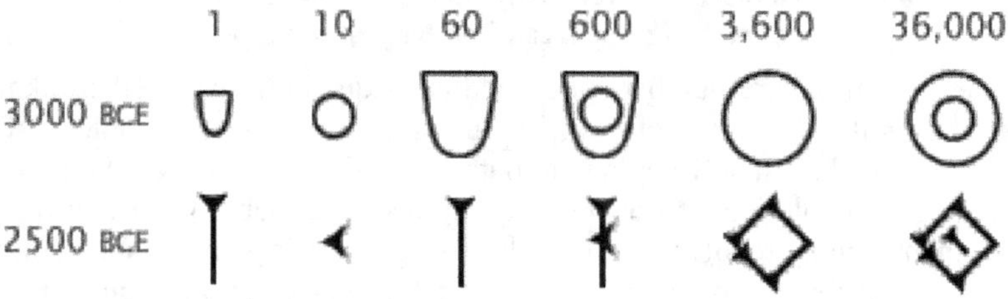

Figure 176 to 179: (Above)

The key is to note the increasing abstractive character of the symbols, moving away from the pictorial.

The story of mathematics

The Babylonians, following on the Sumerians, invented another revolutionary mathematical concept (something that the Egyptians, Greeks and Romans did not have)—a circle character for zero, although its symbol was really still more of a placeholder than a number in its own right.

We have evidence of the development of a complex system of metrology in Sumer from about 3000 BCE, and multiplication and reciprocal (division) tables, tables of squares, square roots and cube roots, geometrical exercises and division problems from around 2600 BCE onwards. Later Babylonian tablets dating from about 1800 to 1600 BCE cover topics as varied as fractions, algebra, methods for solving linear, quadratic and even some cubic equations, and the calculation of regular reciprocal pairs (pairs of number which multiply together to give 60).

One Babylonian tablet gives an approximation to $\sqrt{2}$ accurate to an astonishing five decimal places. Others list the squares of numbers up to 59, the cubes of numbers up to 32 as well as tables of compound interest. Yet another gives an estimate for π of 3 ⅛ (3.125), a reasonable approximation of the real value of 3.1416—pi, in mathematics, is the ratio of the circumference of a circle to its diameter. The symbol for pi is π. The ratio is the same for all circles.

Figure 180: Babylonian Clay tablets from c. 2100 BCE showing a problem concerning the area of an irregular shape

The idea of square numbers and quadratic equations (where the unknown quantity is multiplied by itself, e.g. x^2) naturally arose in the context of the measurement of land, and Babylonian mathematical tablets give us the first ever evidence of the solution of quadratic equations. The Babylonian approach to solving them usually revolved around a kind of geometric game of slicing up and rearranging shapes, although the use of algebra and quadratic equations also appears. At least some of the examples we have appear to indicate problem-solving for its own sake, rather than in order to resolve a concrete practical problem. (Mastin 2010)

Mesopotamian perspective

Ecology: In ancient Mesopotamia, a land of blazing sun and very little rainfall, irrigation was vital for farming. Centuries before the beginning of known history, the Sumerians undertook the stupendous task of building embankments to control the floodwaters of the Euphrates River. Gradually, they drained the marshes and dug irrigation canals and ditches. Large-scale cooperation was needed to build the irrigation works, keep them in repair, and apportion the water. This need gave rise to government and laws.

The rich soil produced abundant crops of barley, emmer (a kind of wheat), beans, olives, grapes, and flax. For the first time, there was a surplus to feed city workers, such as artists, craftsmen, and merchants. This great change in living habits brought about modern civilization—defined as a city-based society held together by economic enterprises. There were no nations then, only small city-states.

Economy of the Gods: Each Sumerian city rose up around the shrine of a local god. As a reflection of a city's wealth, its temple became an elaborate structure. The temple buildings stood on a spacious, raised platform reached by staircases and ramps. From the platform rose the temple tower, called a ziggurat (holy mountain), with a circular staircase, or ramp, around the outside. On the temple grounds were quarters for priests, officials, accountants, musicians, and singers; treasure chambers; storehouses for grain, tools, and weapons; and workshops for bakers, pottery makers, brewers, leatherworkers, spinners and weavers, and jewelers. There were also pens for keeping the sheep and goats that were destined for sacrifice to the temple god.

Horses and camels were still unknown here in the south, but sheep, goats, oxen, donkeys, and dogs had been domesticated. The plow had been invented, and the wheel, made from a solid piece of wood, was used for carts and for shaping pottery. Oxen pulled the carts and plows; donkeys served as pack animals. Bulky goods were moved by boat on the rivers and canals. The boats were usually hauled from the banks, but sails also were in use. Before 3000 BC, the Sumerians had learned to make tools and weapons by smelting copper with tin to make bronze, a much harder metal than copper alone.

Mud, clay, and reeds were the only materials the Sumerians had in abundance. Trade was therefore necessary to supply the city workers with materials. Merchants went out in overland caravans or in ships to exchange the products of Sumerian industry for wood, stone, and metals. There are indications that Sumerian sailing vessels even reached the valley of the Indus River in India. The chief route, however, was around the Fertile Crescent, between the Arabian Desert and the northern mountains. This route led up the valley of the two rivers, westward to Syria, and down the Mediterranean coast.

Agriculture: Early in Sumerian civilization, eighty to ninety percent of those who farmed did so on land they considered theirs rather than communal property. Here, too, the Sumerians were expressing a trend that was common among others.

Another individual effort was commerce. With a growth in commerce, the Sumerians had begun using money, which made individual wealth more easily measured and stored. Commerce required initiative, imagination, an ability to get along with people, and luck. Of course, some merchants were more successful than others. Farming took stamina, strength, good health, good luck and organization. Once again, some farmers were more successful than were others.

Those farmers who failed to harvest enough to keep themselves in food and seed borrowed from those who had wealth in surplus. Those who borrowed hoped that their next harvest would give them the surplus they needed to repay their loan. However, if the next harvests were also inadequate to meet their obligations, they might be forced to surrender their lands to the lender or to work for him. When Sumerians lost their land, they or their descendants might become sharecroppers: working the lands of successful landowners in exchange for giving the landowners a good portion of the crops they grew.

Urban Development: Within a few centuries, the Sumerians had built up a society based in twelve city-states: Kish, Uruk (in the Bible, Erech), Ur, Sippar, Akshak, Larak, Nippur, Adab, Umma, Lagash, Bad-tibira, and Larsa. According to one of the earliest historical documents, the Sumerian King List, eight kings of Sumer reigned before the famous flood. Afterwards, various city-states took turns to become the temporary seat of power, until about 2800 BC, when they were united under the rule of one king—Etana of Kish. After Etana, the city-states vied for domination. The great cemetery at Ur, c. 2650 BCE, represents both dynastic cruelty plus significant, tangible, esthetic creativity. The implicit autocracy of this period and contention for supremacy weakened the Sumerians, and they were ripe for conquest—first by Elamites from the northeast, then by up river, blonde Semitic Akkadians.

Survival: The Sumerians had never been very warlike, and they had only a citizen army, called to arms in times of danger. In about 2340 BC, King Sargon of Akkad conquered them and went on to build an empire that stretched westward to the Mediterranean Sea. The empire, though short-lived, still fostered Sumerian art and literature.

Led by Ur, the Sumerians came back again to spread their rule far westward. During Ur's supremacy (about 2150–2050 BC), Sumerian culture reached its highest development. Shortly thereafter, the cities lost their independence forever, and gradually, the Sumerians completely disappeared as a people. Their language, however, lived on as the language of culture. Their writing, their business organization, their scientific knowledge, and their mythology and law were spread westward by the Babylonians and Assyrians

Social Hierarchy: Sumerian society was thought to have been made up of four social classes: nobles, commoners, clients and slaves. The nobles included the ruling family, priests, warriors and families who owned large areas of land, and who were usually very wealthy. Commoners owned smaller amounts of land. Slaves were usually local criminals, others who had wronged the social system. Wealthy land owners were above the artisan working classes. The status of women was a mixture of respect and secondary status. However, the town of Kish was once ruled not by a king, but by an energetic queen called Ku-baba, "a former tavern keeper, about whom we know nothing else."

Religion: The Assyriologist Jean Bottero notes that "Not only were these gods the originators of the universe and mankind, but they remained their supreme masters and guided their existence and evolution from day to day. For that reason, they were regarded as the promoters and guarantors of all the infinite obligations—positive and negative—that govern human life." All aspects of Mesopotamian existence were imbued with a sense of the divine at work, even in the clothes that they wore. (Penglase 1994)

Education: Education was first established for the purpose of training the scribes, a person who wrote books or documents by hand as a profession, and helped to keep track of records. This was necessary to satisfy the economic and administrative needs of the land—primarily, of course, those of the temple and palace. This continued to be the major aim of the Sumerian school throughout its existence. However, in the course of its growth and development, it came to be the center of culture and learning in Sumer.

The temples were called "edubbas," meaning "tablet house." The teacher, and or priest, was called "ummia," meaning "expert." Sumerians created the first known formal education system (schools). It is unlikely that you could get into scribe school if you were the son of a lowly farmer. You had to attend school for many years to learn the written language (cuneiform), the number system, and the methods and conventions of a scribe. Much of scribe school consisted of memorizing and copying cuneiform texts from one tablet to another. Mathematical learning was exemplified in the Sexagesimal (base 60) numeral system with sixty as its base. This numerical sysem originated with the ancient Sumerians.

Most education took place in a temple, associated with a priest. It was a place of education for wealthy families, studied and worked in from sun up to sun down. Sumerians only focused on studying complex grammar, and practiced writing. Most, if not all Sumerian students were male. The greater part of the students came from the wealthier families; the poor could hardly afford the cost and the time which a prolonged education demanded.

Mesopotamian Music: The Sumerians music system concentrated on bringing order to melody and out of chaos. Until recently, no form of musical notation had been known to arise from Sumerians. However, there is a cuneiform tablet containing a hymn and what has been translated as musical instructions for a performer, making this tablet the oldest known musical notation. There were strict instructions on how to perform music, similar to chord progression today. These instructions also seem to point to a strong desire for musicians to play in tune, with steps in performing, requiring frequent attempts to tune the instruments.

MS 2340 Lexical list of harp strings, 23 types of musical instruments. Sumer, 26th c. BC.
The earliest record of music and instruments in history

Figure 181

Mesopotamian music inscription

Were it not for the eleven stringed instruments recovered at Ur (two harps and nine lyres), we would not have a single, actual stringed instrument from ancient Sumer-Babylonia. The Ur graves also yielded a pair of silver wind instruments and a small number of other types of instruments; a few other prehistoric Mesopotamian sites have given us fragments of bone wind instruments.

The original wooden and stringed instruments found at Ur were richly decorated or overlaid with gold, silver, copper, lapis lazuli, mother of pearl, and other non-wood materials that did not deteriorate in the earth over the millennia. It is these materials that enabled the excavators to establish the forms and dimensions of the instruments. When the ceilings of the ancient tombs fell in, some of the instruments collapsed together so badly that it was difficult to sort out which pieces went with which. As a result of many years of study by archaeologists and museum researchers, we believe we have correctly separated the different and distinct instruments and improved our assessments of their original sizes. Thus, we now have concrete evidence we can use to replicate these unique lyres and harps (see next page).

Figure 182: DETAIL FROM THE PANEL ON THE BULL-HEADED LYRE showing an 8-stringed bovine lyre being played. At the top of the lyre, braided material is wrapped around the crossbar under the tuning sticks. The small fox-like animal facing the front of the lyre holds a sistrum, or rattle. [UPM 817694. Detail of neg. 735-110]

Figure 183: SCENE ON A GOLD CYLfNDER SEAL from a grave in the Ur cemetery (PG L054). In the bottom register are 2 "cymbalists" (figures playing clappers), a dancer, and a seated figure playing a bovine lyre. The top register shows festive banqueters. (U. 11904. From Woolley 1934, pt. 1: fig. 2. C.2650 BCE3) [Musical Instruments from Ur and Ancient Mesopotamian Music By: ***Anne Draffkorn Kilmer]***

Legacy

Evidence of wheeled vehicles appeared in the mid 4th millennium BC, nearly simultaneously in Mesopotamia, the Northern Caucasus (Maykop culture) and Central Europe. The wheel initially took the form of the potter's wheel. The new concept quickly led to wheeled vehicles and mill wheels.

The Sumerians' cuneiform script is the oldest (along with Egyptian hieroglyphs) which has been deciphered (the status of even older inscriptions such as the Jiahu symbols and Tartaria tablets is controversial). The Sumerians were among the first astronomers, mapping the stars into sets of constellations, many of which survived in the zodiac system we know today, and were also recognized by the ancient Greeks. They were also aware of the five planets that are easily visible to the naked eye, as probably were the Cro-Magnons.

They invented and developed arithmetic by using several different number systems, including a mixed radix system with an alternating base 10 and base 6. This sexagesimal system became the standard number system in Sumer and Babylonia. They may have invented military formations and introduced the basic divisions between infantry, cavalry, and archers. The Sumerians developed the first known codified legal and administrative systems, complete with courts, jails, and government records. The first true city-states arose in Sumer, roughly contemporaneously with similar entities in what are now Syria and Lebanon. Several centuries after the invention of cuneiform, the use of writing expanded beyond debt/payment certificates and inventory lists to be applied for the first time, about 2600 BC, to messages and mail delivery, history, legend, mathematics, astronomical records, and other pursuits. Conjointly with the spread of writing, the first formal schools were established, usually under the auspices of a city-state's primary temple.

Finally, the Sumerians ushered in domestication with intensive agriculture and irrigation. Emmer wheat, barley, sheep (starting as mouflon), and cattle (starting as aurochs) were foremost among the species cultivated and raised for the first time on a grand scale.

Figure 184: Reconstruction of the Ziggurat at Ur

Figure 185: Ziggurat at Choghā Zanbīl near Susa, Iran, reflects the widespread influence of the Sumerians. ***Robert Harding Picture Library/Sybil Sassoon.***

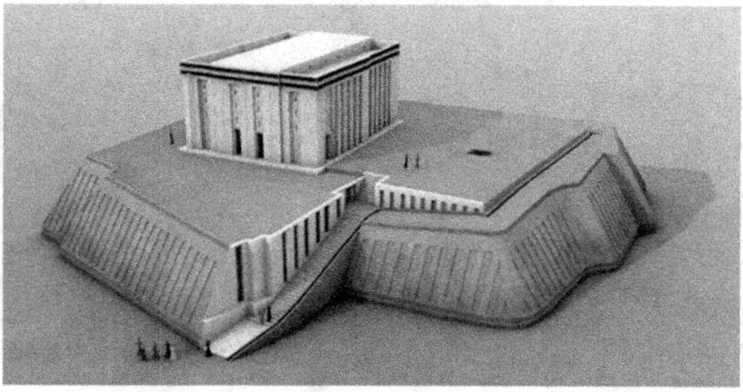

Figure 186: The White Temple and the Great Ziggurat in the Mesopotamian City of Uruk, a modern interpretation

The temple of Inanna—Inanna's House of Heaven, in Uruk—was the greatest of these. This temple is 5000 years old and had been built and rebuilt many times to hold a community of sacred women who cared for the temple lands. The high priestess of Inanna would choose for her bed one she would appoint as shepherd. He would represent Dumuzi, her sacred lover.

Figure 187: Love of Eanna and Dumuzi

Figure 188: Marriage of Eanna and Dumuzi [TangLung published on 16 May 2014]

Figure 189: Ziggurat at Uruk, Contemporary Remains

Figure 190: Queen of the Night: Old Babylonian, 1800-1750 BC—19 inches in height, from southern Iraq. Queen of the Night (Ereshkigal?) perhaps was a symbol or part of a Temple in Ur, devoted to priestly prostitution, a probable continuation of a Sumerian tradition.

This large plaque is made of baked, straw-tempered clay, modeled in high relief. The figure of the curvaceous naked woman was originally painted red. She wears the horned headdress characteristic of a Mesopotamian deity and holds a rod and ring of justice—symbols of her divinity. Her long, multi-colored wings hang downwards, indicating that she is a goddess of the Underworld. Her legs end in the talons of a bird of prey, similar to those of the two owls that flank her. The background was originally painted black, suggesting that she was associated with the night. She stands on the backs of two lions, and a scale pattern indicates mountains.

The figure could be an aspect of the goddess Ishtar, a Mesopotamian goddess of sexual love and war, or Ishtar's sister and rival, the goddess Ereshkigal who ruled over the Underworld, or even the demoness Lilitu, known in the Bible as Lilith. The plaque probably stood in a shrine.

Figure 191: Queen of the Night, detail Of Ereshkigal

The same goddess appears on small, crude, mould-made plaques from Babylonia from about 1850 to 1750 BC. Thermoluminescence tests confirm that the "Queen of the Night" relief was made between 1765 and 1745 BCE (Mark).

The degeneration of mesopotamia

I:196. Such is the equipment of their bodies: and the customs which are established among them are as follows, the wisest in our opinion being this, which I am informed that the Enetoi in Illyria also have. In every village once in each year it was done as follows: When the maidens grew to the age for marriage, they gathered these all together and brought them in a body to one place, and round them stood a company of men: and the crier caused each one severally to stand up, and proceeded to sell them, first the most comely of all, and afterwards, when she had been sold and had fetched a large sum of money, he would put up another who was the most comely after her: and they were sold for marriage. Now all the wealthy men of the Babylonians who were ready to marry vied with one another in bidding for the most beautiful maidens; those however of the common sort who were ready to marry did not require a fine form, but they would accept money together with less comely maidens. For when the crier had made an end of selling the most comely of the maidens, then he would cause to stand up that one who was least shapely, or any one of them who might be crippled in any way, and he would make proclamation of her, asking who was willing for least gold to have her in marriage, until she was assigned to him who was willing to accept least: and the gold would be got from the sale of the comely maidens, and so those of beautiful form provided dowries for those which were unshapely or crippled; but to give in marriage one's own daughter to whomsoever each man would, was not allowed, nor to carry off the maiden after buying her without a surety; for it was necessary for the man to provide sureties that he would marry her, before he took her away; and if they did not agree well together, the law was laid down that he should pay back the money. It was allowed also for any one who wished it to come from another village and buy. This then was their most honorable custom; it does not however still exist at the present time, but they have found out of late another way, in order that the men may not ill-treat them or take them to another city: for since the time when being conquered they were oppressed and ruined, each one of the common people when he is in want of livelihood prostitutes his female children.

Herodotus I:199. *Now the most shameful of the customs of the Babylonians is as follows: Every woman of the country must sit down in the precincts of Aphrodite once in her life and have commerce with a man who is a stranger: and many women who do not deign to mingle with the rest, because they are made arrogant by wealth, drive to the temple with pairs of horses in covered carriages, and so take their place, and a large number of attendants follow after them; but the greater number do thus,--in the sacred enclosure of Aphrodite sit great numbers of women with a wreath of cord about their heads; some come and others go; and there are passages in straight lines going between the women in every direction, through which the strangers pass by and make their choice. Here when a woman takes her seat she does not depart again to her house until one of the strangers has thrown a silver coin into her lap and has had commerce with her outside the temple, and after throwing it he must say these words only: "I demand thee in the name of the goddess Mylitta": now Mylitta is the name given by the Assyrians to Aphrodite: and the silver coin may be of any value; whatever it is she will not refuse it, for that is not lawful for her, seeing that this coin is made sacred by the act: and she follows the man who has first thrown and does not reject any: and after that she departs to her house, having acquitted herself of her duty to the goddess, nor will you be able thenceforth to give any gift so great as to win her. So then as many as* *have attained to beauty and stature are speedily released, but those of them who are unshapely remain there much time, not being able to fulfill the law; for some of them remain even as much as three or four years: and in some parts of Cyprus too there is a custom similar to this.*

Herodotus: The Histories c. 425 BCE

One suspects that Mesopotamia, 5th century BCE, now under the military jurisdiction of the Indo-European Persians, had rigidified in its decline into a state in which the Hellenes' cultural traditions adhere more closely to the status of the female under the Cro-Magnons than the culminating 3000-year Mesopotamian religious controls of the patriarchal priesthood. Now, c. 450 BCE, the Mesopotamian female, even as goddess, is perceived to have only sexual monetary value.

Finale

The Sumerians wandered into the land of the meeting of two rivers, Tigris and Euphrates, c. 4500 BCE. In those days, it was a veritable swamp land. Wherefrom and whose language, we still don't know for sure. But there was something about them that was uniqe. Instead of just living with this great and productive swamp, they undertook its conquering. Gradually, over a period of a thousand years of experimentation and effort, they created something new and powerful—urban civilization. The key period, around 3500 BCE, as writing flourished, was when they could say "wow, what a great idea."

As an academic cultural/historic/religious system of writing and thinking, Sumerian and its tablets long outlived the spoken day to day tongues of the Mesopotamian. This, even after the Chaldean demise. From c. 4000–540 BCE, it was still widely studied. First came Cyrus, the culturally tolerant Persian (see the Jews). Then Alexander and his Macedonians/Greeks, followed by the Seleucid inheritors, who more interested in fostering the cultural integration and military pacifism of conquered states. Finally, came the Persian-speaking (Indo-European) Parthians, with their relatively decentralized and very universally absorptive cultural interests. Yet Sumerian, at the least in the ancient temples, lived on. The last scholars, priests, priestesses and teachers of the Mesopotamian Sumerian heritage are thought to have disappeared from the planet, c. 100 CE, Roman hegemony.

Chapter 11

Sumer 2: First and Greatest Civilization An Interpretation*

Prologue

We will search for originating meanings and momentums, as scholars have uncovered them at the dawning mists of history. Out of the unknown north came the first humans, who had the opportunity and internal impetus to create what we would call a modern, urban civilization.

Sumer arose at the then-extreme southern end of the Tigris/Euphrates Rivers by c. 4500 BCE. Its civilizational discoveries would not have been possible without the biological intelligence that was created for this special taxon of humans, *Homo sapiens sapiens*, perhaps hundreds of thousands of years earlier. As with all animal adaptations created by evolutionary forces impacting on its genetics, time was needed for the creature in question to make full adaptive use of what nature had created. This especially applies to a new form and level of biological intelligence.

External conditions of climate change and ecological opportunity gradually presented themselves. ***Homo sapiens sapiens*** trickled south to take advantage of the new opportunities. Then **commenced** the slow learning to utilize what was inside of this brain and nervous system. It took a few thousand years in these new environs to make the civilizational breakthrough possible.

Indeed, we still know little about this inner nature of ours, a nature that flings itself outward, most often with an unknowing and disastrous exuberance. Then comes nature's inevitable, disciplinary leash. Unquestionably, this explosive investigation of nature's mysteries, and the systematic attempt to master and control them, was first fully demonstrated by the Sumerians.

Making of the great brain

The making and use of tools by the most ancient hominids reflects the basic physical vulnerability of this line of primate mammals. They took the unspecialized route of defense—"stay out of the way"—as a means of survival. Upright, omnivorous scroungers, they had higher than average intelligence, which was derived from a larger brain. But the reward, of mammal domination, was not automatic. Several million years ago, it was merely one of the possible adaptive options for positive natural selection available to the already highly intelligent primates.

Considering the dynamics of environmental and ecological change, a larger brain did work for survival, resulting in this almost **one** billion years of vertebrate perdurance. The tools that humans created were extensions of this extrusive intelligence. Tools went beyond the body in time and space as means **of** obtaining the next meal. This evidences the dawning of the objectivization of experience beyond instinctual automaticity. It argues for a consideration of possibility and predictability. The premeditated fabrication of potentially useful external materials to make a difference in earning this living, was critical for breaking with the instinctual past.

There was more to tool-making than its admittedly crucial, practical usefulness for a physically unspecialized animal. The making of a tool can also give us hints about broader cultural behaviors, even about esthetics. The greatest revolution in cultural behavior coincided with the morphological revolution in brain size and structure represented in the Cro-Magnons of Europe and West Asia. Our theory takes us back hundreds of thousands of years. But the real thing, the fossils and the culture, is first encountered at the time level of c. 45,000 BP. The material results of this as yet hidden biological revolution are then clearly evidenced.

No doubt this revolution in morphology took place much earlier than indicated in the discovered bones and material artifacts. We needed large enough fossil numbers that could signify flourishing communities. Genetic evidence in contemporary northern populations argues for a very small population of ancient founders, which subsequently and slowly spread their genetic imprint throughout the world.

The surprising creation of such a large brain enclosed in an eggshell-thin skull, a *paedomorph*, retaining its infantile characteristics into sexual maturity, constitutes a not rare biological revolution in morphology. The long term consequence in behavior is still reflected in our incapable efforts to come to terms with its nature.

Nature, having negatively selected less adaptive human groups, in competition with the smart ones, allowed enough of the outliers of thin bone and big brain to survive, to procreate. Except for the powerful selective force of language that was rooted in the morphology of tongue, larynx, and brain, modern humans are unspecialized primates. Free of rigid adaptations, our predecessors had to rely on their intelligence, with language, to get them out of whatever vale of annihilation faced them.

A new civilizational threshold

The Cro-Magnons, *Homo sapiens sapiens*, bleached white to help them metabolize Vitamin D from the weak northern sun, were different from other northern Caucasoids. They lacked the heavy bone and, presumably, thick hair, large air-warming nose, **and** short and stubby Eskimo-like adaptations of the

Neanderthals. These Neanderthal characteristics were shaped to ward off the glacial winds. **Physically,** the Cro-Magnons seemed to have originated on the periphery of the northern adaptive zone. The Caucasus region would be one guess as to their origins.

With these humans, our intelligence and our nature were seemingly completed. The cultural and proto-civilizational evidence argues for this climactic evolutionary moment. We need not here repeat the concretia of their cultural achievements: the highly finished tools fabricated for a variety of functions; the painting, sculpture, the sacral settings in which this artistic obsession took place, special caves, most often in interior darkness where only oil lamps provided them the required light.

Even in burial, their connection of thought and wonder about the afterlife unites them with the literate civilizations. They thought no less than the most civilized humans in modern urban settings, about human destiny, birth, death, and immortality. In Russia, in a site dated to c. 28,000 years ago, a male and younger sons were found, dressed in leather, ivory- and bone-decorated clothing as part of a burial ritual, **indicating that they were** clearly regarded with love, wonder, and awe. Other careful burials of young quasi-adolescent women interred with their offspring tell us of the hardships of life, the danger for young child-bearing mothers, and the regard of their families and communities.

The Cro-Magnon, or Upper Paleolithic cultures of Europe and West Asia, extended from c. 45,000 to c. 12,000 years ago. During that time, we see the gradual evolution of their technology, the art, the manner in which they revered the animals they hunted, or with which they were associated. Above all, there is an economic stability, a sense that the food and shelter, the clothing, the life of the small hunting/gathering communities, were in balance with the larger thrust of this new human mind to find external symbolic expression beyond the hunting, scraping tool kit. Art was never detached from the practical—the spear, the dagger, the projectile.

Until the very end of this period, the Magdalenian, the climate generally was glacial, with benign, cool summers, and early spring-like winters that hovered around the freezing level. This allowed for a wide variety of large mammal life to flourish, clearly a magnet for these peoples as they entered Europe from the east.

Their lives, epitomized in their cultural effusions, express our own dawning nature as *Homo sapiens sapiens*. Held in an economic context by this moderate late-Pleistocene glacial ecology, they clearly lived good, stable, disciplined lives.

Across the valley, the Neanderthals crowded at the edge, gradually ceding the better hunting/gathering ranges, no doubt continuously harried, if not ejected, by these aggressive, brainy newcomers. Eventually, the Neanderthals succumbed, disappeared (c. 40,000–27,000 BP), not able to reproduce themselves. No doubt, they left **behind** many hybrid young among the Cro-Magnon dominators.

The Cro-Magnons clearly lived within a context of small numbers in a geographical community, trading with other Cro-Magnon bands. We find amber caches hundreds of miles from their original points of gathering along the northern seas. These humans expressed their wonder of the world around them in chronometry, making incisions on their stone and flint tools to record the rhythms of nature, the phases of the moon, the menarche of the mature female, the geographic settings of the sun at the seasonal climaxes, the comings and goings of migrating animals, **and even** the budding of the flowers. We recognize the sexual energies stimulated by the animal protein, the vulvas that decorated their voluptuous mobiliary stone Venuses. So too, with the decorative sculpturing around the caves.

They apparently were not yet seasoned of mind by social threats to their way of life. Yet, the high intelligence, skills, and esthetics that went into their material culture, discovered after so many thousands of years, tells us much about the civilizational potentialities of *Hss*.

Transition

Eventually, this cool paradise had to end. The economic basis of their proto-civilization was altered, **and there was** a change of climate, **becoming** more continental in nature: wetter, sharper in contrasts. Now, deep snow **created** a foraging crisis for the great mammals: sloths, mammoths, giant elk, saber tooth tigers. The Cro-Magnons do not seem to have been responsible for this wipe-out. Evidence for herding of deer, domestication of the horse **and** the dog, all argue that the Cro-Magnons were inherently "green" in their outlook on nature, conserving and protecting what provided for their living and **which** stimulated their mental/artistic desires.

The semi-sedentary life for which this surplus provided disappeared with the advancing forests of the Mesolithic and Neolithic cultural periods, after c. 12 ,000 BCE. The Cro-Magnons now had to follow the herds and migrations of animals.

In contrast to the Upper Paleolithic, some 35,000 years in length, the transition into the Holocene climatic period (after 12,000 BP), the Mesolithic/ Neolithic cultural traditions were quite dynamic in certain geographies, relatively late in others. Although we see instances of highly developed technology—microliths, painted stones, geometrical and architectural engravings in some cave homes/temples during the Mesolithic—it is probable, at least in the north, that life became much tougher. It was difficult to support the luxuriant high art products of the previous, relatively benign ecological environment. So too, for the sojourners south. The old hunters of the north were forced into a more nomadic way of life, with wandering groups trying to make their living in a variety of lush and desertified contexts of the modern climatic period. {7}

From 12,000–8000 BCE, entirely new economic challenges set in. The old, rich, tundra-like hunting grounds became overgrown with vegetation, and the herds vanished into the forests, some becoming extinct. We see this transition in the cultural detritus from these periods: the Mesolithic, thence the Neolithic. The latter is so named because of the appearance of clay pottery and the more intense hunting, gathering, and herding lifestyle of the south. This gradual change in climate and ecology put pressure on humans heretofore adapted to highland and lowland (Sahara) savannahs. They were being forced to move toward relatively fewer sources of water.

Mesopotamian magnet

It is clear that the movement south of the Cro-Magnons was not the first intrusion of modern types of humans along the Mediterranean littoral and the adjoining river valleys, Danube, Nile, Tigris/Euphrates, Indus. Nevertheless, it is clear from the fossil evidence, burial grounds **and** technology, that the first permanent or semi-permanent communities as found in Palestine, Jericho, in Anatolia, and along the Nile and Tigris/Euphrates reveal a taller, Cro-Magnon-like people displacing the then-dominant Natuftians, people probably hybridized with older human taxons. {8}

Sickles made with flint blades cemented into bone or wood handles can be found from this era. Baskets woven from natural grasses, presumably used to carry the harvested wild grains and naturally occurring vegetables, are also part of the Neolithic economy. Mortars, pestles and grinding slabs suggest knowledge of husking the grains, as well as evidence for roasting, presumably so that they would not sprout. Storage pits, evidently used for the seasonal harvests, argue for awareness of the economic realities of living in a new economic environment. {9}

Petr Charvat sees this general cultural adaptation for life in the south as evidence of these wanderers taking hold in this universal, mid-Eastern context, especially in Mesopotamia. He calls this period, 9000–5000 BCE, the age of inspiration. Here, humans who had been so well-integrated and adapted to the autumnal northern glacial climate, with its meat protein basis, now had to find a balance between agriculture, hunting, and domestication.

Samarra culture

(Left) Figure 192: Female statuette, north Mesopotamian Samarra culture, 6000 BCE, pre-Ubaid

Deep-reaching knowledge of animal life, leading to deliberate and rational hunting practices and subsequently to domestication, preoccupation with plant life, resulting in the protection of and care for the most promising cultigens and, finally in transformation of plant genetic structures to produce new species without wild ancestors-all these results would glorify any of the modern breeding stations and laboratories. These naked “savages” endowed with a spirit of intellectual adventure worthy of any modern discoverer, handled nature with much more responsibility and understanding than modern industrial civilization...Theirs was the age of freedom: never did the contemporary communities allow any of their material creations to chain them to a particular site, landscape or specialized activity. {10}

Agricultural transitions

Toward the end of this period, c. 6500 BCE, the first irrigation channels were being built in Jericho, and along the lower Danube. M. Gimbutis has noted the building of multi-room houses and meeting rooms, as well as fortifications along the lower Danube. {11} Great monuments: megaliths, as in Stonehenge, England, 1800 BCE, also are revealed, as well as those along the Tigris/Euphrates valleys, though not as dramatically preserved. The Mesopotamian climate and the ravages of changing human empires precluded our finding more than hints of their existence.

(Left) Figure 193: Ubaid III pottery, 5300–4700 BCE [Louvre Museum—Pre-Sumerian]

Such projects, Choga-Mami, Tell es Sarwan-Samarra culture, did nevertheless represent a lesson well learnt. The fame of such superhuman feats undoubtedly circulated far and wide. People must have talked them over around the campfires of their large winter congregation sites in the lowlands, where the multitudes gathered to hunt game in the steppes, gather a winter crop, exchange experiences, and perform the necessary rites before the whole community.

"In the following summers, when these large groups (dispersed?) and went to the higher lying grounds that protected them from the heat and offered fresh pastures to their animals, well watered valley bottoms to their peasant minded members and hitherto untapped wild-life resources for every one, many of these experiences were undoubtedly tested and found useful. Thus the memory of them was not lost and generations of free-born and proud men and women challenged the forces of nature equipped with the experiences of their predecessors. This incessant traffic in the plains managed nevertheless to create a certain 'mental environment' common to all lowland communities among which ideas, experiences and manners of artistic expression (e.g. the seals) circulated at large, coalescing into the first cultural 'koine' of this part of the world .... " {12}

These peoples, c. 6500–5300 BCE, were in all likelihood part of a Semitic-speaking migration, as they subsequently remained in the central and northern sections of the Tigris/Euphrates valley, approximately sixty-five miles north of modern Baghdad. Cultural development continued in an internal dynamic.

The cultural artifacts that we find from this period are a composite of the new and the ancient Upper Paleolithic Cro-Magnon proto-civilization, still part of their living mental heritage: "Stone sickles, blades, scrapers, stone ornaments, pendants, beads sown on clothing (belts), alabaster vessels, pottery of different sizes and shapes, figurines of females and animals. Grave goods, semi-precious stones, carnelian and turquoise, the latter probably from Iran, copper pendants." {13}

Wild grains from the fringes of the watercourses of the Zagros and Taurus ranges became a part of their diet, making the transition from a nut-like snack to a main course staple. "These changes [are represented by the transition] from wild two row (chromosome) wheat into einkorn wheat, two row barley, four row wild grasses into domesticated emmer wheat. Soon six row wheat appear and barleys without any known wild ancestors." {14}

Some of these new forms of wheat, with stronger stems and larger berries, were undoubtedly introduced from northern latitudes. These early Neolithic and Chalcolithic cultivators, c. 7000–5000 BCE, were certainly on the move, especially with the introduction of domesticated sheep, goats, **and** oxen. In the northern hills and valleys, they would have noticed different forms of wheat, barley, flax. Then, the experimentation process would have continued in the south in the winter months. In fact, this alertness through migration, trade, or conquest, is what is supposed to have underlain the introduction of southwestern Asiatic forms of grain crops into the Nile Valley. The success of such experiments **also** clearly hastened the agricultural development of Egypt. {15}

Careful analysis by Neolithic peasants, as with all farmers, of the cultivation process made them sensitive to the fact that artificial pollination of certain kinds of plants would yield better results, gradually departing from the wild prototype. "We may envisage protection of plants, sowing them in specially conditioned plots of land, selection of sturdier and especially non-shattering individuals (because of the losses of the shattering varieties releasing their grains easily in the cutting process), and, ultimately, cross-breeding." {16}

Urban origins

Here, close to the village, artificial irrigation was practiced. Village buildings had expanded to **including** up to **ten** rooms. Early hunters in the Upper Paleolithic had long before been accompanied by tame dogs. Neolithic hunters continued this practice by picking out the tamest of the wild goats and sheep that they hunted, gradually domesticating them, breeding out the wildness. Even Bos primigenius, the great wild ox of Roman hunter fame, was gradually transformed into tame cattle, probably by offering domesticated grain crops as lures into corrals and enclosures, c. 5500 BCE. Most early domesticated animals "displayed a high degree of variability and low type stability," clearly for their maximum adaptability to changing external conditions.{17}

While, as Jean Bottero argues, these early domesticators were from the Zagros Mountains and the Piedmonts of Kurdistan, many other groups wandered widely throughout the Fertile Crescent as it gradually drifted climatically into the semi-arid environment that we see described in the Hebrew Bible. "We know that in this new, muddy, fertile territory people seem to have devoted themselves to the growing of grains, besides the husbandry of small animals and cattle. Due to the extreme climate and the lack of rain, it was indispensable to establish a tight network of canals that would connect the two rivers and give the earth an exceptional fertility. This collective enterprise not only enriched the inhabitants enormously but permitted the skilled grouping of workers, first in villages and soon afterwards in larger agglomerations that became the centers of administration and distribution."{18}

> In all spheres of human life we feel the tightened grip on nature. Peasants now round up the range of cultigens yielding nourishment to their communities and settle down to establish not only fields but even such labor intensive cultures as gardens and orchards. Shepherds start keeping more cattle for food, but perhaps also as an energy source for traction and load bearing. Specialists in all sorts of crafts attack the resources of nature, gather new experiences and apply new know how. Sedentarization implies the necessity of transport in this age, regular traffic on the twin rivers of Mesopotamia, most probably using rafts of inflated skins (kelek in Arabic), first becomes a probability. The sedentarization of major social centres now traps whole settlement zones, the inhabitants of which can no longer disperse and congregate at will, being bound by various means to their capitals which constitute the central points of the coordinate systems in which their activities now take place.

Society begins

"This is the end of sometimes ragged and worn but blessed liberty: people now [had] to learn to work together, live together and even die together, this incarceration paying for the increase in living standards. Material seizing of the environment walked hand in hand with a better mental grip on things of the invisible world. The holding of private property now appears on the horizon of historical vision, both by individuals and by whole human communities, as is indicated by the establishment of the first extramural cemeteries, which probably anchored rights to territorial segments claimed by particular communities or rather, community clusters. Socially engineered exchange of material goods, which probably constituted the cementing agent of political alliances, is now institutionalized for the first time in history, most probably in the form of reciprocity."{19

Civilizational revolution: the sumerians

A change occurred in the structure of cultural progress during this period. It is as yet an historically and archaeologically inchoate transition. Still, it merits discussion. Because, while we recognize elements of this Neolithic/Chalcolithic stage of cultural life throughout the Near East—Nile and Danube Valleys, Anatolia, perhaps as far east as Persia and the Indus River Valley—something different had begun to occur in the Mesopotamian south.

It is part of this process of sedentarization mentioned above. The success of these evolved Neolithic semi-stable communities, the fact that they laid claim to a territory essential to their economic and community survival, argues for an eventual surrendering of that ancient joy and trial—the movement of the band with the seasons to new beckoning horizons, sometimes dangerous, sometimes alluring in potential. But always, the rhythm was: anchors away, onto the "high seas" of travel, we are going somewhere new.

The Tigris/Euphrates valley began to rise from its glacial inland sea after c. 6500 BCE. The gradual drying up of the southern world saw the soil rise to emerge from the Persian Gulf, from today's Nasariyah, northwest. It was then still a great freshwater marsh with a rich variety of wild animals, birds, shellfish, vertebrate fish and other water0attracted creatures. The town of Ur was thought to have been founded as a rising island in a great swamp.

Surely, peoples were also attracted to this domain, if not as permanent residents. The mysterious names of the geographical localities reflect the same kinds of reference, as do Amero-Indian names in many of America's rivers and cities; names of an unknown language. There were throughout the subsequent historical period languages of many wandering tribes that we, to this day, cannot decipher or relate to known tongues. The Elamites to the East, in today's Iran (Susa), were a highly developed people, part of the Mesopotamian complex. Their language is undeciphered and not established in its relationship to others. Later, the Gutians, Kassites **and** Hurrians enter our historical purview as conquerors, quickly accommodating their own culture and mysterious (possibly Indo-European) languages to the dominant Mesopotamian civilization.

It is in this context that we meet up with the Sumerians. They transformed what was a wide ranging and developing cultural trend (agriculture, water management, domestication) among all the Cro-Magnon peoples who had descended into the south. Now, the stabilization of secure communities along these water courses allowed for a gradual development into a full-fledged literate civilization, a civilization critically self-conscious of its accomplishments as compared with the then-contemporary level of life among its neighbors.

(Left) Figure 194: Ubaid: Serpent Man, c. 5000 BCE [Ancient Origins.net]

We know little about the origin of the Sumerians. That they were outsiders is hinted in their retention of names for places, flora, and fauna that were not Sumerian, Akkadian (Semitic), or Elamite (Iranian). There seems to have been a cultural change, c. 4500 BCE, evidenced in the type or pottery ware that was being used in southern Mesopotamia and in the trade of the region from that time period. From c. 4000 BCE, we find true urban settings with temple sites that indicate a more intensely settled cultural scene. Eridu, Uruk, Ur **and**

Lagas, important Sumerian cities in their heyday from c. 3000 BCE, are sites that reveal the existence of more ancient temples upon which these towns were built, going back to the dim horizons, evidence of a people branding the terrain with its cultural symbols.

The Sumerians always referred to themselves as the "blackheaded" people, in contradistinction to the light-haired Semites around them. Sargon himself, the Semitic conqueror of the Sumerians, c. 2340 BCE, boasts of his military, but not his cultural suzerainty, over the "blackheaded people." His daughter, Enheduana, sent as high priestess to Sumer at Lagas, writes a hymn to that event: "Urukug , shrine which causes the seed to come forth, belonging to the holy An, called by a good name, Within you is the river of ordeal which vindicates the just man; house of widespread counsel, store house which eternally possesses silver and lapis lazuli, Etarsirsir, from which decisions and me's {arete'=virtues } come forth where the "man" greets (the goddess), Your princess, the merciful princess of the land, the mother of all lands, the lady, the great healer of the dark-headed [Sumerians-not Semites], who determines the destiny of her city, The first born daughter of the holy An, mother Bau, has, oh house of Urukug, placed the house upon your ... has taken her place upon your dais ... The house of Bau in Urukug [Lagas]."{20}

Language, settlement

The Sumerian language was no closer to that of its northern Semitic neighbors, the Akkadians, than is French to Chinese. {21} In its seemingly agglutinative structure, it resembles that of the family of Finno-Ugarit-Altaic languages of Central Asia. And, indeed, considering the wandering needs and tendencies of humans, it is not improbable that this was a tribe that wended its way south instead of east, as many tribes of this language family later did: Kazakhs, Mongolians, Japanese, Koreans.

By 4000 BCE, it is clear that these people were well-settled in the rich, marshy environment of the southern Tigris /Euphrates valley. Perhaps they looked at its potential with new eyes, as foreigners often do. In contrast to the harsh barrenness of the northern steppes, this land teemed with resources that demanded an invitation for disciplined exploitation. A potential garden of Eden, the Sumerians saw opportunity.

(Left) Figure 195: Alabaster with shell eyes, male worshiper from Eshnunna, Sumer, 2750–2600 BCE

Based in the city of Uruk, a thousand-year evolution can be traced—the transformation of cultural life into a full civilization. A companion city, perhaps even older and holier, Eridu, became victim to one of the turns of the wayward river, Euphrates. It was finally abandoned, its achievements buried in the sands as the river moved to another course. A shrine in Eridu, dated to 4500 BCE, home city of Enki/Ea, later "stranded deep in a sandy and rocky desert," was finally transferred to the nearest city, Ur, to stay within the complex of the god, Nanna, patron god of Ur. "No one lived there [Eridu] after the early dynastic period, c. 2600 BCE."{22} This god, Enki, as we will later discuss, became the thematic Sumerian god of myth. As with Apollo of the later Greeks, his activities emulated the youthful spirit of this innovating people.

Uruk: civilization's city

Technology: "Most of the significant developments that were involved in that transformation seem to have taken place in the Uruk period which follows the Ubaid [6000–4000 BCE], monumental architecture, public art, specialization and standardization of industrial production, the invention of writing and greatly expanded trading horizons ... The continuity with the Ubaid period is epitomized in the famous sequence of temples at Eridu, enlarged time and again through the centuries, the latest surviving remnants of the temple's platform are in fact from the Uruk period ... excavations from the Anu ziggurat at Uruk itself has shown that the Uruk period temple on its platform was also built over the site of an Ubaid period temple ... It is fair to conclude that the major cities of the Uruk period are in part at least an indigenous development."{23}

"Technologically, it was a time of rapid and important changes. In metallurgy we see the use of sophisticated casting processes for the first time. In pottery we see the use of the fast wheel; perhaps most significantly of all we see the introduction of the first pictographic writing on clay tablets. This was accompanied by the introduction of the cylinder seal. The Uruk period was arguably the most innovative and important in the history of Mesopotamia and its influence was felt as far as the Mediterranean and the Anatolian plateau."{ 24} " ... extensive irrigation systems, and of techniques of alloying and casting not previously found on the plain itself ... Some may have been brought in by immigrants from the Iranian plateau, for instance, but they all appear within the span of a few hundred years and transform life. No comparable innovations can be attributed to the succeeding Early Dynastic, Agade or Ur III periods [3000–2000 BCE]; they seem to have built on the new techniques and refined them, but fundamentally the technological tool-kit remained unaltered. So too did the range of raw materials available to the craftsmen." {25}

(Left) Figure 196: Mesopotamian king as Master of Animals on the Gebel el-Arak Knife, dated c. 3300–3200 BC, Abydos, Egypt. This work of art suggests early Egypt-Mesopotamia relations, showing the influence of Mesopotamia on Egypt at an early date, and the state of Mesopotamian royal iconography during the Uruk period. (Louvre Museum)

(Left) Figure 197: Similar portrait of a probable Uruk King-Priest with a brimmed, round hat and large beard, excavated in Uruk and dated to 3300 BC. (Louvre Museum)

Uruk period craftsmen and craftswomen embarked upon a course of systematic experiments, some of which succeeded while others failed. Attempts at building in concrete in the Uruk VI period—temple foundations partially made of concrete/cement, 3500–3400 BCE—were abandoned in favor of more traditional material: clay. Of a number of experiments with alloys, the smiths of the period invented the viable formula of tin bronze at the close of the period, 3200–3000 BCE, but in one site only. On the other hand, the invention of the fast wheel for pottery making, c. 3800 BCE, explains the plain, undecorated mass-production pottery of the period.

So, too, was the discovery of iron, which enriched human culture for millennia. Advances in metallurgy, mostly of imported materials, trade with Aratta in the mountains of northern Iran, or again, with Anatolia, is exemplified in smelting copper into wire, sheets, **and** the lost wax casting techniques: single and bivalve moulds. Lead, for the making of bowls, used as a single or compound/alloy, was mostly for repair. As in later Greek and Roman civilizations, much of these advances in metallurgical and other technology were primarily directed to the temple for cultic paraphernalia. Stonework advanced mostly for seals' administration—or **for** works of art.

Trade: In Iran, the city of Susa, bitumen and woolen textiles were exported in exchange for Uruk metal tools. Uruk-type beveled bowls are to be found in Pakistan (Makran Province). Raw materials for chipped stones from Iran, obsidian and metals from Anatolia **and** chert from Canaan are to be found in Uruk sites, alabaster vessels from the Taurus Mountains, pottery from Anatolia.

Uruk-type terracotta cones, used in the decoration of the great temples and painted in different colors, show up in Egypt, in the Nile Delta, presumably transported by ship, or overland on sledges or wheeled vehicles. Domestication of the donkey as a beast of burden took place during this time. The names, Dilmun, Meluhha and Magan, appear in the later written record, signifying distant but long-standing trading partners—Indus River Valley, Persian Gulf oases, Ethiopia in East Africa, and again, more surreptitiously, with the Nile River valley. Uruk farmers continued the experiments of their predecessor domesticators. They attempted to domesticate antelope and/or gazelles. We have discovered illustrated signs that depict the head of such an animal with an ear of corn at its mouth, implying its feeding with grain. 26

(Left) Figure 198: Terracotta cones, painted in bright colors

(Left) Figure 199: 480 × 691

Cultural elan

"With the creation of the Uruk corporate entity [c. 4000 BCE], this essentially prehistoric development, the community as the most important social component, came to its climax ... The Uruk corporate entity, short of introducing any major subsistence innovations, did the trick by realistic analysis, rational administration, common consensus and intellectual attraction. It simply committed no mistakes and there was hardly any escape from the tyranny of its all pervading *success*."

"The Uruk culture managers assessed the economic potential of all their component communities, determined the type of enterprise most suitable to every single one of them and arranged for the distribution of results, maintaining a steady flow of goods throughout the entire community and buffering the impact of unexpected events. Tasks requiring specialized knowledge were carried out by the centres and the products released into the redistribution network. Some of them involved the seeking of alternative solutions by testing several hypotheses, the erroneous ones of which were subsequently abandoned [metalwork]." {27}

"The founding fathers of the Uruk culture must have run their homes precisely on schedule. The ideal of the Uruk culture leadership must have been a society in which no one actually starved, as the best individuals racked their brains to procure for everyone what he or she needed and to release the potential of the natural resources in accordance with a preconceived vision of the external world in which non-economic aspects were not absent." {28}

"The Uruk culture settlement network included large scaled settlement agglomerations which may be termed pristine urban centres and military bases, presumably with deterrent functions. All this went on in a predominantly egalitarian atmosphere without traces of social stratification other than the ability of the centre to mobilize manpower and with evidence for holding assemblies [Lugal= 'big man'] essentially for the decision of common matters ... "{29}

The sociopolitical terminology of leadership is interesting. "Ensi" was the term for head of state during the Early Dynastic period, c. 2900–2600 BCE. This term meant, in the early Uruk period, c. 4000–3500 BCE, "tenant farmer of the god," later, c. 3500–3000 BCE, "the lord who established the foundation of the temple." "Lugal" also meant "great man." First, in the Uruk period, it denoted a leader of the community. Then, in the Dynastic period (during which the city states were in competition with each other), after 3000 BCE, "lugal" became the king who lived in the "lugal," or great house—the palace. {30}

Figure 200: Clay impression of a cylinder seal with monstrous lions and lion-headed eagles, Mesopotamia, Uruk Period (4000 BC–3000 BC). (Louvre Museum)

Uruk, 4000–3000 BCE: "In all spheres of society the principle of universality and equality comes to the fore and struggles with the particular and concrete manifestations of the visible world in order to make way for human advance. In the spiritual sphere, the torrents, twists, and turns of the mythopoeic thinking in symbols embedded in reality are surmounted, if not straightened out, by application of universal principles and rules. The world perceived as a unity in diversity is accordingly organized into a social whole. The natural variability of communities, in most cases linked principally by the factor of co-residence and accompanied by industrial sites and service holdings [is] interspersed by groupings that see themselves as manifestations of divine will, communities divided among elites and commoners is constantly leveled. The material standard of living is equalized by means of redistribution, external threats are eliminated by garrisons posted to ward off any attack."

"Everyone is close to everyone else, people meet in assemblies to discuss and decide matters of common interest with at least some of the resolution put down in writing. All receive the same treatment both in life and in death. The world is an organized place where the economy thrives because all the discoveries and inventions of previous ages are now put into practice and systematically."

"Unity of purpose drives the best brains of the epoch to experiments both economic, some of which fail and social and spiritual, some of which succeed. The flow of goods throughout the community is directed and scheduled ... The world is weighed, measured, disposed of in organized form, from cake baking to manipulations of time and space." {31}

Literacy

From the fourth millennium [3000 BCE] on, what we can call an urban high civilization flourished. It was complex and original, and it was the first in world history. Numerous elements were gathered into this civilization [from 4000 BCE]: social and political organization; the creation of institutions of obligations, and of laws; the production of all goods of use and exchange, procured over-abundantly by a planned effort, and their circulation in the interior of the country as well as abroad; the appearance of superior and monumental art forms, the basis of the scientific spirit [divination and neocromancy, the functions and status of the gods] characterized first of all by a constant urge to rank, to classify, and to clarify the universe.

"And finally, around the year 3000 BCE, came the last great innovation, doubtless the most decisive discovery of an importance that can barely be measured: the establishment of a system of writing. It was at first [4000–3500 BCE] a simple mnemonic device, but in a few centuries it enabled the recording of all that is expressed by the spoken language, and in the way it is expressed by it. This ability of the script allowed people to objectivize knowledge, to organize it in an entirely different way, and to propagate it. Hence knowledge became rapidly more extensive and more profound." {32}

We have tokens from Mesopotamia, Iran **and** Jericho, that date from shortly after 4000 BCE. These are made of clay in the shape of spheres, cones, bi-cones, tetrahedrons, sometimes with incised marks—indications of personal or corporate exchange. People are negotiating and recognizing these marked agreements over time and space. From the mid-fourth millennium, c. 3500 BCE, in Uruk and elsewhere in Sumeria, these tokens contain pictographic symbols: human heads, trees, animals.

Soon, hollow clay balls were devised—bullae—again, to be found widely throughout Mesopotamia, north Syria and Iran (Elam/Susa). These had incised symbols on the outside, to coincide with the clay-enclosed tokens. Such often thumb-sized cylinder seals would be rolled over clay to identify the owner, soon to be followed by pictographic symbols, often denoting the meaning content on small clay tablets. Soon, conventions about meaning had to be established among these pictographs, and more abstract symbols **were** added to clarify rebus and homophonic elements unclear from the pictographs.

The requirements of this urban, economic **and** political revolution necessitated the establishment of clear, accepted legal public agreements: economic/business exchange, long-distance trade between cities, governmental and temple laws.{33}

"The fact that the Uruk tablets, c. 3000 BCE, were located in the enclosure of the great temple of that city, and that the tablet pieces clearly constitute accounts of the movements of goods, listing numbers first in detail and then totaled, make us think that this script was established mainly in order to memorize the numerous and complicated economic operations centered in that temple. The temple was the exclusive or principal owner or redistributors of the products of the labor of the land. In other words Mesopotamian writing did apparently grow from the needs and necessities of the economy and the administration, and therefore any kind of religious, or purely intellectual preoccupation seems to have been excluded from its origins." {34}

The script, created deliberately by Uruk's sages, put at their disposal both an ordered series for things of the visible and invisible worlds which they could manipulate to their hearts' content and an efficient device for the tracing of movements of material goods throughout the complex circuits of the Uruk corporate entity. Thus they rose to the status of masters of the visible world, guardians and managers of its fertility and creators of a cosmic order. Nevertheless, they bowed low before their gods who represented the highest values of Uruk culture/society, reserving for themselves [the gods-temples] all the most precious goods and justifying the mobilization of a considerable amount of manpower for all-community projects.

"The gods had to determine and to decide first of all the destinies of all things, in order to produce and govern the world and the people from day to day. Their orders had to be written down in order to give them substantiality, publicity,and force . Utilizing as pictograms and ideograms the things to come, which they created as needed, they impressed in them the individual words' of their decrees.. Whoever understood the code used by the gods, { cuneiform writing}... could decipher the signs and read in them the irrevocable will of their authors."

"Priests, scribes, and kings could thus associate themselves with a higher order of powers and the future events. The search for meaning in the external world of mostly unique events, beneficial or dangerous, as well as the powers of humans organized to secure their lives, took on a mythological, religious cast. This required of every city its higher commitment to the will of the gods and the temples, as symbols of both dependency as well as the resurgent intelligence of the community under its gods." {35}

Transition: Charvet claims that the Uruk period was unique in its treatment of life and death. "The citizens who came to the shrines where they sometimes left their personal articles as tokens of faith even tolerated an absolute break in the treatment of their dead who departed to the nether world in a manner radically different from those of the preceding periods. No veneration of the dead was now possible with the new policy of mass graves or the incineration of the dead. Thus we find few graves, the disappearance of cemeteries, as compared with earlier periods." {36}

This claim raises an interesting conundrum. For, in the succeeding dynastic period, we find much evidence of cemeteries close to towns, often within living compounds, at the least, of the extended family. There are, of course, the royal burial tombs in Ur, discovered by the British researcher, C. Leonard Wooley. They were discovered, along with a great urban cemetery of c. 2600 BCE, hinting of a radical reversion in cultural practice as represented in the transition between the Uruk and Dynastic periods from c. 3000 BCE.

A supposition about this transition can be made from the Epic of Gilgamesh, a great Homeric tale of adventure **and** misadventure, that is now believed to have been written c. 2700 BCE, but refers to events, such as the great flood, presumed to have taken place c. 3000–2900 BCE. This flood had such an impact on the psyche and history of Mesopotamia and the Near East, that it reappears in a more moral/religious setting nearly 2000 years later in Genesis, of the Hebrew Bible, now featuring Noah and his sons.

It is not at all unlikely that this catastrophic event, and the necessity to rebuild the Sumerian cities, gave rise to a wholly new cultural outlook and destiny for this people.

Sumer's Achievement

Timeline of dominance: Fifteen hundred years, 4500–3000 BCE, approximates the Sumerian adaptation to life amidst the swamps and fields of the southern river system. A vast accommodation had to be made by a people wandering from the mountains and plains of the north. Thence, the maturing cities of Sumer become increasingly self-conscious of their history and covetous of their achievements. Eventually, they became engaged in disputes and conflicts with their neighbors as we enter what is called the "Dynastic Period." Certainly, irrigation rights, and other economic contestations figure highly. But also pride of place should not be discounted, as urban centers such as Ur, Lagas, Nippur, Uruk, Kish, Larsa, **and** Isin flourished.

Umma begin to figure in the documents. Writing, computing, surveying, all are perfected within the cuneiform system during a five-hundred-year-plus period of Sumer's international dominance, c.3000–2400 BCE, a dominance that created civilizational impact well beyond its geographical boundaries. Whereas in Sumer, the development of writing and monumental architecture can be traced back to the fifth millennium BCE, in Egypt, it appears suddenly after 3000 BCE, to be taken up as a passion by the ready Egyptians, "why didn 't we think of that?"

The Sumerian merchants and technicians had clearly visited the Nile kingdoms. The influence of Sumer over the ancient world lay as much in its conception of civilization as it did from **its** material achievements. It reached a peak in Ur, as exemplified in the above mentioned discovery by C. L. Wooley **in the** 1920s, of the royal tombs of that city. Here, artifacts amidst the sacrificial suicides of a king's entourage, c.2600 BCE, first revealed the extraordinary achieved level of craft and esthetic skills of this people.

Over time, the Sumerian cities were a magnet for the gradually infiltrating Semites of the valley and surround**ings.** They increased in proportion and influence. There were no "outside" Sumerians to repopulate the area. The Sumerians were sui generis in Mesopotamia, as compared **to** the Semites, with no other known related kin. About 2340 BCE, Sargon, from Agade **and** an Akkadian Semite, conquered the cities of Sumer to create an empire that extended to the Mediterranean. Culturally and intellectually, he was a Sumerian, even if of a different ethnicity. After one hundred fifty years, his empire was felled by Gutian tribesmen from the north. They seemingly acclimated themselves to this urban culture and ruled for almost one hundred years, thence to be ousted by a new Sumerian dynasty, from Ur III.

This reinvigorated Sumerian dynasty achieved a peak of systematization—a Sumerian forte—in every area of life: political, educational, economic and cultural. Now united by a dynastic force of great brilliance, Sumer's language and heritage flourished amidst this ever-multiplying Semitic demography. It was a relatively brief, century-long nova. Its disintegration, c. 2100 BCE, after another one hundred years plus, was accompanied by a wholly new force of peoples from outside—the Elamites. Once a satellite people to the east, also the Amorites a semi-nomadic Semitic people to the west and north. It was from this latter and seemingly barbarous infiltration, that a new cultural center based in the new city of Babylon (Hammurabi) arose from the Sumerian base.

The belief system

Twenty-five-hundred years separate the Sumerian and the Hellenic renaissance, 3000–500 BCE . Both were civilizations of the city state. Each city was attached to its own tutelary god, which represented the local ethos and **a** protective hand. While the Greeks were in the process of creating a written literature, now including philosophy, science and drama, **that** the Sumerians had not yet developed, both had absorbed ancient mythological systems and literary epics (Iliad, Gilgamesh). These latter explained to their ordinary citizens the meaning of their existence as vulnerable beings upon this earth.

During the Akkadian Semitic dynasty of Sargon (2300 BCE), there was catalogued a series of forty-two hymns, including the Hymn of Enheduana, quoted above, by the daughter of Sargon, high priestess to the Sumerians at Lagas. These were recorded as referring to the Dynastic Period, c. 2600 BCE, and represented thirty-five cities, each one with its indigenous temple and gods. From Ur, Eridu, Lagas in the south, to Kazallu, Marad in the northwest**,** Sippur in the north, **and** Esnunna and Der in the northeast hill region. Although the non Sumerian peoples who came to rule over Mesopotamia throughout its independent history, before the Indo-European Persians and then the Greeks added their own special gods to the pantheon, the original Sumerian deities remained a critically important element in this conjoint**ed** cultural heritage. {37}

The myth of Enki (called Ea by the Semites) represented to the Sumerians a poetic exemplification of the inner values and understandings of their own achievements. Enki, one of the great gods (**the** originating god of the city of Eridu, later transferred to Ur), offspring of Nammu, universal primordial mother, is lazy. He invents humans to do the work of the gods.

"The triumph of Enki was not only to have invented mankind, an enormous technical success, but also in finding a function for all the human types, even the imperfect but usable ones ... (sterile women, the court officer, the blind, a-sexual humans for the court, etc). Thus Enki is at the same time civilizer, inventor, keeper of all cultural values, but also the most intelligent, the most clear sighted, the prototypical technician, the only one capable of overcoming all hurdles, and of adapting everything for his purposes, of molding matter for every possible use." {38}

What Enki/Ea does, what he has to do, is never defined in terms of power, of government, or of political authority, but only in terms of organization, of control, of the promotion of life, and to that goal, of intelligence and technical and practical success—and all this, always, in the supreme interests of the divine society [the stable order of things].

"In the end he was portrayed as being at the head of an enormous and extremelycomplicated mechanism, which he guarded and animated. Beginning with his infallible directives and passing through his 'foramen,' the minor deities, which he charged with a section of this gigantic machine, this mechanism reached all the way down to the 'workmen,' the people who made it actually function. These people were first of all the 'technicians.' Enki/Ea is the only god who, according to the entire tradition, is the patron of all of them and of all technologies, as if he had established all of them: from agriculture and husbandry to writing and exorcism."{39}

"Enki/Ea used the apkallu, the scholars, sages, craftsmen to introduce culture into the history of his country, the great technical advances, the successive elements of high civilization that had made, first Lower Mesopotamia in general, then Babylon the cultural center of the world ... At the side of authority, of power, of efficient command, of commanding appearance, there was indeed need for a clear and profound vision, of intelligence, of wisdom, to give a positive sense to these orders-what we call 'the technical function of power,' eminently incarnated by Enki/Ea." {40}

"The entire civilization of the country, their entire life and their way of living was first of all based, since the beginning of time, on communal work, the extensive production and transformation of usable goods. In such a system, in the end, everything is directed by a spiritual activity that researches, invents, promotes, and perfects procedures, not so much to see better, but to do better." "All knowledge, all intelligence, was thus polarized by production and action, and was materialized equally well in the 'practical judgment' of the artisans, and what we would call 'craft' as in the sagacity of the tactician and his ability to adapt; and as in the 'good sense' and the astuteness of what was known in the old days as an 'upright man'—both on the collective level and in individual achievement ... This type of wisdom was incarnated in a younger god, Enki/Ea as if to stress his capacities not only in knowledge but also in action, in speed and in versatility, which cannot be detached from the 'technical function' or from the 'functions of government' that formed the double aspect of power." {41}

Soft culture

Myths, poetry, a bardic tradition that is imputed in such ancient stories as Enki/Ea, reflect only one aspect of a culture which, by the late Uruk, had found a place in its ideals for leisure, play, song and dance. Illustrations on a carving from the Early Dynastic **period**, c. 2900 BCE, depicts banquet scenes with lyres, flutes **and** cymbals as musical entertainment. There are also stone plaques that show wrestling matches, plus other entertainments, dice, complex gaming boards and their pieces—before the mid-third millennium in Ur, c. 2700 BCE.{42}

"Some material culture items do bear out a non-negligible function of dance (a pin in the form of two nude female dancers. The fact that in the 'Curse of

Agade' [Sargonic period, 2300 BCE], Inanna [goddess] assigns dancing as a historically distinguishing feature for young women could indicate the existence of age grades or age groups as early as the Uruk period (4000–3000 BCE)." {43}

Even love merits the conscious efforts of humans to go beyond mere procreation. Here, we can better understand the sacred sexual/fertility sacraments celebrated on the day of the spring New Year by a high priest or priestess. As with the Greek god Dionysus, whose role in the mysteries at his temple in Eleusis just northwest of Athens, was proclaimed a conjunction of sex, wine, and the sacral. Such invocations were not merely salacious events, **à la** "Sadie Hawkins Day." Rather, they were the ritually inscribed and disciplinary representations of a powerful human drive.

Twenty-five hundred years separate Sumer and Athens. So, too, twenty-five hundred years separates Athens from contemporary twenty-first-century **Rome**. In our own era, Vatican Rome's celebrations are royal and mysterious, invoking the sacredness of non-explicit sexuality supposedly held in check by priestly celibacy. How much further have we advanced in the human understanding, **of** the symbolic and ritual acting out of this unique, human biosocial force?

A famous episode is found in the first two tablets of the Epic of Gilgamesh, c. 2700 BCE. In it is explained how Enkidu, the wild man of the steppe, hairy and barbarous, only intimate with animals, and living a life like them, becomes a man in the full sense of the word: **...** a civilized man, a city man who eats bread, drinks beer, and grooms and dresses himself. This transformation is the work of a courtesan from Uruk who came to look for him in the steppe and who introduced him to love; hence not simple intercourse with a female, but love with a real human. Thus, we have civilized human and refined love, i.e., free love.

"Once he had discovered it and had acquired a taste for it, Enkidu could only follow his teacher to the city, where she taught him to eat, drink and dress, and where she completed his transformation. Thus free love is presented as being the point of access to a life that is truly cultural and human. It is difficult to better indicate its worth and its importance." {44}

Marriage contracts guarantee economic, social position and value. Children from marriages established by elders inherit the economic and social base of settled urban life. Such marriages are proforma, and, thus, not usually based on love or emotion. Man is also privileged to take on concubines, slaves for work and sex, if his economic base allows for it. Love, however, is the real key to civilized life.

Beyond Barbarism

Fundamental to the obsessive Sumerian efforts to free themselves from the chanciness of the old order—a wandering subsistence existence always threatened by nature and man—was their awareness that the good life for humans is a constructed life, built out of the possibilities of individual intelligence and conjoint human social efforts. The struggle was against disorganization, disorder, and the primitive, accidental flow of events. Here, discovered consciously, expressed in both life actions and the myths and temples that gave intelligibility to their efforts, they would attempt to extricate themselves from the unknown. The democratic form of cooperative efforts came naturally, as much a part of this new social and economic system as the ancient Cro-Magnon band or tribal unity that undergirded the hunter-gatherer economy on the northern plains, many thousands of years earlier.

Key to understanding the creativity of the early Uruk period Sumerian achievements, a rich heritage that was mined in subsequent centuries and millennia, was the sense of the possible: the need to push back the mysteries of nature out of which they expected to bring into reality the life vision they hardly understood, but knew they had to find. The end of the Dynastic Period (2600–2340 BCE) is characterized by dominating elites that tried to use the temple as a repository of secular power. The palace and its lugal, or ensi ("big man"), now developed independently of the more community-centered tradition, invested in the ancient temples. Thus ensued a struggle between individual achievement and the ancient Uruk community classless traditions.

Yet, "this age, terminated by the Akkadian (Semitic) conquests of the forties and thirties of the twenty-fourth pre-Christian century [2340 BCE], saw the emergence of the principle of statehood, the embryonic forms of political democracy, but also the foundations of modern thought, including mathematics, metrology and the first ethical categories, when such notions as freedom or justice but also guilt and sin found their way into human literary culture. One of the most creative periods of Mesopotamian history, this epoch (the Dynastic-signifying kings and competitive Sumerian states, 2900–2340 BCE) enriched human civilization with achievements that affect the lives of humankind up to this very day."{45}

The Sumerian heritage of the free citizen and the secular culture was epitomized in the fact that rarely did Kings try to transform their secular power achievements into theological ordinations, as gods. Narum-Sin of the Semitic Sargonic era, 2290 BCE; Shulgi, c. 2060 BCE, a Sumerian of the Ur III reestablishment; Ishbi-Erra of !sin, c. 1950 BCE, possibly an Amorite, are the few that attempted thus to anoint themselves. Considering that Mesopotamia had its share of autocrats and warrior kings over a period of almost 2,500 years before Cyrus, the Persian, he of religious tolerance conquered Babylon/Chaldea in 539 BCE, this civic restraint was remarkable. These kings, if not gods, however, felt the need to proclaim that they were "called" by the gods to rule. {46}

What strikes as the gift of Sumer, extended throughout each succeeding dynastic empire, was the commitment to preserving, copying, and cataloging the heritage accumulated over the millennia. Thus, the sanctified role of the scribe, the scholar, the scientist/medical doctor, never waned in Nineveh (Assyria) as well as in Babylon.

Assurbanipal, the last ruler of Assyria, had himself been trained as a scribe and scholar of both Akkadian and Sumerian. It was he who spent fortunes in directing his scholars throughout his realm, 668–637 BCE, to bring the history of

Mesopotamian scholarship from Sumeria to Assyrra, to his Museum in Nineveh. It was here, in the 19th century, that these tablets were discovered and our knowledge of these millennia of Mesopotamian history immensely fructified.

The last king of Babylonia/Chaldea, Nabonidas, under whose rule the Israelites lamented in Nippur, sent his eldest daughter to be high priestess in one of the restored Sumerian temples in the south (550 BCE).

The Mesopotamians always feared that descent and disintegration could come from the wilderness both within and without the nation.

In the time of the rule of Su Sin, Sumerian king during the Ur III reestablishment (c. 2030 BCE) , after the Aggade dynasties of Sargon and the Gutian hill people, a description was written of the Amorite nomads of the western desert, then beginning to overrun the walls along the Euphrates: "A tent dweller ... wind and rain, ... who digs up truffles from the hills, but does not know how to kneel, who eats raw meat; who has no house during the days of his life, and is not buried on the day of his death .... Since that time the Amorites, a ravaging people, with the instincts of a beast ... the sheepfolds like wolves, a people which does not know grain .... "{47}

A few generations later (c. 1900 BCE), the widely disseminated myth of the marriage of Martu made its rounds in the ancient cities: Martu, the god of the Amorites (city of Mari, along the northwestern Euphrates River region), arrives among the southern city dwellers. He engages in a series of wrestling matches with the king's heroes. Martu wins. The king offers him gold and silver as a reward for his achievements. He refuses, wanting instead the king's daughter in marriage. Her friends warn her away, "He is a tent dweller, eats raw food, has no house ... ." She goes ahead and marries Martu.{48}

These Amorites, whose West Semitic dialect (also Hebrew) could not be understood by the Tigris/Euphrates, East Semites, became the Babylonians of Hammurabi: he who introduced new concordats of laws, created a rich and glorious city, to continue and amplify this powerful civilizational tradition.

Bottero sees the living Uruk tradition modifying and limiting the excesses in class and inter-city warfare until c. 2600 BCE of Sumerian dynastic experience.{49} By the time of the Amorite takeover of the ancient civilization, c. 1900 BCE, Sumerian was no longer spoken, even in the ancient cities of the Uruk period—Ur, Lagas, Uruk, Kish, Nippur. It was, however, the language of the scribes, taught in the urban academies, and was the secondary official language of the cuneiform tablets.

" ... Sumerian remained, until the very end, the scholarly and liturgical language, as Latin was to us in the Middle Ages. That is undoubtedly the best proof of the intellectual preponderance of the Sumerians in the Sumero-Akkadian cultural complex." {50}

In giving a postscript to the achievements of the Sumerians, Postgate sees the inner character of the Sumerian advance as constituted of literacy, urbanization, and bureaucracy. This civilizational pattern was subsequently disseminated to the rest of Mesopotamia and the ancient world. { 51}

Chronology—Sumeria/early Mesopotamia

40,000–12 ,000 BCE	Upper Paleolithic, Cro-Magnon, modern *Homo sapiens sapiens*, Eur-Asian peoples
12,000–5000 BCE	Mesolithic-Neolithic, small lowland communities, ceramics , mixed agriculture—proto-Semites
5000–4500 BCE	Possible arrival of Sumerians in Mesopotamia
4500–4000 BCE	Formation of first towns Eridu, Uruk
4000–3000 BCE	Domination of Uruk culture, writing
3000–2350 BCE	Dynastic period, various Sumerian cities
2350–c. 2200 BCE	Sargon, Semitic Dynasty of Akkad
2200–2100 BCE	Gutians, unknown people from north, possibly Indo-Europeans—possible eastern migrant Indo-European Tokharians of Xinjiang Province, c. 2000 BCE
2100–2000 BCE	Sumerians at Ur III dominate, flourishing culture
2000–1800 BCE	Breakdown, lsin /Larsa, Semitic Amorites, also, Elamites, unknown language; Sumerian spoken language in decline
1800–1600 BCE	Semitic Babylonians, Hammurabi, c. 1750 BCE
I 600–1150 BCE	Population movements, Kassites rule, unknown language, possibly Indo-Europeans. West Semitic Aramaeans migrate into the river valleys.

Maps

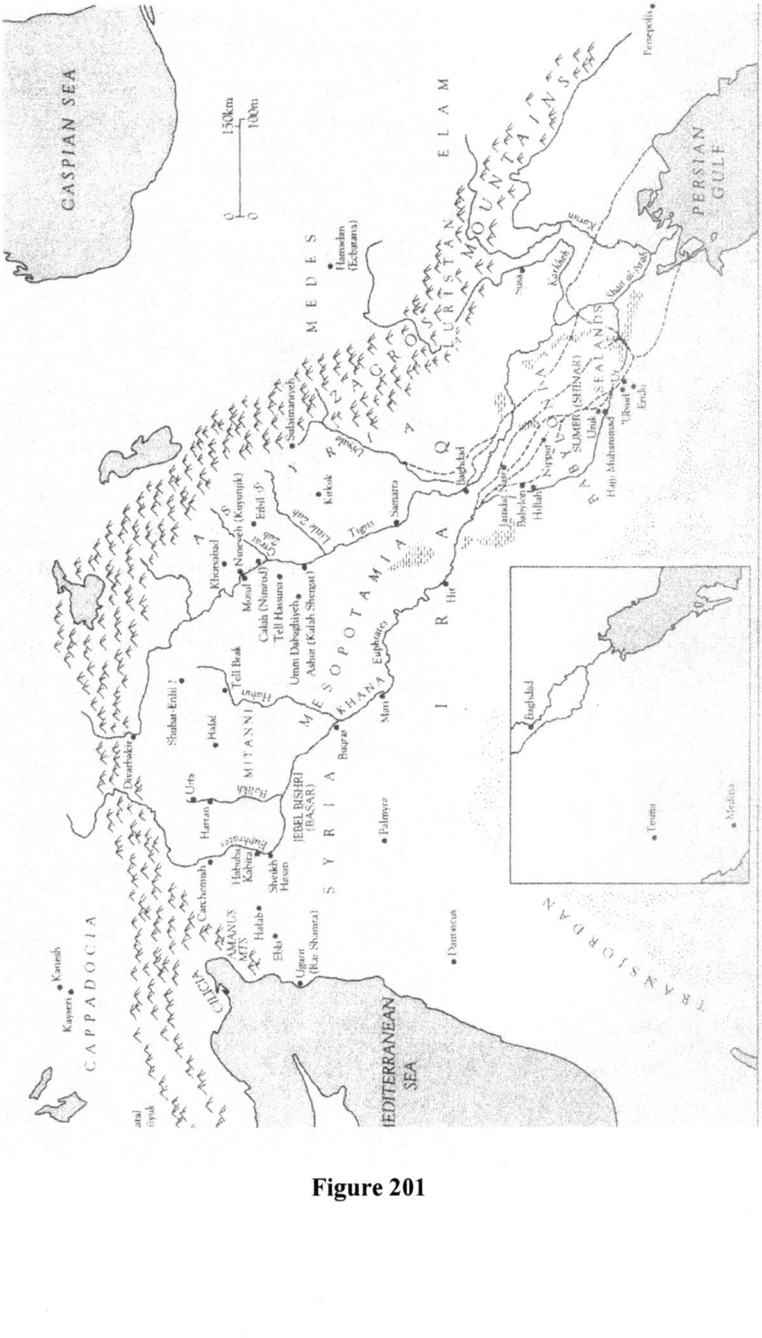

Figure 201

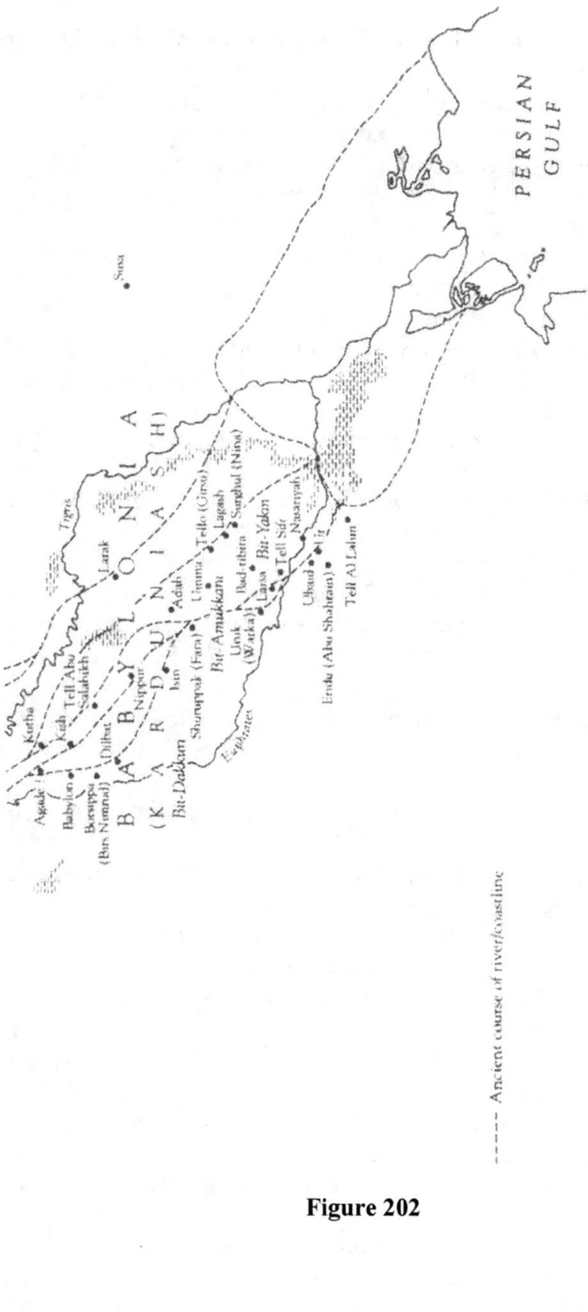
PERSIAN GULF
BABYLONIA
(KARDUNIASH)
Tigris
Euphrates
Susa
Agade
Babylon
Kish
Kutha
Borsippa (Birs Nimrud)
Dilbat
Tell Abu Salabikh
Nippur
Isin
Adab
Larak
Shuruppak (Fara)
Umma
Tello (Girsu)
Lagash
Surghul (Nina)
Bit-Amukkani
Bit-Dakkuri
Bit-Yakin
Uruk (Warka)
Larsa
Bad-tibira
Tell Sifr
Nasariyah
Ubaid
Ur
Eridu (Abu Shahrain)
Tell Al Lahm
Ancient course of river/coastline

Figure 202

Bibliography

Barker, J. W. *Justinian and the Later Roman Empire*. Madison: University of Wisconsin Press, 1966.

Bottero, J. *Mesopotamia*. Chicago: University of Chicago Press, 1992.

Bowra, C. M. *Periclean Athens*. New York: The Dial Press, 1971.

Charvat, P. *Mesopotamia before History*. New York: Routledge, 2002.

Chiera, E. *They Wrote on Clay*. Chicago: University of Chicago Press, 1938.

Cottrell, L. *The Quest for Sumer*. New York: Putnam, 1965.

Crawford, H. *Sumer and the Sumerians*. Cambridge, England: Cambridge University Press, 1991.

Diamond, J. *Guns, Germs, and Steel*. 1997. p. 111.

Durant , W. *The Age of Faith*. New York: Simon and Schuster, 1950.

Frankfort, H. *The Birth of Civilization in the Near East, Garden City*. New York: Doubleday Anchor, 1956.

Gimbutas, M. A. *Goddesses and Gods of Old Europe, 6500–3500 BC*, London: Thames and Hudson, 1982.

Gimbutas, M.A. *The Language of the Goddess*. New York: Harper and Row, 1989.

Grant. M. *The World of Rome*. New York: Mentor, 1960.

Gurney, O. R. *The Hittites*. Baltimore: Penguin, 1964.

Haldar, A. *Who Were the Amorites?* Leiden: Brill, 1971.

Herodotus. *The Histories*. Translated by Aubrey de Selincourt. Baltimore: Penguin, 1954.

Jones, T. B. *The Sumerian Problem*. New York: John Wiley and Sons, 1969.

Kramer, S. N. *History Begins at Sumer, Garden City*. New York: Doubleday Anchor, 1959.

Kramer, S. N. *The Sumerians, Their History*. Chicago: University of Chicago Press, 1963.

Laessoe, J. *People of Ancient Assyria*. Translated by F. S. Leigh-Browne. London: Routledge and Kegan Paul, 1963.

Lyber, A. H. *The Government of the Ottoman Empire in the Time of Suleiman The Magnificent*. Cambridge: Harvard University Press, 1913.

Mallowan, M. E. L. *Early Mesopotamia and Iran*. New York: McGraw-Hill, 1965.

Morrison, M.A., and Owen, D. I., ed., *Nuzzi and the Hurrians*. Winona Lake, Indiana: Eisenbrauns, 1981.

Oded, B. *Mass Deportations and Deportees in the Neo-Asyrian Empire*. Wiesbaden: Reichert Verlag, 1979.

Popovic, A. *The Revolt of African Slaves in Iraq, in the 9th Century*. Princeton, New Jersey.: Marcus Wiener, 1999.

Postgate, J. N. *Early Mesopotamia*. London: Routledge, 1992.

Saggs, H. W. F. *Babylonians*. Berkeley: University of California Press, 2000.

Smith, P. 'People of the Holy Land from Pre-history to the Recent Past,' ' p. 65. 1995.

Bar Yosef. "Earliest Food Producers, Pre-pottery Neolithic (8000–5500 BCE)." In Levy, T., ed., *The Archaeology of Society in the Holy Land.* London: Leicester University Press, 1995.

Thomson, M. L. *The Sumerian Language.* Copenhagen: Akademisk Forlag, 1987.

Thucydides. *The Peloponnesian War.* Translated by Rex Warner. Baltimore: Penguin, 1987.

Toynbee. A. *Greek Civilization and Character.* New York: Mentor, 1953.

Toynbee, A., ed. *Greek Historical Thought.* New York: Mentor, 1952.

Wooley, C. L. *The Sumerians.* Oxford: Clarendon Press, 1930.

Yoshiwara. R. *Sumerian and Japanese, a Comparative Language Study.* Japan: Chiba, 1991.

Endnotes, Appendix

1 Bar-Yosef, O., "The Role of Western Asia in Modern Human Origins," in Aitken, et al., eds. *The Origin of Modern Humans and the Impact of Chronometric Dating* (Princeton, New Jersey: Princeton University Press, 1993), pp. 132–147.

2 Tattersall, I., *The Last Neanderthal*, (New York: Macmillan, Illustrated. 129, 1995), pp. 186–187.

3 Guthrie, R. D., "Mosaics, Allelochemics, and Nutrients,'" in R. S. Martin and R. G. Klein, eds., *Quaternary Extinctions: A Prehistoric Revolution* (Tucson, AZ: University of Arizona Press, 1984), pp. 259–298.

4 Trinkaus, E., and Zilhao, *Proceedings of the National Academy of Sciences, Washington, D.C.* (2000); Wilford, J. N., "Discovery Suggests Man Is a Bit Neanderthal," *The New York Times*, April 25, 1999.

5 Conrad, N., "Thirty Thousand Year Old Carved Figurines," *Nature*, December 18, 2003; Clottes J., and Courtin, J., *The Cave beneath the Sea* (New York: Abrams, 1996).

6 Leakey, R., *The Making of Mankind* (New York: Dutton, 1981), pp. 193–196; Clottes, J., *Chauvet Cave: The Art of Earliest Times* (Salt Lake City: University of Utah Press, 2003).

7 Schenk, G., *The History of Man* (Philadelphia : Chilton, 1961), pp. 158–164.

8 Smith, P., in T. Levy, ed. *People of the Holy Land: From Pre-History to the Recent Past* (1984), p. 65; Bar-Yosef, O. "Earliest Food Producers, Pre-Pottery Neolithic (8000–5500 BCE)," in Levy, T., *The Archaeology of Society in the Holy Land* (London: Leicester University Press, 1995).

9 Ibid.

10 Charvat, P., *Mesopotamia before History* (New York: Routledge, 2002), pp. 235–237.

11 Gimbutas, 1989, p. xix.

12 Charvat, op. cit., pp. 235–237.

13 Charvat, op. cit., pp. 22–27.

14 Charvat, op. cit., p. 26.

15 Personal communication from Dr. Thomas Boyle, Professor of Botany, University of Massachusetts, Amherst.
16 Charvat, op. cit., p. 26.
17 Charvat, op. cit., p. 27.
18 Bottero, J., *Mesopotamia* (Chicago: University of Chicago Press, 1992), p. 47.
19 Charvat, op. cit., pp. 25–27.
20 Postgate, J. N., *Early Mesopotamia* (London: Routledge, 1992), p. 26.
21 Bottero, op. cit., p.48.
22 Postgate, op. cit., p. 299.
23 Postgate, op. cit., p. 24.
24 Postgate, op. cit., p. 170.
25 Crawford. H., *Sumer and the Sumerians* (Cambridge England: Cambridge University Press, 1991), pp. 118–119.
26 Charvat, op. c:it., pp. 120–131.
27 Charvat, op. cit., pp. 158–159.
28 Charvat, op. cit., p. 131.
29 Charvat, op. cit., pp. 235–237.
30 Mallowan, M. E. L. *Early Mesopotamia and Iran* (New York: McGraw-Hill, 1965), p. 88.
31 Charvat, op. cit., pp. 158–159.
32 Bottero, op. cit., pp. 47–48.
JJ Saggs, H. W. F. *Babylonians* (Berkeley: University of California Press, 2000), pp. 46–51.
34 Bottero. op. cit., p. 70.
35 Bottero, op. cit., p. 101.
36 Charvet, op. cit., pp. 150–152, 237.
37 Postgate, op. cit., p. 26.
38 Bottero, op. cit., p. 240; also see Postgate, op. cit., p. 299.
39 Bottero, op. cit., p. 246.
40 Bottero, op. cit., pp. 248–249.
41 Bottero, op. cit., p. 250.
42 Crawford, op. cit., pp. 118–119; Saggs, op. cit., Illus. 38, p. 64.
43 Charvat, op. cit., pp. 158–159.
44 Bottero, op. cit., p. 193.
45 Charvat, op. cit., p. 239.
46 Saggs, op. cit., p. 136.
47 Postgate, op. cit., p. 84.
48 Saggs, op. cit., pp. 92–93.
49 Bottero, op. cit., p. 238.
50 Bottero, op. cit., pp. 47–48.
51 Postgate, op. cit., p. 302

[*Adapted from, Appendix, *Rebuilding Western Civilization, Beyond the 21st Century Collapse* by Seymour W. Itzkoff. Ashfield, MA. Paideia Publishers, 2005]

Chapter 12

The Civilizations of *Hss*—Egypt, China

Egypt: Mesopotamian heritage?

There are certain elements in Egypt's Early Dynastic Period which seem to betray unmistakable Sumerian influence. Egyptian hieroglyphic writing may be one. Another is the so-called "paneled-façade"-type of architecture found in Egyptian tombs, from the First to the Third Dynasty. The most remarkable evidence of cultural connection, is that shown in the architecture of the early Dynastic tombs of Egypt and Mesopotamian seal-impressions show almost exactly similar building techniques.

"For the Sumerian region, intelligible written records begin at about 3000 BC. From these, and from archaeological research, it is evident that even at this early period there were large cities with splendid temples and elaborately planned houses. Stone-carving was well-developed, also metal-working and the fashioning of jewelry. Extensive foreign trade contributed lapis-lazuli from Afghanistan, shells from the Persian Gulf and rare stones such as calcite, obsidian and diorite, none of which are found in southern Mesopotamia. But in the early Egyptian Dynastic Period there was no unified state. Sumer, unlike Egypt had become unified by 3200 BC." (Cottrell 1952)

In 1988, Günter Dreyer and his team discovered an exceptionally large and well-equipped tomb at Umm el-Qa'ab/Abydos (tomb U-j). Its central position in Cemetery U, as well as the discovery of an ivory sceptre inside, led to the conclusion that tomb U-j was the final resting place of a Pre-Dynastic ruler, or at least a member of the highest elite of the region. It was often assumed that Egyptian writing was invented under the stimulus of the Mesopotamian system, developed in the late fourth millennium BC (Petrie 1920, 49–50; **Pope 1966**, 17–23; **Brewer 2005**, 119ff). A variety of artistic and architectural evidence for contact between Mesopotamia and late Pre-Dynastic Egypt has in fact been found, but none of it can be dated precisely in relation to tomb U-j. The resemblance of the repertory of signs on earlier cylinder seals (Naqada II–IIIA1; c. 3800–3300 BC) with early Sumerian/Elamite pictograms, has been used to add weight to the Sumerian/Elamite influence argument (Kaiser 1990, p. 299; Smith 1992, p. 235; Hill 2004, p. 95–104; Honoré 2007, p. 31–45). However, it cannot be proved that these Pre-Dynastic seals already reflected a proper writing system, such as that detected among the U-j labels (Kahl 1994, p. 154; Morenz 2004, p. 40; Regulski 2008b, p. 993–994; Stauder 2010). These seals probably belong to an earlier form of notation that was later replaced by the Egyptian writing system. The earliest solid evidence of Egyptian writing differs in structure and style from the Mesopotamian and must therefore have developed independently. The possibility of "stimulus diffusion" from Mesopotamia remains, but the influence cannot have gone beyond the transmission of an idea (Ray 1986; Bard 1992, p. 297–306; Vernus 1993; Dreyer 1998, p. 181–182; Baines 2004, p. 175).

Figure 203: Closeup of Kaaper's wooden statue, 5th or 4th dynasty of the Old Kingdom, from Saqqara, c. 2500 BCE, ethnic Egyptian prototype.

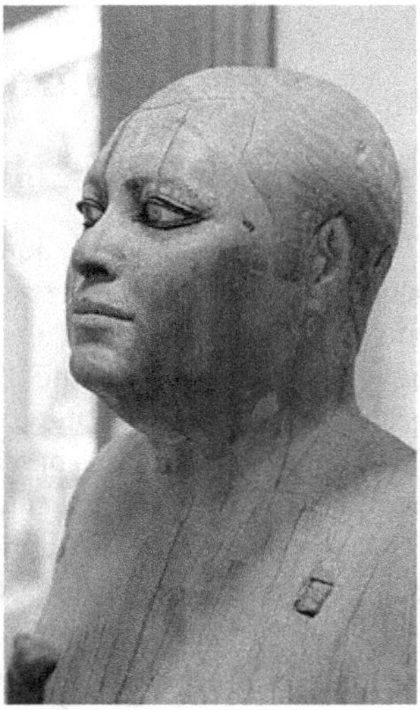

Figure 204: Bust of an unknown Sumerian; Are the Egyptians and Sumerians cousins?

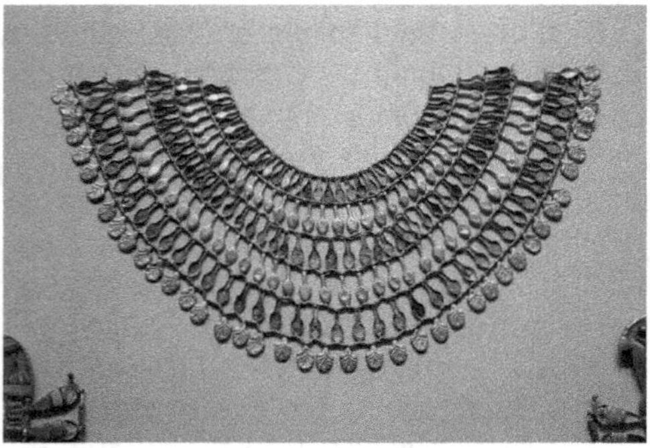

(Left) Figure 205: Egyptian Gold Broad Collar

Historical/Cultural notes

Egyptian ethnic continuity was based upon the Nile River, and prodigious agricultural and food possibilities. Invaders (Hyksos from the north, and Nubians/Kush from the south) were either repelled or absorbed.

Ahkenaten and monotheism was an adaptation of the fiery war god of the people of the desert, the Habiru, Moses. c. 1326–1200 BCE.

Below, paintings of Egyptian women in a soiree with musicians, food, drink, and gossip, possibly influenced by Minoan civilization, c. 1500 BCE. In all civilizational prototypes, there is a search for middle-class stability, family, steady income, modern housing, leisure for music, dance, choral entertainment by professionals.

Figure 206 & 207: Exemplification of Upper-Class Egyptian Women Enjoying a Musical Soiree

Figure 208: The ancient Egyptians maintained a rich cultural heritage, complete with feasts and festivals accompanied by music and dance

Figure 209: Goddess Hathor embracing King Thutmosis III (See Sumerian images of gods embracing, Chapter10)

(Left) Figure 210: Menkaura (Mycerinus) and queen, Old Kingdom, Dynasty 4, 2490–2472 BC. The formality of the pose is reduced by the queen's arm round her husband.

"If thou wouldest be wise, provide for thine house, and love thy wife that is in thine arms. Fill her stomach, clothe her back; oil is the remedy of her limbs. Gladden her heart during thy lifetime, for she is an estate profitable unto its lord. Be not harsh, for gentleness mastereth her more than strength. Give to her that for which she sigheth and that toward which her eye looketh; so shall thou keep her in thy house." (Ptah-hotep c. 3550 BCE; Brown 1923)

Figure 211: Queen Nefertari plays Senet, c. 1250 BCE—Wife of Akhenaten

Hebrew Bible: Genesis Chapter 41:46
"And Joseph was thirty years old when he stood before Pharaoh king of Egypt—And Joseph went out from the presence of Pharaoh, and went throughout all the land of Egypt.

47 And in the seven years of plenty the earth brought forth in heaps.

48 And he gathered up all the food of the seven years which were in the land of Egypt, and laid up the food in the cities; the food of the field, which was round about every city, laid he up in the same.

49 And Joseph laid up corn as the sand of the sea, very much, until they left off numbering; for it was without number.

And the seven years of plenty, that was in the land of Egypt, came to an end.

54 And the seven years of famine began to come, according as Joseph had said; and there was famine in all lands; but in all the land of Egypt there was bread.

55 And when all the land of Egypt was famished, the people cried to Pharaoh for bread; and Pharaoh said unto all the Egyptians: 'Go unto Joseph; what he saith to you, do.'

56 And the famine was over all the face of the earth; and Joseph opened all the storehouses, and sold unto the Egyptians; and the famine was sore in the land of Egypt.

57 And all countries came into Egypt to Joseph to buy corn; because the famine was sore in all the earth."

Moses and the exodus—what happened to Akhenaten?

Within a hundred years of Moses and the Exodus, a weird blip happens on the ancient Egyptian radar. Amenhotep III dies, and is succeeded by his second son, Amenhotep IV (c. 1335–1315 BCE). This son is a visionary who turns away from the state pantheon of gods, and instead, turns Egypt to the worship of a single deity known as the Aten—the disk of the sun.

"Amenhotep IV changes his name to Akhenaten—"he who is beneficial to the Aten"—and shuts down the polytheistic, religious capital at Thebes. Akhenaten builds a new, monotheistic capital 180 miles north at Amarna, and there, serves as high priest to the Aten with his beautiful wife, Nefertiti."

Pharaoh Akhenaten stripped everything from the polytheistic priesthood—hundreds of years of power, prestige, and wealth—and declared the Aten as the one, supreme god.

The art of the period goes from over-the-top idealism—depicting pharaohs and priests as youthful, muscular, and god-like—to in-the-raw naturalism—depicting Akhenaten himself as potbellied, thick-lipped, and effeminate.

It seems Akhenaten reigned as Pharaoh a bit under 20 years, and upon his death, everything returned to Egyptian "normal"—the single god concept was again replaced by the pantheon of deities … and pharaohs, once again, took on larger-than-life imagery.

What happened to Akhenaten? What motivated him to revolutionize centuries of Egyptian culture and religion? Could it be that experiencing the death of the first born of all the males in Egypt some decades prior was such a motivation? (Biblical Archaeological Society, 2018)

Amenophis IV, Naphu(`)rureya, Akhnaten

(Left) Figure 212: ** Statue of Akhenaten in the early Amarna style

Most authorities propose that Moses left Egypt with his "corvee rabble," between fifty and one hundred years after the death of Akhenaten.

Mathematics

The introduction of writing in Egypt in the predynastic period (c. 3000 BCE) brought with it the formation of a special class of literate professionals—the scribes. By virtue of their writing skills, the scribes took on all the duties of a civil service: record keeping, tax accounting, the managemen of public works (building projects and the like), even the prosecution of war through overseeing military supplies and payrolls. Young men enrolled in scribal schools to learn the essentials of the trade, which included not only reading and writing but also the basics of mathematics.

Figure 213: Slab stela of Old Kingdom princess Neferetiabet (dated 2590–2565 BC) from her tomb at Giza, painting on limestone, now in the Louvre. The king's daughter, Neferetiabet, is shown with an offering of 1000 oxen, bread, beer, etc.

One of the texts popular as a copy exercise in the schools of the New Kingdom (13th century BCE) was a satiric letter, in which one scribe, Hori, taunts his rival, Amen-em-opet, for his incompetence as an adviser and manager. "You are the clever scribe at the head of the troops," Hori chides at one point. "A ramp is to be built, 730 cubits long, 55 cubits wide, with 120 compartments—it is 60 cubits high, 30 cubits in the middle … and the generals and the scribes turn to you and say, 'You are a clever scribe, your name is famous. Is there anything you don't know? Answer us, how many bricks are needed?'" Let each compartment be 30 cubits by 7 cubits. This problem, and three others like it in the same letter, cannot be solved without further data. But the point of the humour is clear, as Hori challenges his rival with these hard, but typical, tasks.

What is known of Egyptian mathematics tallies well with the tests posed by the scribe Hori. The information comes primarily from two long papyrus documents that once served as textbooks within scribal schools. The Rhind papyrus (in the British Museum) is a copy, made in the 17th century **BCE**, of a text two centuries older still. In it, is found a long table of fractional parts to help with division, followed by the solutions of eighty-four specific problems in arithmetic and geometry.

The Golenishchev papyrus (in the Moscow Museum of Fine Arts), dating from the 19th century **BCE**, presents twenty-five problems of a similar type. These problems reflect well the functions the scribes would perform, for they deal with how to distribute beer and bread as wages, for example, and how to measure the areas of fields as well as the volumes of pyramids and other solids. (Encyclopedia Britannica)

Figure 214: Measuring and recording the harvest is shown in a wall painting in the tomb of Menna, at Thebes (Eighteenth Dynasty, c. 1500 BCE).

Figure 215: Depiction of craft workers in ancient Egypt

Social classes

The ancient Egyptians viewed men and women, including people from all social classes except slaves, as essentially equal under the law, and even the lowliest peasant was entitled to petition the vizier and his court for redress. Although slaves were mostly used as indentured servants, they were able to buy and sell their servitude, work their way to freedom or nobility, and were usually treated by doctors in the workplace.

Both men and women had the right to own and sell property, make contracts, marry and divorce, receive inheritance, and pursue legal disputes in court. Married couples could own property jointly and protect themselves from divorce by agreeing to marriage contracts, which stipulated the financial obligations of the husband to his wife and children, should the marriage end. Compared with their counterparts in ancient Greece, Rome, and even more modern places around the world, ancient Egyptian women had a greater range of personal choices and opportunities for achievement. Women, such as Hatshepsut and Cleopatra VII, even became pharaohs, while others wielded power as Divine Wives of Amun. Despite these freedoms, ancient Egyptian women did not often take part in official roles in the administration, served only secondary roles in the temples, and were not as likely to be as educated.

Population growth

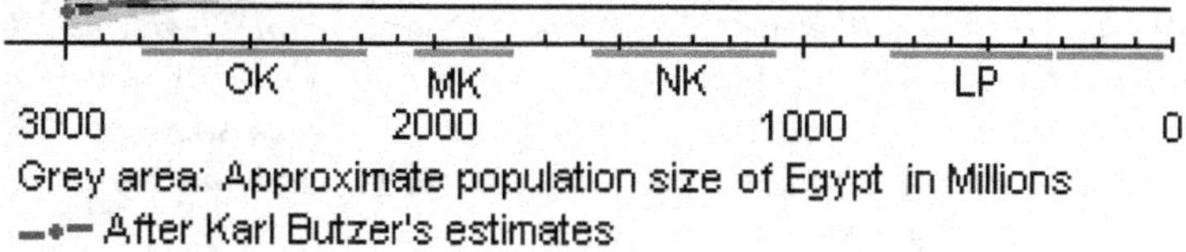

(Left) Figure 216: MK: Middle Kingdom NK: New Kingdom LP: Late Period (Left)

It has been estimated that at the beginning of the pharaonic period, the population count was about one million, and that at time of the Roman conquest, about 5 million people lived in Egypt, give or take a million or so. The sizes of the dips during the Intermediate Periods are arbitrary. These 4–5 million in the 1st century CE, provided about 3–4 months of the grain needs of the city of Rome, c. one million people. **Jump to 2016 CE: Egypt's population=92 million**

Herodotus, histories, c. 425 BCE

The Egyptians in agreement with their climate, which is unlike any other, and with the river, which shows a nature different from all other rivers, established for themselves manners and customs in a way opposite to other men in almost all matters:

for among them the women frequent the market and carry on trade, while the men remain at home and weave; and whereas others weave pushing the woof upwards, the Egyptians push it downwards: the men carry their burdens upon their heads and the women upon their shoulders:

the women make water standing up and the men crouching down: they ease themselves in their houses and they eat without in the streets, alleging as reason for this that it is right to do secretly the things that are unseemly though necessary, but those which are not unseemly, in public:

no woman is a minister either of male or female divinity, but men of all, both male and female:

to support their parents the sons are in no way compelled, if they do not desire to do so, but the daughters are forced to do so, be they never so unwilling.

The priests of the gods in other lands wear long hair, but in Egypt they shave their heads:

among other men the custom is that in mourning those whom the matter concerns most nearly have their hair cut short, but the Egyptians, when deaths occur, let their hair grow long, both that on the head and that on the chin, having before been close shaven:

other men have their daily living separated from beasts, but the Egyptians have theirs together with beasts: other men live on wheat and on barley, but to any one of the Egyptians who makes his living on these it is a great reproach; they make their bread of corn,[l] which some call spelt: they knead dough with their feet and clay with their hands, with which also they gather up dung:

and whereas other men, except such as have learnt otherwise from the Egyptians, have their members as nature made them, the Egyptians practice circumcision:

as to garments, the men wear two each and the women but one:

and whereas others make fast the rings and ropes of the sails outside the ship, the Egyptians do this inside:

finally in the writing of characters and reckoning with pebbles, while the Hellenes carry the hand from the left to the right, the Egyptians do this from the right to the left; and doing so they say that they do it themselves rightwise and the Hellenes leftwise: and they use two kinds of characters for writing, of which the one kind is called sacred and the other common.

(Herodotus)

Historical Issues

"Southern boundary, made in the year 8, under the majesty of the King of Upper and Lower Egypt, Khekuer, who is given life forever and ever; in order to prevent that any Negro should cross it, by water or by land, with a ship, (or) any herds of the Negroes; except a Negro who shall come to do trading in Iken, or with a commission." (Semna stela in the year 8 [c. 1870 BCE] of Senusret III)

While the Hyksos, c.1600 BCE (whose name simply means "foreign rulers"), were hated by the Egyptians, they introduced a great many improvements to the culture such as the composite bow, the horse, and the chariot, along with crop rotation and developments in bronze and ceramic works.

The treaty of Kadesh 1276 BCE

Signed in ancient Syria between Egypt Ramses II and the Hittites (Indo-European Kingdom plus Semite allies), itself contains more than twenty principles and obligations for both sides. However, some of the key points are the following:

"The third obligation is that neither side will attack the other, and is in force till the end of time. Neither the Egyptians, nor the Hittite should and could pass the land of the other nation: There shall be no hostilities between them, forever. The great chief of Kheta shall not pass over into the land of Egypt, forever, to take anything therefrom. Ramses-Meriamon, the great ruler of Egypt, shall not pass over into the land of Kheta, to take anything] therefrom, forever."

Aside from ending the war between the two empires, the treaty also forged an alliance between the two sides in future wars with a third enemy. Obligation No.5 from the treaty states "If another enemy come against the lands of Usermare-Setepnere (Ramses II), the great ruler of Egypt, and he shall send to the great chief of Kheta, saying 'Come with me as reinforcement against him,' the great chief of Kheta shall [come], and the great chief of Kheta shall slay his enemy. But if it be not the desire of the great chief of Kheta to come, he shall send his infantry and his chariotry, and shall slay his enemy."

The treaty also regulates whether prisoners from the one country could ask for exile in the other country. According to the treaty, no man could flee from Egypt to the land of the Kheta (Hittite territory) and vice versa. This obligation is No.11 and it states "if any great man shall flee from the land of Kheta, [and he shall come to] Usermare-Setepnere, the great ruler of Egypt, (from) either a town or a district, or [any region of] those belonging to the land of Kheta, and they shall come to Ramses-Meriamon, the great ruler of Egypt, then Usermare-Setepnere, the great ruler of Egypt, shall not receive them, (but) Ramses-Meriamon, the great ruler of Egypt, shall cause them to be brought to the great chief of Kheta. They shall not be settled." (Mishkov 2016)

Figure 217: Ramesses II at The Battle of Kadesh

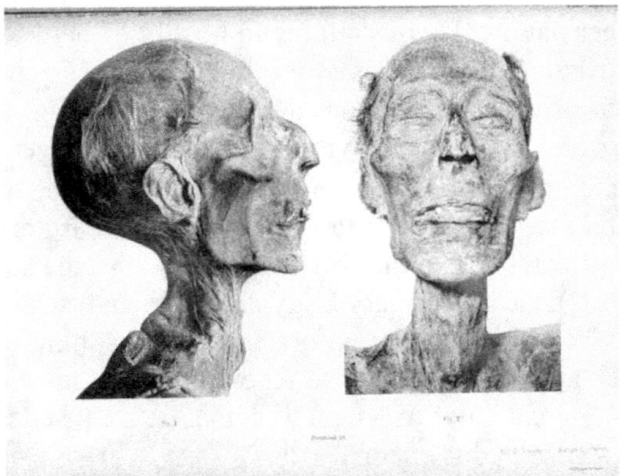

Figure 218: Mummy of Ramses II, who had red hair

The Merneptah Stele c. 1205 BCE

The conclusion reads, in part: "The princes prostrate themselves, saying, 'Peace!' Not one of Nine Bows dares raise his head; Tehenu is plundered while Hatti is peaceful, Canaan is seized by every evil, Ashkelon is carried off and Gezer is seized, Yenoam is made as that which never existed, **Israel** is wasted without seed, Khor is made a widow of Egypt, All the lands are at peace. Everyone who travels has been subdued by the King of Upper and Lower Egypt." (Merenptah was a son of Ramses II), (Yenoam=Lybia).

Egyptian Jewelry

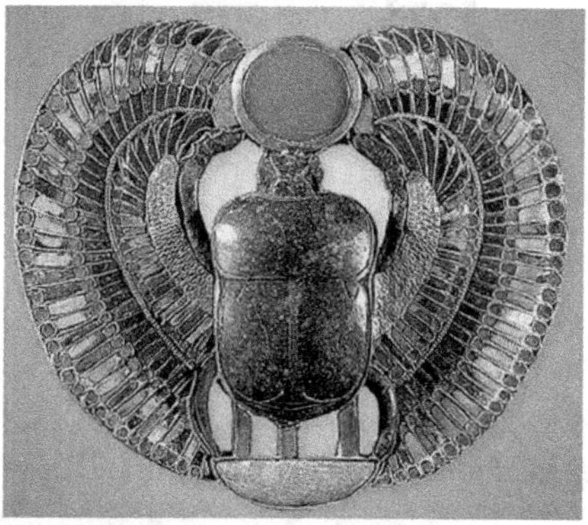

Figure 219: © Zepfanman.com - Pectoral Scarab of KingTuteankhman

Figure 220: Gold amulet with stone inserts

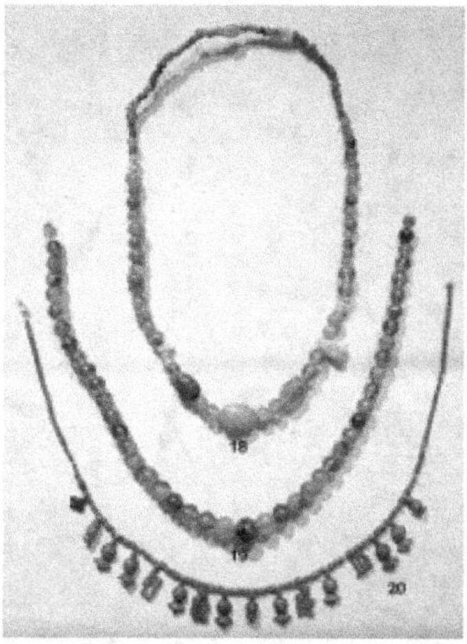

Figure 221: Simple necklaces were often created using tiny beads of lapis, malachite, turquoise, silver and gold. Typical of more proletarian enhancements. (Ancient Egypt Online)

Figure 222: Group of stones collected in Egypt showing the range of colors and textures available to the ancient artists (Kahn Academy).

Figure 223: An Egyptian burial chamber mural, from the tomb of Khnumhotep and Niankhkhnum dating to around 2400 BCE, showing wrestlers in action: Wrestling was another popular sport in ancient Egypt. The Pharaohs would watch this sport very often. (Peluso)

Figure 224: Aboved: c.2000 BCE, Mesopotamian plaque, boxing with a rhythm section

China

Early Dynasties

The Shang dynasty, 1700–1200 BCE, is the earliest dynasty of traditional Chinese history supported by archaeological evidence. Excavation at the Ruins of Yin (near modern-day Anyang), which has been identified as the last Shang capital, uncovered eleven major royal tombs and the foundations of palaces and ritual sites, containing weapons of war and remains from both animal and human sacrifices. Tens of thousands of bronze, jade, stone, bone, and ceramic artifacts have been found. The Bronze Age in China began with the Xia dynasty, c. 2200–1750 BCE. Examples from this period have been recovered from ruins of the Erlitou culture, in Shanxi, and include complex but unadorned utilitarian objects. In the following Shang dynasty, more elaborate objects, including many ritual vessels, were crafted.

(Left) Figure 225: Shang dynasty bronze ritual vessel

Writing

Patricia Buckley Ebrey expresses the consensus on this: exactly when writing was first used in China is not known, since most writing would have been done on perishable materials like wood, bamboo, or silk. Symbols or emblems inscribed on late neolithic pots may be early forms of Chinese graphs. Early Shang bronzes sometimes have similar symbols cast into them. The earliest evidence of full sentences is found on the oracle bones of the late Shang. From these divinatory inscriptions, there can be no doubt that the Shang used a language directly ancestral to modern Chinese, and moreover used a written script that evolved into the standard Chinese logographic writing system still in use today. (Ancient History Encyclopedia)

(Right) Figure 226: Oracle bone writing, late Shang, c. 1300 BCE

Civilization

"Many of the achievements of mankind date back to the Chinese civilization. China became a birth-place of gun powder and paper. It is here where the Great Silk Road sprang from, becoming a sui generis linking bridge between the East and the West. Starting from Sian, the former capital of ancient China, the Great Silk Road ended at the Mediterranean seashore. It is this road, by which China exported paper, silk, gun powder, precious stones, oriental spices, to Europe. The Great Silk Road played an important role in formation of Chinese culture, because it was the route, by which Buddhism, one of the main religions of Chinese civilization, found its way to the Heavenly Empire." (Adventour)

Hybridization—Xian, 210 BCE

Figure 227 & 228: Terracotta army of Emperor Qin Dynasty, Shi Huang, c. 210 BCE in Xian, on the Yellow River.

These soldiers exemplify an emperor's desire to recognize great achievements without necessarily sacrificng the lives of his soldiers. In ethnicity, at a distance of well over a thousand years in time from the origin of the Shang dynasty, c. 1700 BCE, these soldiers still reflect in their visage and height, a hybrid people, partly Caucasoid, partly of Mongoloid/Australid racial heritage, very similar to the indigenous people of Kazakhstan

The Tokharians—c.2000 BCE–900 CE

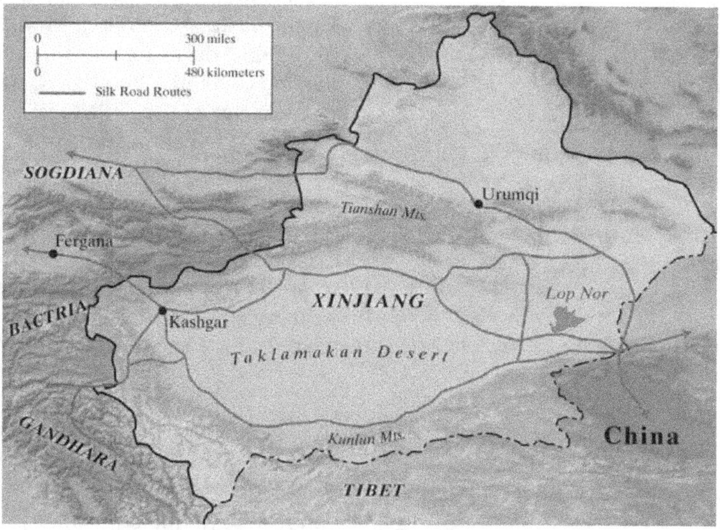

Figure 229: Ancient homeland of Tokharians

"One of the most fascinating, and controversial, discoveries made to date in the history of world archaeology is the exceptionally well-preserved 'mummified' human remains—men, women, and infants—around the Tarim Basin of northwestern China. (Although they are not actual mummies but simply bodies naturally preserved through desiccation in the arid desert environment, I will follow usual convention and use the term 'mummy' in this paper.) The well preserved state of these Tarim mummies surpasses any of those coming from ancient Egypt … many of them also appear western European in origin, with reddish-blond or brown hair, deep-set round eyes, aquiline nose, and, among the adult males, prolific facial hair. The identification of the mummies with Europeans has led to considerable historical and political controversy in China and around the world." (Kaufman **2011)**

Figure 230: The "Beauty of Kroran (Loulan)," the actual mummy and an artist's reconstruction based on hair color and facial features. (www.ancientworlds.net)

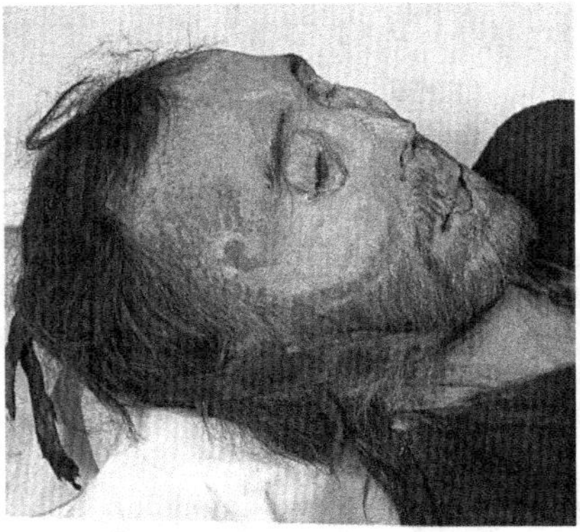

(Left) Figure 231: Perhaps the most famous mummy discovered in the Tarim Basin: "Cherchen Man," (c. 1000 BCE) excavated in the southern Tarim region of Cherchen. Left: Solid as a warrior of the Caledonii (Scotland) tribe, the man's hair is reddish brown flecked with grey, framing high cheekbones, a long nose, full lips and a ginger beard. When he lived three thousand years ago, he stood six feet tall, and was buried wearing a red twill tunic and tartan leggings. He looks like a Bronze Age European. In fact, he's every inch a Celt. (Coonan 2006)

(Left) Figure 232: A Celtic stone head excavated near Mšeké Žehrovice in the modern Czech Republic. (www.ceramicstudies.me.uk)

Archaeological findings suggest initial settling of the Tarim oases by peoples from the west, in the early second millennium BCE. "There is ample archeological and historical evidence that chariots and fortified towns came to China from the West" (Lubotsky 1997, p. 382). Findings suggest that there was a time of major transition in the region to semi-nomadic pastoralism, horseback-riding, the manufacture of colorful textiles (including plaid-twill), iron-making in the late second millennium BCE (Thornton and Schurr 2004, p. 98), and the use of wheeled vehicles, including wagons and chariots, in the Tarim.

The use of bronze began in the Tarim Basin as early as 2000 BCE, while it did not begin in China until c. 1500 BCE (Romgard 2008, p. 32). "The introduction of both bronze and iron tools seems to have traveled from west to east" (id., p. 30). Metal analyses done in China indicate a possible direct link to the Urals (id.). Mud-brick construction and irrigation techniques also made their appearance in the Basin around 2000 BCE (id., pg. 38). Since archaeological and genetic evidence indicates that the Han Chinese did not arrive in the Tarim until the late first millennium BCE (Thornton and Schurr 2004, p. 98), these cultural innovations likely correlate with the arrival of the Tokharians in the region.

It was here in the Eurasian steppe, where domesticated sheep and cattle first appeared by 5000 BCE, horseback-riding by 3500 BCE, and wagons by 3000 BCE (Anthony 2007). The earliest evidence of lightweight spoked wheels dates to c. 2000 BCE and is possibly an outgrowth "of experimenting with bending wood for making a more efficient bow" (Barber 1999, p. 203–204) and occurs at sites in the steppe area just east of the southern Ural Mountains along the Ob and Ural Rivers.

An examination of the evidence, however, does lead us to the conclusion that the Tokharians were of Indo-European stock, and furthermore, were of western European origin, perhaps related to those who became known as the "Celts" in Europe. As Romgard states, "the idea of a European entry either directly from the West or from the steppe cultures in the north (into the Tarim Basin) is today the prevailing theory" (2008, p. 33).

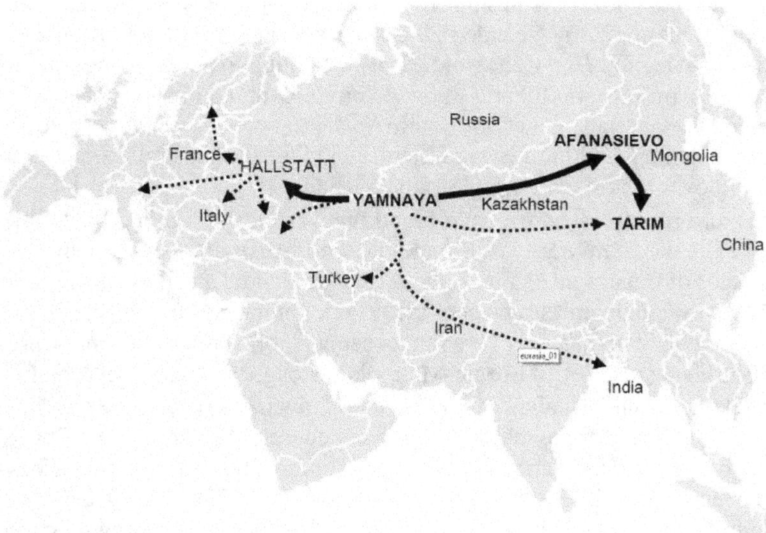

Figure 233: Map of Eurasia with bold solid lines showing probable Proto-Tokharian and Proto-Celtic migrations east and west from the Indo-European (IE) homeland in the Caucasus Mountain region (Yamnaya horizon) and dotted lines showing migrations of other IE groups, including Celts from the Hallstatt region (modern Austria) and Indo-Iranians heading into South Asia as well as toward the western Tarim. (Blank map from [www.prattmuseum.org] Hallstatt culture, c.1000 BCE [see also== "Prehistoric 'Europeans' in Xinjiang? A Case for Multiple Interpretations" Schurr, T. 2005—Academia; Dickens, M.]

"Everything You Always Wanted to Know about Tocharian….." [**www.oxuscom.com/eyawtkat.htm**][; Amy Quinn, "The Tarim Mummies: The Use and Abuse of History" The Trinity College Dublin Histories and Humanities Journal, 2012] **Figure 234 (Right)**

Discussion: who are the Chinese?

As stated above with regard to the terracotta soldiers, 210 BCE, in Xian, some 1400 years after the establishment of the Shang dynasty in that area of modern

China, the Chinese are a hybrid people. Today, they largely understand and accept this fact. The earliest human migrants out of Africa must have occurred some two million years ago as evidence from one of its descendants, the Flores miniatures in Indonesia. The Flores represent a form of hominids mid-way between the Australopithecines and *Homo*. Over the millennia, these Australid sub-species migrants from Africa flowed in large numbers into the east.

The Australid (Melanesian) sub-species of *Homo* dominated the landscapes of India to the east into Siberia, down into the the Indonesian islands and then New Guinea, Australia, and Tasmania. Here, we include China, Korea, and Japan. The Mongoloids came later in smaller numbers, perhaps a version of the Capoids (Bushman ancestors), the Neanderthal Caucasoids, and most lately from 200,000 BP, the Cro-Magnon *Hss*, Caucasoids. It is interesting to note that the mysterious Denisovians of Eastern Siberia show a mixture of ethnic origins, including Australid (Melanesian) genes. In South America, scientists have also found an admixture of racial/ethnic heritage, here including the Australid genes carried with them over the frozen Bering Straits, c. 15,000 BP.

In more recent times, we have found that practically all the nationalities of East Asia have a mixture of Mongoloid and Australid genes with varying percentages of Caucasoid elements, both Neanderthaloid and Cro-Magnon. In the map below, the blue areas represent the original homeland of the Shang and the to-be-Han ethnic dynasties. The red occupants of southern China were the hybridized speakers of Australo-Mongoloid tongues, today represented by Cambodia, Laos and Thailand. It is also clear, given the recent research into the history of the Tokharians, that the Shang conquerors themselves, c. 1700 BCE, were an offshoot of the various Cro-Magnon/Indo-European-speaking tribes that had their impact as far south as Egypt, the Hyksos c. 1700–1300 BCE (Semites and Indo-Europeans). For example, when the Joshua-led Hebrews arrived in ancient Palestine, c. 1200 BCE, the towns such as Jerusalem were inhabited by Indo-European people under the protective wing of the Egyptians.

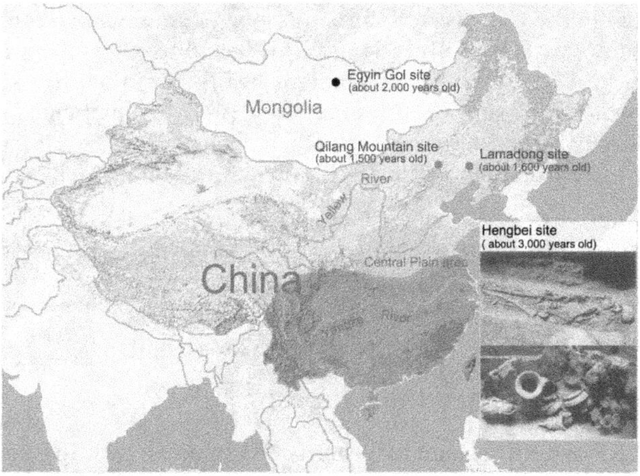

Figure 235: [Zhao, Y-B, et al, "Ancient DNA Reveals,,,,," *PLOS]*

The issue of the hybridization of the Han Chinese is only as important as the message it sends to our own time—that the hybridization of *Hss* is as yet incomplete. Destiny tells us that at some time in the future, humanity will create a super-species—international—that will render the ancient separation of races an irrelevant dimension of a current human dynamic.

Great art of the Han Chinese

Figure 236: Li Zhaodao's early 8th century CE Emperor Minghuang's Journey into Shu

Note the poetry, philosophy that must always accompany a visual scene.

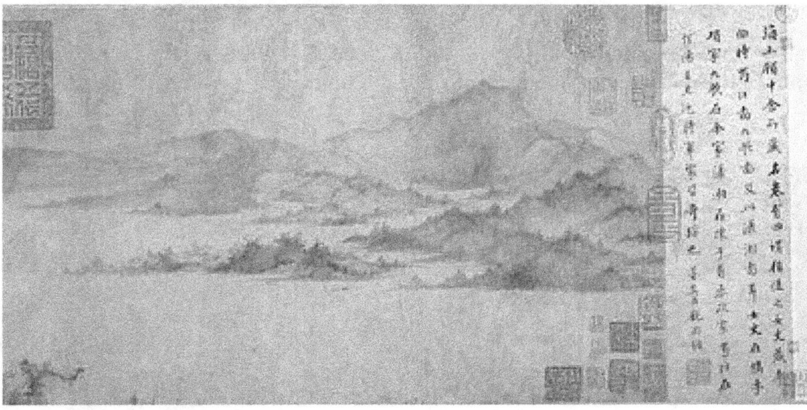

Figure 237: Part of Eight Views of Xiao-xiang, an imaginary tour through Xiao-xiang by Li Shi (李氏), 12th-century CE scroll, 30 x 400 cm. Ink on paper. (Tokyo National Museum)

Figure 238: A wooden Bodhttva from the Song dynasty (960–1279 CE)

Figure 239: Chinese jade Cup with dragon handles, 12th century CE

Figure 240: Traditional Chinese portrait Hanging scroll portrait on silk

Figure 241: Blue underglaze statue of a man with a conch, Jingdezhen porcelain, Ming dynasty (1368–1644 CE)

Figure 242: Tiger from Shang dynasty, c. 1500 BCE Bronze

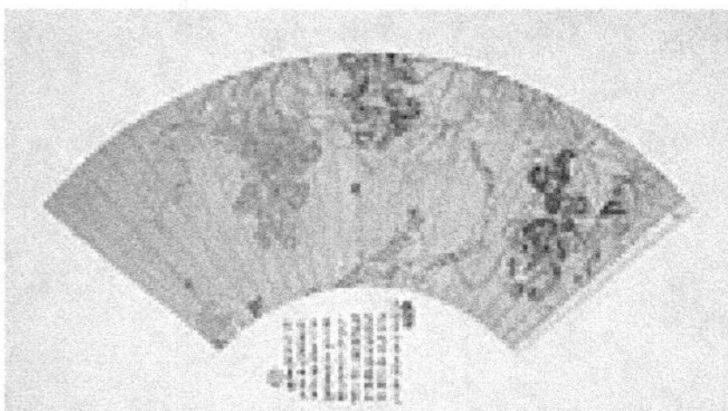

Figure 243: *Grapes and Mantis*, Ma Quan, 18th century CE (Hunan Museum)

The inner kingdom withdraws

Zheng He, Wade-Giles romanization Cheng Ho, **whose** original name **was** Ma Sanbao, later Ma He (born **c**. 1371 CE, Kunyang, near Kunming, Yunnan province, China—died 1433 in Calicut [now Kozhikode], India), **was an** admiral and diplomat who helped extend the maritime and commercial influence of China throughout the regions bordering the Indian Ocean. He commanded seven naval expeditions, almost a century before the Portuguese reached India by sailing around the southern tip of Africa. Zheng He was from a Hui (Chinese Muslim) family. His father was a hajji, a Persian Muslim who had made the hajj (pilgrimage) to Mecca. His family claimed descent from early Mongol invaders.

When Ma He was about **ten**, the Ming army invaded Yunnan to take it back from the Mongols and bring it under Ming control. The Ming soldiers killed Ma He's father in the fighting and captured Ma He. As was customary with juvenile captives, they castrated him by cutting off his testes and penis with a sword. He survived this trauma and was handed over to be a servant in the household of the emperor's fourth son, Zhu Di. The prince whom Ma He served, Zhu Di, was only 11 years older than He. They were based in Beijing, in the north of China, near Mongol territory, and they spent a lot of time together campaigning on horseback on the Mongolian steppe. Ma He grew unusually tall and strong and became a skilled fighter and brave leader.

Figure 244: Zheng He

Zhu Di, now emperor Yongle, made his companion Ma He the director of palace servants (similar to a chief of staff), and changed Ma's name to Zheng He, in commemoration of his role in battles to win the throne. (Zheng was the name of Yongle's favorite warhorse.) Yongle ruled from 1402 to 1424. Since Yongle was determined to control trading in the Indian Ocean, one of his first acts was to commission the construction of 3,500 ships, with Zheng He supervising the construction and then commanding the fleet. Zheng He's nine-masted flagship measured about 400 feet long; for comparison, Christopher Columbus's *Santa Maria* measured just 85 feet.

On the first voyage, from 1405 to 1407, sixty-two nine-masted "treasure ships" led the way, followed by almost 200 other ships of various sizes, carrying personnel, horses, grain, and 28,000 armed troops. On his seventh and final voyage, from 1431 to 1433, Zheng He apparently died at sea and was likely buried off the coast of India. After these voyages, the Chinese were too interested in internal affairs to look beyond their world, until Christopher Columbus and his fellow Europeans descendants dragged them into the modernity of the 19th and 20th centuries.

Ethnic women—illustrated

Figure 245: Chinese woman

Figure 246: Korean woman

Figure 247: Japanese woman

Figure 248: Russian woman

Above is a random comparison of women from four different ethnicities. The difference between the Russian face and the three Asiatic faces tell us about the relative unity of the Asiatic ethnicities. Arriving in their new geographies in the time frame between 200 BCE and 200 CE, the Korean and Japanese lands received a new influx of genes. These shaped their cultural, linguistic, and historical position. Being influenced by an already powerful Chinese Han cultural scene, these new ethnicities, having an originating migratory hegira from central Asia, and sporting an Altaic Mongolian/Siberian linguistic armament, possibly even rooted further back in time to the Uralic linguistic heritage of the Finns and the Hungarians, these newcomers were already heavily Caucasoid in genetics and culture.

Certainly, with the exception of the Samurai of the Japanese feudal military world, these were heavily esthetic civilizations, continually influenced by what was happening in China. The art and philosophy/religion was deeply abstract and poetic, suffused with ethical motifs. We may not see such a powerfully insular civilization again in the history of *Hss*. The status of these three ethnicities in the dynamics of the great nations of our time, is testimony enough of their capacity to turn from an earlier abstract envisionment of cultural symbolism to what is taking place in this high-tech 21st century.

The eugenic factor

It is true, the European colonialist took over these nations from the 18th century, when they could not muster the economic/military/technological power to withstand the hurricane emanating from the West. The assumption in the West was that these Asiatic ethnicities were permanently weak. The West was fascinated by the esthetics of these cultures, but disregarded the implicit power of the Middle Kingdom's cultural levels. The power, intellect and organizational adeptness that we see now had to have important socio-biological roots.

Forms of concubinage are universal in recorded history. Many thinkers believe that in China, the practiced went far beyond the royal house or the intellectual bureaucrats, to the more well-off secular classes, as far down as economics would allow for the support of concubines and their children. One of the off-shoots of all these arrangements has been the social disenfranchisement of the lower class male. In a modern state as in the Arab world, the children produced by both plural wives and concubines have created a potentially large ruling class, creating a socio-biological group of able, potential rulers. Today in China, after the coming of Mao and communism, 1949, the strict outlawing of plural marriage and concubinage has led to a plurality of males, one wife, one child, and an increase in the value of educated females, as well as a temporary social crisis for the millions of men without legal relations with a woman.

The taking of concubines was always highly selective in terms of their persona as well as social background, yet possibly inferior to the political elite, and highly wealthy commoners. So, you now have the talents of the able female of slightly lower social heredity and the many children only of elite males. Consider that such arrangements took place over a period of several thousand years, and over the intellectual potency of a stable nationality, the inner kingdom, with clear pathways to power and wealth for most. An objective outside observer could predict the potentiality of a nationality endowed with ever higher layers of *Hss* intelligence.

Another element in the incipient eugenic uplift of the masses of Chinese, Japanese and Koreans is the issue of literacy. Critical to the advancement of any society from implicit civilizational capacity, to the full use and creative involvement of the citizenry in advanced abstract skills and vocations, is writing. It showed itself in China soon after the Shang arrived, possibly stimulated by written Tokharian. No matter the origin, the approach of the Chinese literati in their advance from pictographic symbols of meaning, to more abstract, agreed-upon symbols of trade, mathematics, and even ethnic mythology, was the form of writing.

The Chinese written symbols, followed in time by Korean and Japanese forms, are called "logographs, symbols of meaning without a spoken, phonetic equivalent, 'ideograms/graphemes.'" Children and adults have to learn the meaning of hundreds of these symbols and their modifiers. This was unlike the simplifying attempt of western Mediterranean or Egyptian Semite workers, c. 1900–1600 BCE, then passed on to Phoenecian traders who secured the simplification of uniting sound and sight in language—the alphabetic system—then sending it on to the Greeks and the West.

The Chinese resisted, and possibly for social class reason, attempted to maintain a scribal monopoly, because residents of the south of China, who could not understand the dialect of those in the north, would have to depend on the official Mandarin system. This system was, in theory, translatable into the home dialects. For those who could not master such an intellectual, highly abstract symbolic system of communication (so said the great linguist Edward Sapir), it would be tough luck—a future of unskilled physical labor, illiteracy.

"*Who would want to marry and have children with such a person.*" Thus, another route for the encephalization of the society, and the permanent separation and suppression of a possible Han underclass.

The moral issues are many. In India especially, this has led to a bifurcated society, as we shall explore below. It can work in societies which are ethnically homogeneous, and is much more difficult in nations which are multi-ethnic. It would be valuable to compare such societies, those allowing for concubinage/polygany, here also Korea and Japan, with nations, such as in Christianity, where the moral imperative ostensibly enforces monogamy as the only moral and legal relationship between man and woman. Of course, below the surface of such a moral interdict, there could and does exist a subterranean world of prostitution, hidden mistresses, serial monogamy, etc. within the social structure.

Thus, the challenge for any nationality to continue to exist, and flourish no less, is the challenge of maintaining the moral status of the female, while at the same time preserving the intellectual capital of the nation. What we observe today in China is the historic product of questionable moral and social traditions that yet produced a great abstract, intellectual, and artistic civilization, now growing into modernity, deeply capable in science, technology, social and cultural power for all its citizens of both genders.

Figure 249: 19th century photograph of royal concubines

Chapter 13

Civilizations of *Hss*—India

Indus River civilizations

The Indus Valley Civilization is often separated into three phases: the Early Harappan Phase, from 3300 to 2600 BCE, the Mature Harappan Phase, from 2600 to 1900 BCE, and the Late Harappan Phase, from 1900 to 1300 BCE.

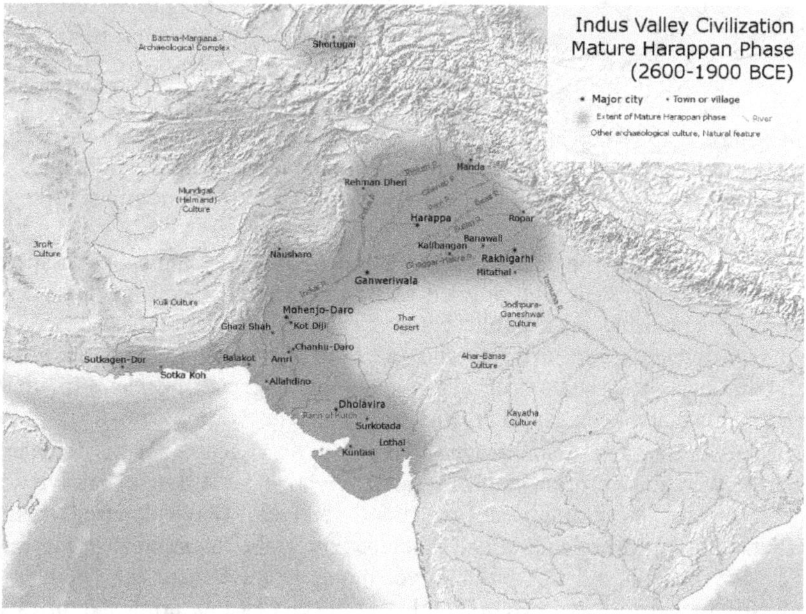

Figure 250: This map shows the extent of the Indus Valley Civilization during the Mature Harappan Phase, 2600–1900 BCE. Civilization is highlighted in brown in the area of modern-day Pakistan and northern India. The rest of the map is green and is a partial map of India and the area northwest of Pakistan.

Mohenjo-daro is thought to have been built in the twenty-sixth century BCE; it became not only the largest city of the Indus Valley Civilization, but one of the world's earliest major urban centers. Located west of the Indus River in the Larkana District, Mohenjo-daro was one of the most sophisticated cities of the period, with advanced engineering and urban planning.

Below: The Great Bath at Mohenjo-daro. Amid the brick ruins of a 3rd-millennium BCE city, stairs descend on two sides into a large, rectangular, brick-lined pit. Wooden stakes and wire encircle the perimeter, preventing entry by modern-day tourists.

(Left) Figure 251: Archaeological remains at the lower town of Lothal. The bricks are uniform in size and are gray-brown colored. They are in a field of dead grass and are bordered by low-lying green trees and shrubs. Fire-baked bricks will hold up to moisture, making them suited to building baths and sewers.

Harappa was a fortified city **that still** exists in modern-day Pakistan (15,000 population). that is believed to have once been home to as many as 23,500 residents living in sculpted houses, **in** flat roofs made of red sand and clay. The city spread over 150 hectares—370 acres—and had fortified administrative and religious centers of the same type used in Mohenjo-daro.

The remains of the Indus Valley Civilization cities indicate remarkable organization; there were well-ordered wastewater drainage and trash collection systems and possibly even public baths and granaries—storehouses for grain. Most city-dwellers were artisans and merchants grouped together in distinct neighborhoods. The quality of urban planning suggests efficient municipal governments that placed a high priority on hygiene or religious ritual.

Ancient Harappa demonstrated advanced architecture, with dockyards, granaries, warehouses, brick platforms, and protective walls. These massive walls likely protected the Harappans from floods and may have deterred military conflicts. Unlike Mesopotamia and Ancient Egypt, the inhabitants of the Indus Valley Civilization did not build large, monumental structures. There is no conclusive evidence of palaces or temples—or even of kings, armies, or priests—and the largest structures may be granaries. The city of Mohenjo-daro contains the Great Bath, which may have been a large, public bathing and social area.

The trade relationship during the later 3rd millennium was a direct one: ships from Meluhha (Sumerian designation of the Indus) docked in Mesopotamian ports; some Meluhhans settled in Sumer; and there is a seal belonging to a Mesopotamian, whose job it was to act as an interpreter of the Meluhhan language. (McIntosh) There is extensive presence of Harappan seals and cubical weight measures in Mesopotamian sites. Specific items of high-volume trade are timber and specialty wood, such as ebony, for which large ships were used. Luxury items also appear, such as lapis lazuli mined at a Harappan colony at Shortigai (modern Badakhishan in northern Afghanistan), which was transported to Lothal, a port city in Gijarat in western India and shipped from there to Oman, Bahrain and Sumer.

At this point, our knowledge of the Indus Valley civilization, now in Pakistan, is impeded by our inability to decipher the various writings (inscriptions) thus far found amidst the ruins of their cities. The probability is that the Indus Valley peoples came down from the north at the time of the Mesopotamian Semites, and the Hamites of Egypt. The Elamites of the Iranian plateau were northeast of the later arriving Sumerians.

The coming of the Elamites and Harappans, c.10,000–6,000 BCE.

Further east along the Indus River, these migrants from the north created their own versions of an incipient civilization, thence were quietly absorbed by the oncoming Indo-Europeans c. 2200–2000 BCE. The Gutian (proto-Tokharian) conquest of Akkad/Sumer, c. 2100 BCE, is a possible parallel.

To the east of the Indus, was the sub-continent proper of India where the proto-Dravidian peoples, a probably hybrid amalgam of Australid (Melanesian) racial/ethnic peoples and wandering early Cro-Magnon genes, from c. 100,000 BP was creating its own civilizational profile. Thus, as with the Elamites, the Indus River cultures, the Dravidians in India proper remain enigmatic civilizational preludes to the history and civilizations of the Indian sub-continent.

(Left) Figure 252: The Dancing Girl, a prehistoric bronze sculpture made in approximately 2500 BCE in the Indus Valley Civilization city of Mohenjo-daro.

Below: The Great Bath at Mohenjo-daro: Amid the brick ruins of a 3rd-millennium BCE city, stairs descend on two sides into a large, rectangular brick-lined pit. Wooden stakes and wire encircle the perimeter, preventing entry by modern-day tourists.

Figure 253: The Great Bath of the Indus Valley civilization

The Great Bath is one of the well-known structures among the ruins of the ancient Indus Valley Civilization at Mohenjo-daro in Sindh, Pakistan. Archaeological evidence indicates that the Great Bath was built in the 3rd millennium BCE, soon after the raising of the "citadel" mound, on which it is located.

(Below) Figure 254: Excavated ruins of Mohenjo-daro, with the Great Bath in the foreground and the Buddhist Stupa in the background.

Dravidians and Aryans

Reich et al. (2009) discerned two major ancestral components in India, namely the *Ancestral North Indians* (ANI), who are "genetically close to Middle Easterners, Central Asians, and Europeans," and the *Ancestral South Indians* (ASI) which are clearly distinct from ANI and "not closely related to groups outside the subcontinent." Basu et al. (2016), discerned two additional components, Ancestral Tibeto-Burmese (ATB) and Ancestral Austro-Asiatic (AAA), noting that the ASI and the AAA were early settlers of India, who differentiated after their arrival in India. The ANI and ASI mixed in India between 4,200 and 1,900 years ago (2200 BCE–100 CE), whereafter a shift to endogamy took place, possibly by the enforcement of "social values and norms" by the "Hindu Gupta rulers." Northern Indians and higher castes are more related to West Eurasians, while southern Indians and lower castes are less related to West Eurasians.

Map: Dravidian languages of Indian sub-continent

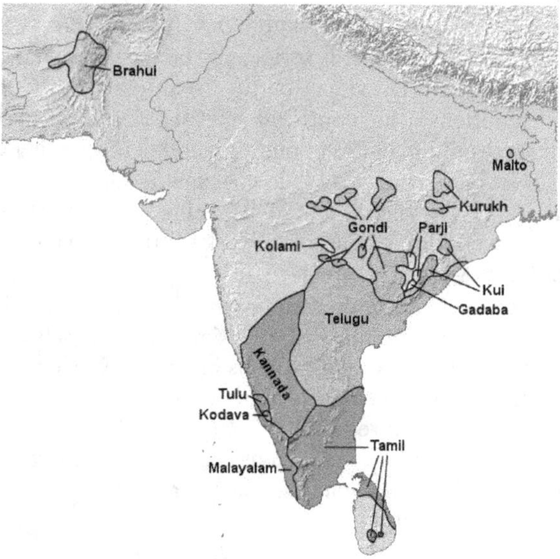

(Left) Figure 255: Early language divisions in India

Though some scholars have argued that the Dravidian languages may have been brought to India by migrations in the fourth or third millennium BCE, or even earlier, the Dravidian languages cannot easily be connected to any other language family, and they could well be indigenous to India.

Proto-Dravidian and onset of diversification

As a proto-language, the Proto-Dravidian language is not itself attested in the historical record. Its modern conception is based solely on reconstruction. It was suggested in the 1980s that the language was spoken in the 4th millennium BCE, and started disintegrating into various branches around 3rd millennium BCE. According to Krishnamurti, Proto-Dravidian may have been spoken in the Indus civilization, suggesting a "tentative date of Proto-Dravidian around the early part of the third millennium." Krishnamurti further states that South Dravidian I (including pre-Tamil) and South Dravidian II (including Pre-Telugu) split around the eleventh century BCE, with the other major branches splitting off at around the same time. Kolipakam et al. (2018) estimate the Dravidian language family to be approximately 4,500 years old.

The Indus Valley civilization (3,300–1,900 BCE), located in Northwestern Indian subcontinent, is often understood to have been Dravidian. Cultural and linguistic similarities have been cited by researchers Henry Heras, Kamil Zvelebil, Asko Parpola and Iravatham Mahadevan as being strong evidence for a proto-Dravidian origin of the ancient Indus Valley civilisation. The discovery in Tamil Nadu of a late Neolithic (early 2nd millennium BCE, i.e. post-dating Harappan decline) stone celt allegedly marked with Indus signs has been considered by some to be significant for the Dravidian identification.

Yuri Knorozov surmised that the symbols represent a logo-syllabic script and suggested, based on computer analysis, an underlying agglutinative Dravidian language as the most likely candidate for the underlying language. Knorozov's suggestion was preceded by the work of Henry Heras, who suggested several readings of signs based on a proto-Dravidian assumption.

Linguist Asko Parpola writes that the Indus script and Harappan language are "most likely to have belonged to the Dravidian family." Parpola led a Finnish team in investigating the inscriptions using computer analysis. Based on a proto-Dravidian assumption, they proposed readings of many signs, some agreeing with the suggested readings of Heras and Knorozov (such as equating the "fish" sign with the Dravidian word for fish, "min") but disagreeing on several other readings. A comprehensive description of Parpola's work until 1994 is given in his book, *Deciphering the Indus Script*.

Although in modern times, speakers of the various Dravidian languages have mainly occupied the southern portion of India, in earlier times they probably were spoken in a larger area. After the Indo-Aryan migrations into north-western India, starting c. 1800 BCE, and the establishment of the Kuru kingdom c. 1100 BCE, a process of Sanskritisation started, which resulted in a language shift in northern India. Southern India has remained majority Dravidian, but pockets of Dravidian can be found in central India, Pakistan, Bangladesh and Nepal.

Aryans in India

Indo-Aryan migrations started approximately 1,800 BCE, after the invention of the war chariot, which also brought Indo-Aryan languages into the Levant and possibly Inner Asia. It was part of the diffusion of Indo-European languages from the Proto-Indo-European homeland at the Pontic steppe, a large area of grasslands in far Eastern Europe. This started in the 5th to 4th millennia BCE, followed by the Indo-European migrations out of the Eurasian steppes, which started approximately 2,000 BCE.

Linguistic studies have shown that as the Aryans migrated into the Indian subcontinent, and the Aryan language gained ascendency over the local languages. They also gradually adopted an agricultural lifestyle, which was widely established by around 1000 BCE. Despite the fact that we have no historical records of this period, we have a mythical record in the form of religious literature known as the Vedas, one of the most important religious books in Indian history. Stories about military conflicts and other hostilities are recorded in the Vedas, but the historical reliability of these episodes is uncertain. (Violatti)

The Aryan influence, some scholars claim, gave rise to what is known as the Vedic Period in India (c. 1700–150 BCE), characterized by a pastoral lifestyle and adherence to the religious texts known as the Vedas. Society became divided into four classes (the ***Varnas***) popularly known as "the caste system" which were comprised of the ***Brahmana*** at the top (priests and scholars), the ***Kshatriya*** next (the warriors), the ***Vaishya*** (farmers and merchants), and the ***Shudra*** (labourers). The lowest caste was the ***Dalits***, the untouchables, who handled meat and waste, though there is some debate over whether this class existed in antiquity. At first, it seems this caste system was merely a reflection of one's occupation but, in time, it became more rigidly interpreted to be determined by one's birth; one was not allowed to change castes nor marry into a caste other than one's own. This understanding was a reflection of the belief in an eternal order to human life dictated by a supreme deity.

While the religious beliefs which characterized the Vedic Period are considered to be much older, it was during this time that they became systematized as the religion of Sanatan Dharma (which means "Eternal Order") known today as Hinduism (this name deriving from the Indus [or Sindus] River where worshippers were known to gather, hence, "Sindus," and then "Hindus"). The underlying tenet of Sanatan Dharma is that there is an order and a purpose to the universe and human life and, by accepting this order and living in accordance with it, one will experience life as it is meant to be properly lived. While Sanatan Dharma is considered by many a polytheistic religion consisting of many gods, it is actually monotheistic in that it holds there is one god, *Brahma* (the Self), who, because of his greatness, cannot be fully apprehended save through the many aspects which are revealed as the different gods of the Hindu pantheon.

It is *Brahma* who decrees the eternal order and maintains the universe through it. This belief in an order to the universe reflects the stability of the society in which it grew and flourished as, during the Vedic Period, Governments became centralized and social customs integrated fully into daily life across the region. Besides the Vedas, the great religious and literary works of the Upanishads, the Puranas, the Mahabharata, and the Ramayana all come from this period. (Mark)

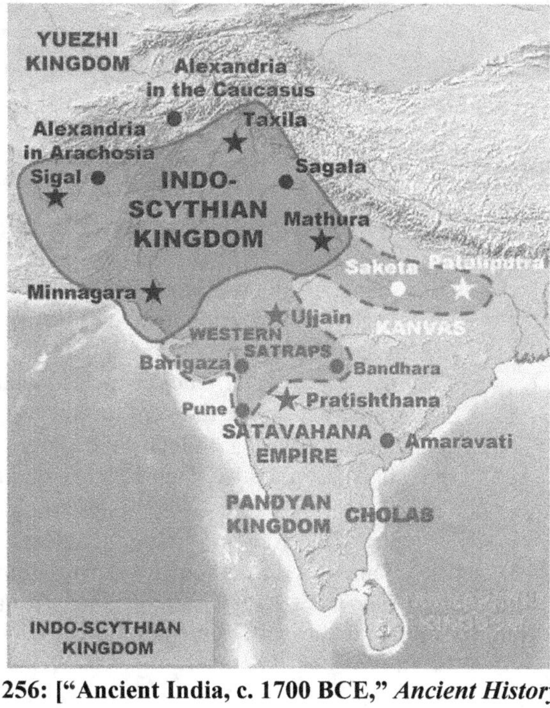

(Left) Figure 256: ["Ancient India, c. 1700 BCE," ***Ancient History Ecyclopedia*****]**

(Top) Figure 257: Bodhisattva Head, Gandhara

(Left) Figure 258: Ceramic Chalcolithic Malwa culture goblet from Navdatoli, c. 1300 BCE

Historians have analysed the Vedas to posit a Vedic culture in the Punjab region and the upper Gangetic Plain. Most historians also consider this period to have encompassed several waves of Indo-Aryan migration into the Indian subcontinent from the north-west. The peepal tree and cow were sanctified by the time of the Atharva Veda. Many of the concepts of Indian philosophy espoused later, like dharma, trace their roots to Vedic antecedents.

Early Vedic society is described in the Rigveda, the oldest Vedic text, believed to have been compiled during 2nd millennium BCE in the northwestern region of the Indian subcontinent. At this time, Aryan society consisted of largely tribal and pastoral groups, distinct from the Harappan urbanisation (Indus River Civilization)which was abandoned c. 1900 BCE.

Figure 259: An ancient temple in Tamil Nadu, India.

Figure 260: Mallas defending the city of Kusinagara, as depicted at Sanchi. Malla was an ancient India republic (Gaṇa sangha) that constituted one of the *solasa* (sixteen) Mahajanapadas (great realms) of ancient India as mentioned in the *Anguttara Nikaya*. Mallas c. 600 BCE.

Ancient India: an evaluation

The pre-Aryan Harappan civilization in the Indus valley seems to have had many feminine and egalitarian qualities, but unfortunately without writing little is known of their history and beliefs. Floods and over-grazing may have made them more vulnerable to conquest. The invasion of light-skinned Aryan conquerors of the dark natives in India is documented in the *Vedic* scriptures of the Hindus. Powerful religious beliefs, similar to the Iranians', were used to justify the establishment of a caste system based on skin color and occupations. Hindu society was dominated by the Brahmin priests and Kshatriya warrior-kings, supported by artisan, merchant, and farming Vaishyas, all of whom exploited the labor of the natives. Aryan ways were patriarchal and violent.

Yet, somehow in India, the western religion combined with the eastern methods of yoga and meditation to develop a remarkable spiritual philosophy and ascetic way of life based on inner awareness and renunciation of the world. The sages of the *Upanishads* left teachings that written, would inspire millions with their mystical wisdom. The doctrines of karma and reincarnation explained how spiritual justice transcended one lifetime, and the mystical methods offered seekers a path of liberation from the cycle. An ethical life of nonviolence was only the first step in such an awesome endeavor, while renouncing worldly success made the society more inward than other materialistic cultures.

The practice of nonviolence by Parshva was developed into a major religion by the noble Mahavira, whose extraordinary ascetic disciplines and spiritual awareness attracted devoted followers. Adding chastity to the ethical disciplines of nonviolence, truthfulness, not stealing, and freedom from possession, Mahavira established a religious community that spread Jainism. Yet, the extremity of the asceticism, which some believed required nudity, did not become as popular as a similar but more moderate religion founded in the same era by Siddartha, the *Buddha*.

The life of Siddartha Gautama, and his teachings as the *Buddha*, have inspired millions of people seeking peace and enlightenment to live more ethically. His renouncing of princely wealth and power to become an ascetic, only to discover that a moderate path between the extremes of strictness and luxury was the most successful approach to spiritual awareness, is an archetypal story of great significance.

The four noble truths of the Buddha are: 1) life is painful; 2) pain is caused by craving; 3) stopping craving stops pain; and 4) the way to stop craving is by correct understanding: intention, speech, action, livelihood, attention, concentration, and meditation. The Buddha, by his counseling, prevented a war between the Kolyas and the Shakyas. The Buddha refused to discuss speculative and metaphysical questions as irrelevant to ending suffering and finding enlightenment. He overcame attempts by Devadatta to cause a schism in the Buddhist community and refused to harm him even after Devadatta tried to kill him.

The teachings of the Buddha offered a practical way to reduce social harm as well as personal suffering. The Buddha diagnosed the psychological causality of attachment in his theory of dependent origination. Ethical behavior is an important part of the Buddhist quest for enlightenment. The Buddha's leadership of the community that formed around his teachings set an example of wisdom. His teachings were passed on orally and then in writing in numerous dialogs analyzing human consciousness and ethical conduct.

One of the greatest Buddhist works on ethics is the poetic *Dhammapada*, which emphasized compassion, self-mastery, and awareness. *The Questions of King Milinda* has the Hellenistic Bactrian king, who converted to Buddhism, ask many difficult questions, which are answered by Nagasena.

Thousands of people were profoundly influenced by Buddha's teachings in his own lifetime, and Buddhism spread throughout India and then into China, Japan and southeast Asia in the next five centuries, influencing the policies of kings as well as individual seekers. Although the injustices of war and the caste system were certainly not eliminated, there can be little doubt that efforts to practice Buddhist compassion by so many greatly improved the ethics of Indian society. (Beck)

Science, mathematics, technology

The great 7th-century Indian mathematician and astronomer, **Brahmagupta,** was the first to give rules to compute with ***zero*** i.e. ***When zero is added to a number or subtracted from a number, the number remains unchanged, a number multiplied by zero remains zero.*** Secondly, he is also known for dealing with **positive (fortune) and negative (debt)** numbers i.e. ***debt (negative) minus zero is debt (negative), similarly fortune (positive) minus zero is*** fortune ***(positive),*** etc.

Notation generally implies a set of well-defined representations of quantities and symbolic operators. It is believed that a mathematical notation to represent counting was first developed at least 5,000 years ago. Arabs called it the *hind* numerals, which were also adopted by western countries centuries later.

Ancient Indian Sage and Philosopher, Kanad (who lived between 6th century and 2nd Century BCE), who was also known as Kashyap, Kananda, and founded the Vaisheshika (one of the six orthodox schools of Vedic Systems) school of Indian philosophy, invented the theory of the atom, an indestructible particle of matter. He also used invariance arguments to determine properties of the atoms.

Notation

Vedic Indian Scholar Pingala Invented Binary Number System in 2nd Century BCE, in his book ***Chandaḥśāstra***. Pingala (in Chandaḥśāstra 8.23) has assigned the following combinations of **zero** and **one** to represent various numbers, much in the same way as the present day computer programming procedures.

Aryabhatta, the first major ancient mathematician from classical ancient mathematics age happens to be between 476–550 CE. He played a vital role in place value system and zero, approximation of π, trigonometry, indeterminate equations, and algebra.

Figure 261: Brahmagupta (598–668 CE)

Brahmagupta's invention in mathematics

In his work on arithmetic, Brahmagupta explained how to find the cube and cube-root of an integer and gave rules facilitating the computation of squares and square roots. He also gave rules for dealing with five types of combinations of fractions. He gave the sum of the squares of the first n natural numbers as $n(n + 1)(2n + 1)/6$ and the sum of the cubes of the first n natural numbers as $(n(n + 1)/2)^2$.

Metals

As we know, India is still full of mysteries and discoveries which are untouched, or **are** a work in progress. India also invented various metals like Wootz Steel (**a** pioneering steel alloy matrix developed in India). Smelting of zinc (India was first to smelt zinc **at** Zawar in the Tiri valley of Rajasthan. It is the world's first known ancient zinc smelting site) and **a** seamless celestial globe was made in Kashmir by Ali Kashmiri ibn Luqman in the reign of Emperor Akbar, 1556–1605.

Medicine

Figure 262

As written in The Sushruta Samhita by Sushruta in 6th Century BC, which is an ancient Sanskrit text on medicine and surgeries which played a vital role at that time. The Sushruta Samhita mentions many treaties, including plastic surgery of the nose practiced at that time, also known as rhinoplasty. The first cataract surgery also performed by Sushruta in the same time span. Therefore, plastic surgery and cataract surgery can be said to have been invented by ancient Indians.

Chess

Figure 263

Chess is believed to be derived from the Indian game chaturana, some time before the 7th century. Chaturanga is also the likely ancestor of the Eastern strategy games—xiangqi, janggi, and shogi. Chess reached Europe by the 9th century, due to the Muslin Umayyad conquest of Hispania. The pieces assumed their current powers in Spain in the late 15th century; the modern rules were standardized in the 19th century.

Vedic abstraction

As far back as possibly 500 BC, the Jyotish Vedanga used sophisticated methods of calculation to fix the position of the new and full moon and other astronomical inferences. A group of 16 sutras, or word formulae, were used widely in Vedic times to solve arithmetic and algebraic problems, and attempts are now being made to resurrect this science under the rubric of "Vedic math."

In later times, Hindu astronomers and mathematicians, of whom the most famous were Aryabhata I (5th century), Brahmagupta (6th century), Mahavira (9th century) and Bhaskara (12th century), made groundbreaking contributions to the development and elaboration of mathematical concepts, unknown to the West until the Renaissance or even later.

Aryabhata I, for instance, had calculated that the earth revolves around the sun about a thousand years before Galileo was persecuted for the same claim. It is well known that the concept of the zero, called the shunya, and the decimal system, originated in India, and reached the West through the Arabs, who, for long, called mathematics Hindsat, the "science of India."

The Syrian astronomer-monk Severus Sebokht wrote with awe in the 7th century of the rational system of mathematics of the Hindus, "and of their method of calculation which no words can praise strongly enough." A. L. Basham, the world-renowned historian of ancient India, writes that the "unknown man" who devised the decimal system "was from the world's point of view, after the Buddha, the most important son of India." **(Varma 2016)**

The Caste Systerm

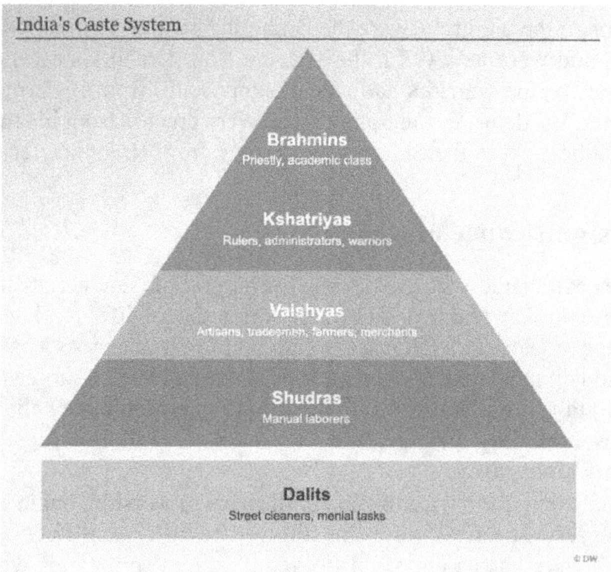

Figure 264

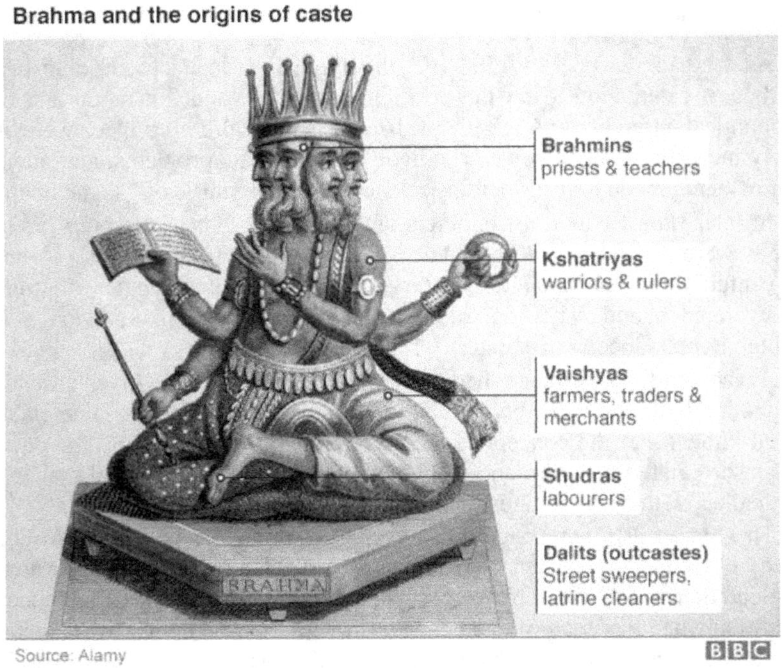

Figure 265

"India's caste system is among the world's oldest forms of surviving social stratification." (BBC)

At the top of the hierarchy were the Brahmins, who were mainly teachers and intellectuals, and were believed to have come from Brahma's head. Then came the Kshatriyas, or the warriors and rulers, supposedly from his arms. The third slot went to the Vaishyas, or the traders, who were created from his thighs. At the bottom of the heap were the Shudras, who came from Brahma's feet and did all the menial jobs.

Daily significance of caste

Practices associated with caste varied through time and across India, but all shared some common features. The three key areas of life dominated by caste were marriage, meals and religious worship. Marriage across caste lines was strictly forbidden; most people even married within their own sub-caste or *jati*. **In terms of religion, the priestly class, Brahmins presided over religious rituals and services including preparation for festivals and holidays, as well as marriages and funerals.**

The Kshatrya and Vaisya castes had full rights to worship, but in some places, Shudras (the servant caste) were not allowed to offer sacrifices to the gods. Untouchables were barred entirely from temples, and sometimes were not even allowed to set foot on temple grounds. People who violated social norms could be punished by being made "untouchables." This was the lowest caste--they and their descendants were condemned, completely outside of the caste system.

Untouchables were considered so impure that any contact with them by a caste member would contaminate the member. The caste-person would have to bathe and wash his or her clothing immediately. The Untouchables did work that no-one else would do, like scavenging animal carcasses, leather-work, or killing rats and other pests. **Untouchables could not even eat in the same room as caste members and could not be cremated when they died. (Quorum)**

Inter-caste marriage

The caste system in India prohibits marriage outside the caste. However, inter-caste marriages have gradually gained acceptance due to increasing education, employment, middle-class economic background, and urbanization. According to a survey in 2014, about 5% of marriages are inter-caste in India.

The caste system is thought to have prevailed for more than 2,000 years, long enough to have impressed a subtle genetic signature on the population, according to a team of American and Indian biologists.

Although marriage within one's own caste is strongly preferred, women are occasionally allowed to marry men of higher caste, with their children assuming their father's status. This practice gives rise to a rigidly stratified population in which women have slight upward mobility but men none.

Sampling DNA from people of various castes in Andhra Pradesh, a province in southern India, the biologists reported in a recent issue of *Nature* that there was indeed a significant genetic distance, as measured by the number of differences in the DNA, between women in the higher and lower castes. The differences can be explained by a low level of gene flow carried by women moving one way in a firmly layered system. If women moved both ways, or not at all, there would be little difference between the castes.

Dr. Michael J. Bamshad of the University of Utah, a co-author of the study, said that he and his colleagues at Andhra University in India hoped to extend their methods to explore the population history of India and the role of different groups in the caste system. "There has been a greater Asian influence in the lower and middle castes, and a greater Caucasian contribution to the upper castes," he said, a finding that probably reflects the Indo-European invasion of India. (Wade 1998)

What Bamshad and his Indian colleagues are saying referring to the "Asia" influence is of course not the Mongoloid/Caucasoid/Australid hybrids of Northeast Asia, but rather the large Australid/Caucasoid aboriginal populations that the Indo-European tribes encountered as they made their way into this vast sub-continent. The Dravidian civilization that the Indo-Europeans encountered as they entered the lands of India/Pakistan proper was not the Indus River civilization at the western edge of today's Pakistan, which was probably destroyed, c. 1900 BCE, by the invading rough-necks from the northwest. The eastern sub-continent's existing population, c. 1500 BCE was probably an amalgam of early Cro-Magnon settlers in the sub-continent and the hybridization of these few in numbers with the existing indigenous Australid tribes.

This Dravidian civilization was in all likelihood a mixed entity. The now, c. 1500 BCE Indo-European (Sanskrit) invaders, pushed the Dravidians south in repugnance, but also interbred with them, male genes into vulnerable females. This is where Dr. Bamshad and colleagues gets it right. And this phenomenon of females marrying, breeding, entering into concubinage or polygamy upwards in social rank is a universal characteristic of the earliest civilizations, including, as noted earlier, the Sumerians, Egyptians, and Chinese, plus where ever else one could look and find.

The result for India was the religious rationalization of the separation of these peoples with the light haired and light skinned invaders from the north. But only for awhile, as the usual male weakness for a plurality of sex partners made its impact on the appearance of most Indians, soon a well but unofficially hybridized people; as the invaders attempted to maintain their socio-economic status, turning the semi-skilled Dravidians, back then, from 1500–1000 BCE, into a lesser subservient caste.

The irony is that in the south of India in recent times, a selective process of marriage and position has been going on, such that many parts of the south of India are in the vanguard of economic modernization, perhaps the inevitable dilution of once iron-clad caste separations.

Figure 266: Gandhi visiting Madras (now Chennai) in 1933, on an India-wide tour for Dalit (he used Harijan) causes. His speeches during such tours and writings discussed the discrimination-against "lower" castes of India

(Above) Figure 267: Yakshi Bracket Figure. Easttorana of the Great Stupa. At Sanchi, 1st Century BCE

(Left) Figure 268: The Great Stupa at Sanchi, c. 273–232 BCE (Mauryan Empire), enlarged c. 150–50 BCE Shunga Dynasty

Irony, India Accomplished?

Given the great historic accomplishments both in religion/ethics and in math and science, the entire Indian sub-continent, yet, was taken over in the 18th century as a colony by the East Indian Company of England, only to be voluntarily freed as one nation in 1947 CE. Then it was to be broken up in wars, now into three independent nations; cost, one million deaths.

The great intellectual achievements of Indian migrants all over the world as well as in today's advancing technology, in the home country, ostensibly secular but historically Hindu India, is testimony to the original insular retention by the higher castes of historic India of their genetic Cro-Magnon capital. However, now having been officially released from the bindings of this exploitative system. it remains to seen how deep the illicit sharing of this genetic potential has infiladed into the still unabsorbed masses of Indians. They still await entry into the modern world of the abstract.

What the Ancient Civilizations Tell Us

Taxonomy: This is a term used in the biological sciences to classify, relate organism to each other in terms of pre-determined criteria. In the taxonomy of human evolution the term *Homo sapiens* has recently been used to incorporate a great diversity of fossil humans into our closest relatives in ancient times, before 100k BP.

Whether or not we call these fossils early forms of *Hs,* it has resulted in satisfying the socio-political interests of peoples and nations around the world wherever the remains of these ancient humans have been found. They covet their heritage of sapiency. However it does not clarify the cultural elements of human diversity, even as the morphological distinctions between fossil humans has been blurred.

In this essay we have noted that most scientist see verbal communication as a key to the entrance of humans into modernity. And they do not especially confirm that the Neanderthals, for one, or other ancient humans had the capacity to express explicit, directive, descriptive comments about themselves and the world in modern linguistic exposition.

The great human revolution seems to have taken place in eastern Europe/western Asia some time between 400-200k BP, the so-called 'genetic blockade', what we have here called '*the funnel effect*', small groups of *paedomorphic* humans (the retention of infantile characteristics into sexual maturity) breaking off from an ancient northern Neanderthal type of advanced *Homo erectus,* of light Caucasoid skin and features, and spreading out in small groups, around the continents.

We have evidenced cultural signs of the impact of this enormous change in morphology and the consequent change in behavior, noting the fragmentary genetic/cultural evidence found in southern Africa (Boskopoid, Khoisan, Bushman) the art, the three pointed fishing spears, found in the Negroid areas of central Africa, similar to Magdalenian tools in Europe at 15k BP, but there in Africa, c. 100k BP. Cro-Magnon like tools have been found in East Asia, and certainly give evidence of the spread and hybridization of humans throughout the continents, Denisovans in Asia, Amero-Indians in North and South America.

Below See Sitting Bull, (evidence of three ancient races).

Figure 269: Sitting Bull, 1831-1890 CE, Dakotas, USA

Cultural Question

Considering the lack of secular philosophical and medical/scientific knowledge of our universe, rule by emperors, priests, bureaucrats, small populations vulnerable to marauding tribal primitives, how did the middle classes of the ancient civilizations live in comparison with ourselves in the 21st century? Could they have described the fulfillments of their civilization's lifestyle as a rich experience as they looked back on their own three or four score years of life?

Doesn't this have to be a critical question as we look back on a world that knew less than we do—this, a contemporary prejudice of every generation that has lived on this planet. What does this knowledge of the civilizations of the past tell us about our human nature that will be useful for future generations.

One thing is sure. These civilizations had a "half life," colloquially meaning, they eventually disappeared from the face of this earth, and with it their practical envisionment of the good middle class life. Can we here find meaning? Next chapter.

Chapter 14

Greece

Perspective on civilization

Each of the civilizations, derivative from the Cro-Magnon migrations south and east, had their day in the sun. They, all in one way or another, after revealing such different yet fascinating cultural/historical variations on the *Homo sapiens sapiens,* the *Hss* theme, faded from history. They told us about all the different literary possibilities of high intelligence and the consequent intellectual advances which created real middle class lives for their citizens, even under autocratic political suzerainty.

The fading from history was most directly attributed to the stagnant religion/mythological structures of understanding, belief, and the consequent political forms that impeded change and progress. The gods and their monarchs stood in the way of a breakout into a new world of understanding. This, in culture and awareness of the nature of our human world of symbols. Akhenaten's attempt at religious reform in Egypt was the most critical exemplification of a civilization going nowhere.

Yes, the Greeks come on the scene later in time. They may have gained perspective from the stagnant world views of the old timers along the Nile, and Euphrates, and their consequent weaknesses. In reality, this perspective was more of an indigenous Hellenic difference in place, geography/ecology that sent the Greeks off in a different direction of cultural development and power.

The point is, the Greeks gave us modernity and a belief perspective that has been passed down to our own time. At this moment in the history of our world-wide technological civilization, we might not appreciate that we and the Greeks are closely aligned in our views of reality, deeply embedded in our potential for renewal, but over time.

It is important to note the special historical circumstances and the nature of this unique people who, like the Sumerians, exploded a wholly new way of understanding and acting on the nature of the human cultural/historical experience.

The making of a people

Minoans: The earliest proto-Greeks were the Minoans of the island of Crete. Sometime before 2000 BCE, they descended, probably from the northeast (DNA), a mystery of origins similar to the Indus River civilization. Protected to a great extent by their island isolation from the migratory currents of Europe proper, they developed an indigenous way of life, highly influenced by their close proximity to the great Egyptian civilization in North Africa along the Nile.

But the ethos of this people was very different from that of the Egyptians. In spite of signs of military capability, the main cities seemed to have been unwalled. The material culture was light and deeply involved with the esthetics of living in a rich, physical socio-economic style. They had an intrinsic, probably proto-Indo-European language, and their Linear A written language deviates radically from the Egyptian's. In fact, it was adapted into recognizable Greek in the Minoan syllabic style in the Linear B of the land-based Mycenaeans (c. 1500 BCE).

The Minoan civilization itself was probably overrun by the migratory comings of the Mycenaeans and the sea peoples, c. 1500–1200 BCE. It was a world on the move in those centuries. Egypt was attacked several times: by the Hyksos from the northern lands, by the Philistines and other Aegean peoples, the Etruscans, and the tribe of Dan. The Israelites also seem to have been involved in these tumultuous times. The Minoan peace passed into history as we focus on a new incipient civilization on the Greek mainland.

Crete was an important centre for the Minoan civilization.

Figure 270: The palace at Knossos, the capital of the Minoan civilization on Crete, c. 1800 BCE, was an important centre for the Minoan civilization

(Top) Figure 271: Reconstruction of a fresco on Knossus showing Minoan women, c. 1800 BCE

(Left) Figure 272: Minoan Snake Goddess, Knossos, c. 2000–1700 BCE

Mycenaeans: Moving into Greece from somewhere in the northeast, speaking an outlier language of the Indo-European family (closer to Persian and Hindi), these people learned quickly the seafaring life and the building of great architectural palaces/fortresses, a rich culture defintiely influenced by the now extinct Minoan vision, c. 1200 BCE. It is probable that the semi-romantic Hellenic extolling of Homer and this bardic tradition derives from memories of the Mycenaean warrior tradtion. Certainly, throughout the world of the discovery of literacy and civilization, bloody war even amongst first cousins was the rule of the day. But so was poetry and the re-discovery of the individual and his/her place in the community.

(Below) Figure 273: Mycenaean City Gate

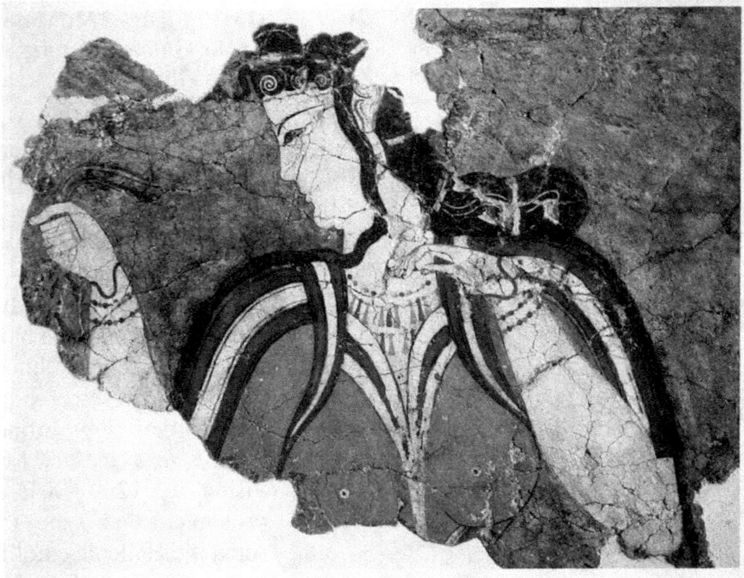

Figure 274: Fresco depicting a female figure in the acropolis of Mycenae, 13th century BCE.

The Mycenaeans established the first advanced civilization in mainland Greece, c. 1300 BCE.

The Mycenaean civilization flourished in the late Bronze Age, from the 15th to the 13th century BCE and extended its influence not only throughout the Peloponnese in Greece but also across the Aegean, in particular, on Crete and the Cycladic islands. The Mycenaeans were influenced by the earlier Minoan civilization (2000–1450 BCE) which had spread from its origins at Knossos, Crete, to include the wider Aegean. Architecture, art and religious practices were assimilated and adapted to better express the perhaps more militaristic and austere Mycenaean culture.

“Major Mycenaean centers included Mycenae (traditional home of Agamemnon), Tiryns (perhaps the oldest centre), Pylos (traditional home of Nestor), Thebes, Midea, Gla, Orchomenos, Argos, Sparta, Nichoria and probably Athens. Beyond trading relations, the exact political relationship between these centers is not clear. However, there were many shared cultural features such as architecture, frescoes, pottery, jewelry, weaponry, and of course, the Greek language and writing in the form of Linear B (an adaptation of the Minoan Linear A).” (Mark 2013)

Dorians

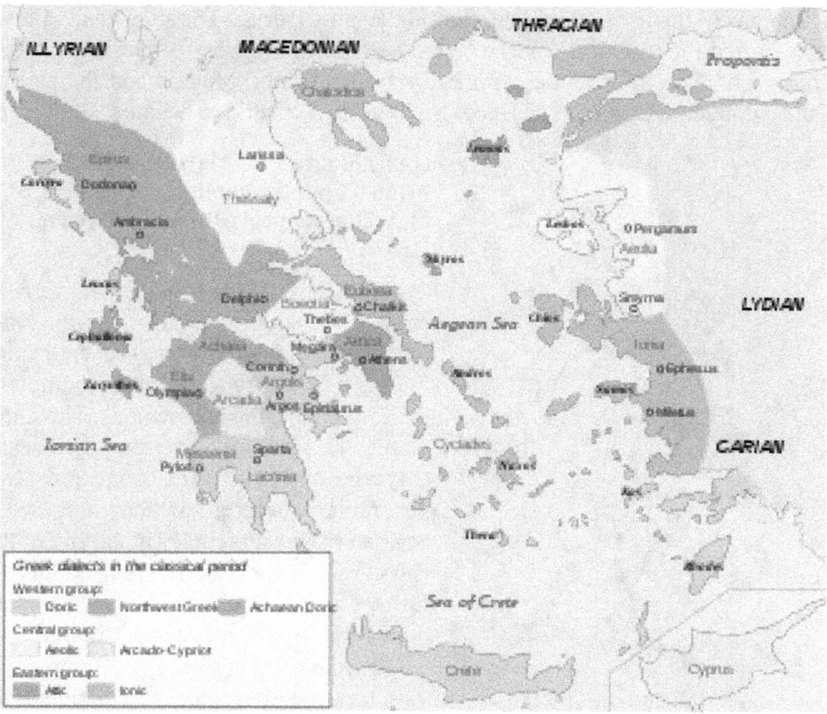

Figure 275: Distribution of Greek dialects in Greece in the classical period.

Western group:	**Central group:**	**Eastern group:**
Doric proper	**Aeolic**	**Attic**
Northwest Doric	**Arcad**	**Ionic**
Achaean Doric		

Accounts vary as to the Dorians' place of origin. One theory, widely believed in ancient times, is that they originated in the northern mountainous regions of Greece—ancient Macedonia and Epirus—and obscure circumstances brought them south into the Peloponnese, to certain Aegean islands: Magna Graecia, Lapithos and Crete. Mythology gave them a Greek origin and eponymous founder, Dorus son of Helen, he, the mythological patriarch of the Hellenes.

Though most of the Doric invaders settled in the Peloponnese, they also settled on Rhodes and Sicily, in what is now Southern Italy. In Asia Minor existed the Dorian Hexapolis (the six great Dorian cities): Halikarnassos (Halicarnassus) and Knidos (Cnidus) were in Asia Minor; Kos, Lindos, Kameiros and Ialyssos were on the island of Rhodes. The six cities would later become rivals with the Ionian cities of Asia Minor. The Dorians also invaded Crete. These origin traditions remained strong into classical times.

Attic (Athenian) historian, Thucydides, saw the Peloponnesian War in part as "Ionians (Athenians and allies) fighting against Dorians (Spartans and allies)" and reported the tradition that the Syracusans in Sicily were of Dorian descent. Other such "Dorian" colonies, originally from Corinth, Megara and the Dorian islands, dotted the southern coasts of Sicily from Syracuse to Selinus.

(Left) Figure 276: Marble statue of a helmed Spartan hoplite (5th century BC) at Archaeological Museum of Sparta, Greece

Tyrtaeus, a lame Athenian warrior-poet, became advisor of the Lacedaemonians (Spartans) in their mid-7th-century war to suppress a rebellion of their neighbor, the Messenians. The latter were a remnant of the Achaeans (Mycenaeans) and were conquered "two generations before," which suggests a Spartan rise to supremacy at the end of the Dark Age, rather than during and after the fall of Mycenae. The Messenian population was reduced to serfdom.

Tyrtaeus: "For 'tis a fair thing for a good man to fall and die fighting in the vain for his native land, ... let us fight with a will for this land, and die for our children and never spare our lives."

In the earliest period of their national formation, the Dorian Spartans were noted for their material culture, their poets: not too dissimilar to other Greek states, if not superior. However, after the decision of the leadership was made to preserve their superiority on the Peloponnese, they decided to put this cultural egotism to the test, to militarize and put its neighbors to the sword of helotry, peasant feudal dependency and helot labor.

Thucydides: “For a short time the league (after the Persian wars) held together, till the Lacedaemonians (Dorian Spartans) and Athenians quarrelled, and made war upon each other with their allies, a duel into which all the Hellenes sooner or later were drawn … the real cause I consider to be ... the growth of the power of Athens and the alarm which this inspired in Lacedaemon.”

Sparta was the warrior society in ancient Greece, that reached the height of its power after defeating rival city-state, Athens, in the Peloponnesian War (431–404 BCE). Spartan culture was centered on loyalty to the state and military service. At age seven, Spartan boys entered a rigorous, state-sponsored education, military training and socialization program. Known as the Agoge, the system emphasized duty, discipline and endurance. Although Spartan women were not active in the military, they were educated and enjoyed more status and freedom than other Greek women. Because Spartan men were professional soldiers, all manual labor was done by the slave class, the Helots. Despite their military prowess, the Spartans’ dominance was short-lived: in 371 BC, they were defeated by Thebes at the Battle of Leuctra (Epaminandas), and their empire went into a long period of decline.

Athens

(Right) Figure 277: Caryatids of the Erechtheion, Acropolis, c. 440 BCE

According to legend, Athens was formerly ruled by kings, a situation which may have continued up until the 9th century BCE. From later accounts, it is believed that these kings stood at the head of a land-owning aristocracy, known as the *Eupatridae* (the “well-born”), whose instrument of government was a Council that met on the Hill of Ares, called the Areopagus, and appointed the chief city officials, the archons and the polemarch (commander-in-chief).

Before the concept of the political state arose, four tribes, based upon family relationships, dominated the area. The members had rights, privileges, and obligations. During this period, Athens succeeded in bringing the other towns of Attica under its rule. This process of *synoikismos*—the bringing together into one home—created the largest and wealthiest state on the Greek mainland, but it also created a larger class of people excluded from political life by the nobility. By the 7th century BC, social unrest had become widespread, and the Areopagus appointed Draco to draft a strict new code of law (hence the word "draconian"). When this failed, they appointed Solon, with a mandate to create a new constitution (in 594 BCE).

All of Greece was going through an economic and cultural expansion at the the end of the 7th century BCE, 650–600 BCE. The Spartans had earlier sent out a number of colonies to the west in Sicily (Syracuse), which forgot their militaristic origins for commerce and culture. Syracuse did defeat Athens, c. 415 BCE, in their home waters during the Peloponnesian Wars. The nearby Akragas stayed neutral and raised up an enormours Doric ensemble of temples along their shoreline, to give testimony to this new wealth.

Then, there was Croton on the mainland of Italy, where Pythagoras of Samos (Ionian) Islands migrated from, died c. 495 BCE. Neapolis, further north, was a great stimulator to the surrounding Latin tribes. In the east, the Ionian islanders and the Ionian Greek mainlanders, on the now west coast of Turkey, were even richer in their economic interactions. Miletus, the wealthiest of these cities, spawned an indpendent school of naturalistic thinkers, headed by Thales, c. 545 BCE.

Thales is recognized for breaking from the use of mythology to explain the world and the universe, and instead explained natural objects and phenomena using theories and hypotheses, in a precursor to modern science. Almost all the other pre-Socratic philosophers followed him in explaining nature as deriving from a unity of everything, based on the existence of a single ultimate substance, instead of using religio/mythological explanations. Aristotle regarded Thales as the founder of the Ionian School and reported that Thales' hypothesis of the originating principle of nature, and the nature of matter, was based on a single material substance: water.

One hundred and forty-six years later, 399 BCE, Plato (*Apology*) wrote that Athenian citizen Socrates still had to acknowledge the existence of the Greek gods. Socrates distinctly separated their existence from the dialectical method of searching for truth, with or without the assistance of the gods. The Athenian democracy killed him for his impudence.

The reforms that Solon initiated in Athens, c. 594 BCE, dealt with both political and economic issues. The economic power of the *Eupatridae* was reduced by forbidding the enslavement of Athenian citizens as a punishment for debt (debt bondage), and by breaking up large landed estates and freeing up trade and commerce. This allowed the emergence of a prosperous urban trading class.

Politically, Solon divided the Athenians into four classes, based on their wealth and their ability to perform military service.

The poorest class, the *Thetai* (Ancient Greek *Θῆται*), who formed the majority of the population, received political rights for the first time and were able to vote in the *Ecclesia* (Assembly). But only the upper classes could hold political office. The four aristocratic tribes that dominated the Areopagus continued to exist, but its powers were reduced. The great historical Solonnian contribution was the innovation of eliminating "blood guilt"—that children must pay for the crimes of their fathers.

The population of the city of Athens proper, 600 BCE, could not have been more than c. 50,000. It is interesting to note that the Sumerian cities, c. 3000 BCE, at this population level were making the transition from limited democracy to unlimited monarchy.

The new Solonnian system laid the foundations for what eventually became Athenian democracy, but in the short-term, it failed to quell class conflict. After twenty years of unrest, the populists led a revolutionary tyrant into power. These tumultuous political times were taking place all over Hellas. Indeed, the Greek word *tyrannos* does not mean a cruel and despotic ruler, but merely one who takes power by force. Here came Peisistratus (power c. 566 BCE, died c. 528 BCE), in fact a very popular ruler, who helped to make Athens wealthy, powerful, and a center of culture (Pan-Athenaic games). He also instituted Athenian naval supremacy in the Aegean Sea.

Peisistratus was sensitive to events throughout Hellas, economic and political. He stimulated the development of an olive oil and wine industry, supporting the planting of trees and vines, but also made the port enclave of Piraeus a center of trade, where new traders and artists poured the wine and olive oil into beautiful packaging: amphorae, designed by these non citizen residents (perioikoi). Finally, Peisistratus initiated a great religious, nationalist program to build a city of painted, wooden temples high up on the Acropolis. These, the Persian king, Xerxes, 480 BCE, burnt in the second Persian War.

The political, economic and intellectual ferment in Athens during this century was not only unmatched throughout Hellas, but also could be felt throughout the civilized world. Buddha, c. 500 BCE; Confucius c. 500 BCE; Lao Tzu, c. 500 BCE; Zoroastrianism (flourished in Persia, c. 600 BCE), Jeremiah, Jews, c. 600 BCE.

After Peisistratus' death in 528 BCE, and the chaos of the rule of his sons, another relative of the Solon/Peisistratus clan, Cleisthenes, took power under a new populist movement, 508 BCE. The reforms of Cleisthenes replaced the traditional four Ionic "tribes" (*phyle*) with ten new ones, named after legendary heroes of Greece and having no class basis. These acted as electorates. Each tribe was in turn divided into three *trittyes* (one from the coast, one from the city and one from inland) divisions, while each *trittys* had one or more *demes*—depending on their population—which became the basis of local government.

(Left) Figure 278: Coinage of Athens at the time of Cleisthenes. Effigy of Athena, with owl and AΘE, initials of "Athens," c. 510–500/490 BCE.

The tribes each selected fifty members by lot for the *boule*, the council which governed Athens on a day-to-day basis. This wildly turbulent democratic public opinion of newly enfranchised voters could and was influenced by political satires written by the comic poets and performed in the city theaters. The Assembly, or *Ecclesia*, was open to all full citizens and was both a legislature and a supreme court, except in murder cases and religious matters. These became the only remaining functions of the *Areopagus*. Most offices were filled by lot, although the ten *strategoi* (generals) were elected.

Persian Wars

By c. 560 BCE, the Indo-European kings of Persia had completed their control over the Mesopotamian river system, plus most of the Anatolian Plateau (modern Turkey). The interior of this great expanse still contained large remnant populations of the ancient Indo-European Hittite Empire. In c. 539 BCE, after the Persian conquest of the Chaldean Babylonians, Emperor Cyrus sent back the Semitic Jews who wished to return to Jerusalem, to rebuild their temple. Jerusalem was once occupied by populations of Indo-European Jebusites, descendants of the Hurrians and Mitanni, before the conquest by David, c. 1000 BCE.

Cyrus, who was killed in battle, c. 530 BCE, and the Persians were easy on religious diversity, themselves being a devotee of monotheistic Zoroastrianism. Darius became emperor in 522 BCE.

The western coastline of Anatolia and the Aegean Islands were lined with increasingly independent, wealthy and ethnically proud Ionian Greeks. Inevitably, in c. 500 BCE, their independence from Persia stimulated revolution and war, in which Athens became involved. What followed was not only revolutionary warfare mostly put down by Persia, but an awakened awareness by Darius, son of Cyrus. The latter set his eyes on the over-the-seas instigator of this tumult—Athens.

Athens and Plataea confronted a landed army and armada headed for Athens, 490 BCE. The ensuing Marathon defeat of Darius sent the Persians back to Asia. Ten years later, his successor, Xerxes, mounted and even vaster invasion. In 480 BCE, Athens was completely evacuated and burned. But the sea battle that developed around this now deserted city, in the Saronic Gulf between the Piraeus and the evacuee island of Salamis, saw this vast, united Persian fleet destroyed (see illustration of the large Triremes the Athenians mustered against the numerically larger allied [Greek] Persian fleet).

(Right) Figure 279: Greek *Triremes* (warships), symbolic of the Athenian victory over Persia and allies at Salamis, 480 BCE (Encyclopedia of Ancient History)

The Persian allied armies retreated to Plataeae, where a great united Greek army, under the leadership of the Spartan Pausanius, defeated a disorganized Persian offensive, killed its leader, Mardonius, and saw Xerxes quickly fleeing back into Asia (see below Herodotus' description of Pausanius' educational view of the results).

Herodotus Histories Ch.9—After the victory of the Greek allies over the Persians plus their Macedonian and Greek allies, Plataea, 379 BCE

> Then Pausanias [lead general, Sparta] made a proclamation, that no man should touch the spoil, and bade the helots gather all the stuff together. They, scattering all about the camp, found their [Persian] tents adorned with gold and silver, and couches gilded and silver-plated, and golden bowls and cups and other drinking-vessels; and sacks they found on wains, wherein were seen cauldrons of gold and silver; and they stripped from the dead that lay there their armlets and torques, and daggers of gold; as for many-coloured raiment, it was nothing regarded. Much of all this the helots showed, as much as they could not conceal; but much they stole and sold to the Aeginetans; insomuch that the Aeginetans thereby laid the foundation of their great fortunes, by buying gold from the helots as though it were bronze.
>
> 81: Having brought all the stuff together they [Greek allies] set apart a tithe for the god of Delphi, whereof was made and dedicated that tripod that rests upon the bronze three-headed serpent, nearest to the altar; another they set apart for the god of Olympia, ((Olympics)whereof was made and dedicated a bronze figure of Zeus, ten cubits high; and another for the god of the Isthmus, whereof came a bronze Poseidon seven cubits high; all which having set apart they divided the remnant, and each received according to his desert of the concubines of the Persians, and the gold and silver, and all the rest of the stuff, and the beasts of burden. How much was set apart and given to those who had fought best at Plataea, no man says; but I think that they also received gifts; but tenfold of every kind, women, horses, talents, camels, and all other things likewise, was set apart and given to Pausanias.
>
> 82: This other story is also told. Xerxes in his flight from Hellas, having left to Mardonius [killed in battle] his own establishment, [encampment] Pausanias, seeing Mardonius' establishment with its display of gold and silver and gaily-coloured tapestry, bade the [Persian] bakers and the cooks to prepare a dinner in such wise as they were wont to do for

> Mardonius. They did his bidding; whereat Pausanias, when he saw golden and silvern couches richly covered, and tables of gold and silver, and all the magnificent service of the banquet, was amazed at the splendour before him, and for a jest bade his own servants prepare a dinner after Laconian [Spartan] fashion. When that meal was ready and was far different from the other, Pausanias fell a-laughing, and sent for the generals of the Greeks. They being assembled, Pausanias pointed to the fashion after which either dinner was served, and said: 'Men of Hellas, I have brought you hither because I desired to show you the foolishness of the leader of the Medes[and Persians]; who, with such provision for life as you see, came hither to take away from us ours, that is so pitiful.' Thus, it is said, Pausanias spoke to the generals of the Greeks.

The interregnum between the two great wars that Athens fought was a series of struggles with both Persia and Sparta, the Persians never to return in victory to Greece. In 478 BCE at the height of Athens power, the Delian league was instituted. A treasury was contributed to by over 150 Greek states for the protection by Athens and its navy. In 454 BCE, Pericles, then a "strategos" general of the Athenians (from 461 BCE), removed the treasury from Delos to Athens, using it to commence their enormous investment of wealth in the building of the marble Acropolis (high city) to commemorate their history and present position of leadership in the arts and architecture.

Pericles and the Peloponnesian Wars

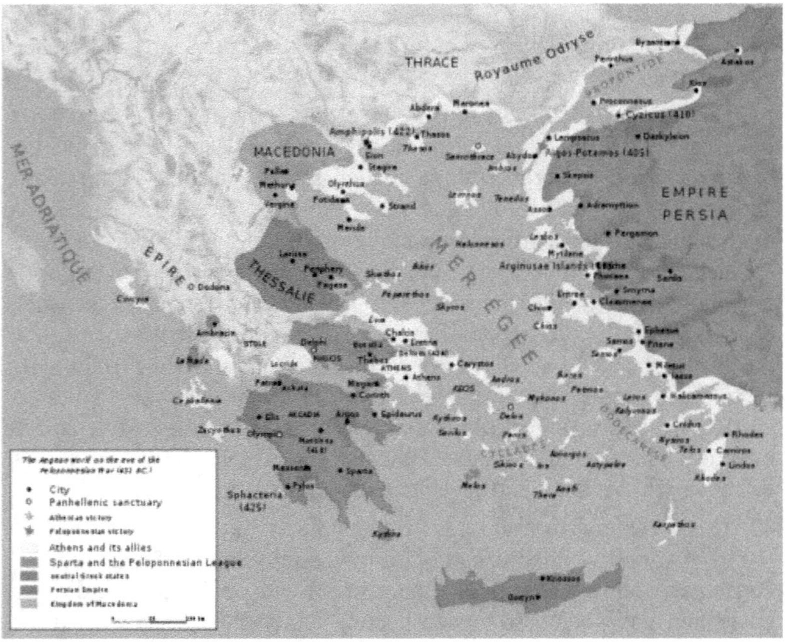

Figure 280

Peloponnesian War

The historian Thucydides, in his sober and brilliantly interpreted history of the Peloponnesian war between the two great powers, Athens and Sparta and their numerous allies (431–404 BCE), saw Pericles as the visionary of democracy, the arts and philosophical knowledge (the Sophists). He bemoaned the untimely death of Pericles in the plague that beset Athens after it called the Attic population into its walls for protection, 429 BCE.

Thucydides' "The Peloponnesian Wars," paraphrased Pericles' memorial speech, 431 BCE, for the first fatalities in the conflict with Sparta and its allies. He speaks of the self-conscious contribution of Athens to Hellas and the world.

> If we look to the laws, they afford equal justice to all in their private differences ... if a man is able to serve the state, he is not hindered by the obscurity of his condition. The freedom we enjoy in our government extends also to our ordinary life. There, far from exercising a jealous surveillance over each other, we do not feel called upon to be angry with our neighbor for doing what he likes … We throw open our city to the world, and never by alien acts exclude foreigners from any opportunity of learning or observing, although the eyes of an enemy may occasionally profit by our liberality … advancement in public life falls to reputations for capacity, class considerations not being allowed to interfere with merit ... our ordinary citizens, though occupied with the pursuits of industry, are still fair judges of public matters ... at Athens we live exactly as we please, and yet are just as ready to encounter every legitimate danger.
>
> In short, I say that as a city we are the school of Hellas; while I doubt if the world can produce a man, who, where he has only himself to depend upon, is equal to so many emergencies, and graced by so happy a versatility as the Athenian ... for the Athens that I have celebrated is only what the heroism of these and their like have made her ... none of these men allowed either wealth with its prospect of future enjoyment to unnerve his spirit, or poverty with its hope of a day of freedom and riches to tempt him to shrink from danger. No, holding that vengeance upon their enemies was more to be desired than any personal blessings, and reckoning this to be the most glorious of hazards, they joyfully determined to accept the risk ... Thus, choosing to die resisting, rather than to live submitting, they fled only from dishonor ...
>
> Therefore, having judged that to be happy means to be free, and to be free means to be brave, do not shy away from the risks of war.
>
> If then we prefer to meet danger with a light heart but without laborious training, and with a courage which is gained by habit and not enforced by law, are we not greatly the better for it? Since we do not anticipate the pain, although, when the hour comes, we can be as brave as those who never allow themselves to rest; thus our city is equally admirable in

peace and in war. For we are lovers of the beautiful in our tastes and our strength lies, in our opinion, not in deliberation and discussion, but that knowledge which is gained by discussion preparatory to action. For we have a peculiar power of thinking before we act, and of acting, too, whereas other men are courageous from ignorance but hesitate upon reflection. And they are surely to be esteemed the bravest spirits who, having the clearest sense both of the pains and pleasures of life, do not on that account shrink from danger.

Athens' constitution is called a democracy because it respects the interests not of a minority but of the whole people. When it is a question of settling private disputes, everyone is equal before the law; when it is a question of putting one person before another in positions of public responsibility, what counts is not membership of a particular class, but the actual ability which the man possesses. No one, so long as he has it in him to be of service to the state, is kept in political obscurity because of poverty.

… the admiration of the present and succeeding ages will be ours, since we have not left our power without witness, but have shown it by mighty proofs; and far from needing a Homer for our panegyrist, or other of his craft whose verses might charm for the moment only for the impression which they gave to melt at the touch of fact, we have forced every sea and land to be the highway of our daring, and everywhere, whether for evil or for good, have left imperishable monuments behind us.

(Thucydides)

(Left) Figure 281: Pericles (495–429 BCE)

With his early death, shortly after having given this funeral oration, Athenian democracy seemingly fell apart. Their decisions, whether tactical or strategic, led to one military disaster and another. Whether the blame can be laid upon the oligarchic (*hoi oligoi*) wing always able to muster power, or the Assembly of the *hoi poloi* It led to such genocidal episodes as the killing of the one thousand leaders of the island of Lesbos, in revolt against the Athenian control of the Delian league's wealth, or the even more terrible genocide of the entire male population of the island of Melos, when they attempted to assert their neutrality in the Peloponnesian war, and the subsequent selling into slavery of Melos' female population. Then came the introduction into this once-prosperous island, Melos, of Athenian settlers to repopulate it.

The period during which Pericles led Athens, roughly from 461 to 429 BC, is sometimes known as the "Age of Pericles," though the period thus denoted can include times as early as the Persian Wars, or as late as the next century.

The Periclean building program

(Left) Figure 282: The Parthenon

Pericles promoted the arts and literature; it is principally through his efforts that Athens acquired the reputation of being the educational and cultural center of the ancient Hellenic world. He started an ambitious project that generated most of the surviving structures on the Acropolis (including the Parthenon). This project beautified and protected the city, exhibited its glory, and gave work to the people. Pericles also fostered Athenian democracy to such an extent, that critics call him a populist.

He sponsored, at the age of twenty or so, the earliest surviving tragic drama by Athens' and Greece's first master of that evergreen theatrical genre, *The Persians of Aeschylus*, first staged in the Theater of Dionysus at the foot of the Athenian Acropolis in early 472 BCE. Pericles was intimately connected with the building program on the top of the Acropolis, that witnessed the construction preeminently of the Parthenon (447–432). He hobnobbed with leading intellectuals of the day, both Athenian and foreign. His private life—living with a foreign Greek woman to whom he could not legally be married, thanks to a law that he had himself sponsored in 451—was a scandal that contemporary writers of comic drama considered a gift.

Above all, so far as posterity is concerned, Pericles made an enormous—and favorable—impression on Herodotus' principal successor as a writer of broadly based Hellenic history, Thucydides of Athens (c. 472–400). Thucydides nearly promoted his memory as the uncrowned monarch of Athens, and was thus stimulated to write his history (until c. 411 BCE) of the Atheno-Peloponnesian War (431–404). Thucydides' vision centered on the Athenians' adherence to or failure to fulfill the policies and strategies advocated, so persuasively, by Pericles—as Thucydides understood and presented them.

Democratic incompetence and defeat

An example of growing Athenian incompetence and desperation in a war that Pericles thought should have been won on a dime, was the Arginusa, a naval battle near the island of Lesbos, 406 BCE. Athens was victorious, the Spartan fleet under Callicratidas lost 70 ships, while the Athenians lost 25 ships. But, due to bad weather, the Athenians were unable to rescue their stranded crews or to finish off the Spartan fleet. Despite their victory, these failures caused outrage in Athens and led to a controversial trial. The trial resulted in the execution of six of Athens's top naval commanders.

Symptomatic of a modern invitation to disaster, Joseph Stalin's suspicions, the subsequent killing of his general staff in 1938 CE, prelude the Soviet collapse at the invasion of the Nazis in the summer of 1941 CE and again the consequent unjustified execution of a great part of the left over Soviet general staff in that year.

Athens saw the Spartan allied fleets, Corinth, and Thebes quickly sailing to blockade the Dardanelles food suppliers to now overpopulated Athens.

Facing starvation and disease from the prolonged siege, Athens surrendered in 404 BC, and its allies soon surrendered as well. The surrender stripped Athens of its walls, its fleet, and all of its overseas possessions. Corinth and Thebes demanded that Athens should be destroyed and all its citizens should be enslaved. However, the Spartans announced their refusal to destroy a city that had done good service at a time of greatest danger to Greece, and took Athens into their own system. Athens was "to have the same friends and enemies" as Sparta.

While the walls connecting the port of Piraeus with upland Athens were being taken down, the supervising Spartans engaged an orchestra of maidens to accompany the Spartan efforts with flute music. Put into power by the Spartans, were a "quisling" group of oligarchs, soon to be overthrown by the democratic populists, the Spartans apparently washing their hands of Athenian affairs. The *hoi polloi,* now in power, took up by trial, and guilt by association, Socrates' numerous wealthy young friends (oligarchs). Socrates himself was condemned to die by the democratic Assembly, 399 BCE, not withstanding Plato's attempt to free him on bond. (Plato's *Apologia*)

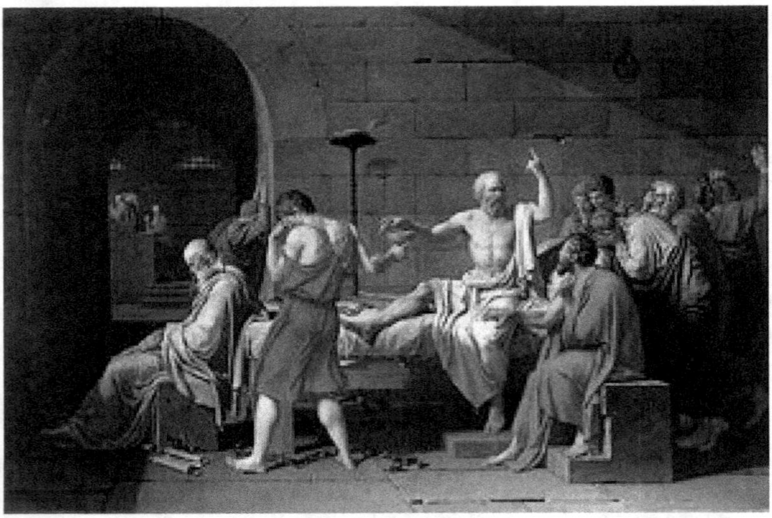

Figure 283: ***The Death of Socrates by*** **French painter Jacques-Louis David in 1787**

Socrates

(Left) Figure 284: A marble head of Socrates 469-399 BCE, in the Louvre

Epilogue

It is always difficult to make civilizational comparisons over broad swaths of time. The approximately three-thousand-year difference between the hay days of Sumer and Athens gives us only a blurry picture of the Sumerian rise and fall. Yet the similarities intrigue. Sumer, consisting of small city states, were semi-democratic in governance, individualistic and enormously creative of our paradigm civilizational mode. Then, *hybris*: the decadence of wealth and success, infighting and wars between the states. Finally, an outside force hungry and willing, Sargon the Semite Akkadian takes them over, as did Sparta to Athens, but with respect.

Sumer makes a comeback later, but it is short lived. Sumer goes down in Mesopotamian history as a holy originator, important as a memory and model, ultimately, until recently, forgotten. Greece and Athens still had a political history after 404 BCE. It is in many ways an ironic history, before a new Sargon, Philip, the Macedonian gulps down all of Greece at Chaeronea, 338 BCE. Athens and company again are defeated near the city of Thebes.

Now, Thebes was an ally of Xerxes in the final Persian invasion, and suffered mightily for this "quisling" action. In the Peloponnesian War, Thebes sided with Sparta, who had obviously forgiven the city. After the fall of Athens, the usual bickerings between the cities for dominance seem to always reveal a new combination of allies. For a good part of the period between 404 BCE and 338 BCE, a number of wars were fought in which Athens and Thebes combined to fight Sparta and its allies.

A name which lives in history, Epaminandas, the leader of a revitalized Thebes, is considered to be one of the greatest military minds in history, perhaps equaling Pericles, Themistocles, Pausanius, Lysander—all Athenian and Spartan greats. In the battles of Leuctra, 371 BCE, and Mantinea, 362 BCE, Epaminandas designed a whole series of tactical strategies, which even the Macedonians, Philip and Alexander, later took up. The Spartans were defeated in both battles. Epaminandas, who was at the front of the line, as is wont in Greek warfare, was killed at Mantinea.

The Thebans had long had a regiment called the "sacred band," male lovers who wished to die or win alongside their hearts. Epaminandas was one of them. He led the combined Theban and Athenian forces to the point where Sparta itself collapsed as an important military, political entity. The story is told that at the end of one of the battles, as the surviving Spartans straggled back to their town, the Spartan women, emancipated as no other Greek women were, because of their eugenic proclivity in bearing healthy young sons, emerged to confront the retreating, defeated warriors. They heckled: "*You were supposed to return in victory, alive, with your swords and spears, else lying and borne in heroic death upon the shield. Here you straggle home with nothing.*" Royal visitor: "Why are you Spartan women the only ones who can rule men?" Answer: "Because we are also the only ones who give birth to men." |source=Gorgo, Queen of Sparta and wife of Leonidas I| ?

Greek Hss: civilizational fulfillment

Under Pericles, he a relative of the political dynasty from Solon to Cleisthenes, Athens entered her golden age. An unusual number of great thinkers, writers, and artists flourished in the city. Herodotus, the 'father of history,' lived and wrote in Athens. Socrates, the 'father of philosophy,' taught in the marketplace, the Parthenon was built on the Acropolis and the Temple of Zeus at Olympia. Democritus envisioned an atomic universe. Aeschylus, Euripedes, Aristophanes, and Sophocles wrote their famous plays. This legacy would continue as, later, Plato would found his Academy outside the walls of Athens in 385 BCE and, further, Aristotle's Lyceum would be founded in the city centre.

Tragedy

Although plays were specifically commissioned for competition during religious and other types of festivals, many were re-performed and copied into scripts for "mass" publication. Those scripts regarded as classics, particularly written by the three great Tragedians, were even kept by the state as official and unalterable state documents. Also, the study of the "classic" plays became an important part of the ancient school curriculum.

There were, however, new plays continuously being written and performed, and with the formation of actors' guilds in the 3rd century BCE and the mobility of professional troupes, the genre continued to spread across the Greek world, with theatres becoming a common feature of the urban landscape from Magna Graecia to Asia Minor. (Cartwright 2013)

Below are examples of theatre genres and the corresponding, notable playwrights:

Tragedy: Phrynicos **(**511 BCE**)**, Aeschylos **(**472 BCE**),** Agathon (c. 448 BCE**)**, Sophocoles **(**442 BCE**)**, Euripides **(**438 **BCE),** Thespis (c. 320 BCE)

Comedies: Susarion of Megara **(c.** 580 BCE**), Epicharmus of Kos (c. 540 BCE), Aristophanes (425 BCE)**

Figure 285: Seating of the Theater at Epidaurus, in Peloponnese

Greek Philosophy

Figure 286: *The School of Athens* (1509–1511) by Raphael, depicting famous classical Greek philosophers in an idealized setting inspired by ancient Greek architecture

Ancient Greek philosophy arose in the 6th century BC and continued throughout the Hellenistic period, one in which Greece and most Greek-inhabited lands were part of the Macedonian and then the Roman Empire. Philosophy was used to make sense out of the world in a secular, non-religious way. It dealt with a wide variety of subjects, including astronomy, mathematics, political philosophy, ethics, metaphysics, ontology, logic, biology, rhetoric and aesthetics.

The first of these philosophers are called "Pre-Socratics" which designates that they came before Socrates. The Pre-Socratics were from either the eastern or western regions of the Greek world. Athens—home of Socrates, Plato and Aristotle—is in the central Greek region and was late in joining the philosophical game. The Pre-Socratic's most distinguishing feature is emphasis on questions of physics; indeed, Aristotle refers to them as "Investigators of Nature," Thales, Miletus, c. 600 BCE.

Their scientific interests included mathematics, astronomy, and biology. As the first philosophers, though, they emphasized the rational unity of things, and rejected mythological explanations of the world. Only fragments of the original writings of the pre-Socratics survive, in some cases merely a single sentence. The knowledge we have of them derives from accounts of philosophical writings, such as Aristotle's *Physics and Metaphysics, The Opinions of the Physicists*, by Aristotle's pupil, Theophratus, and Simplicius, a Neoplatonist who compiled existing quotes.

Philosophy was first brought into connection with practical life by Pythagoras of Samos (about 582–504 BCE), from whom it received its name: "the love of wisdom." Regarding the world as perfect harmony, dependent on number, he aimed at inducing humankind likewise to lead a harmonious life. His doctrine was adopted and extended by a large following of Pythagoreans.

"For, men were first led to study philosophy, as indeed they are today, by ***wonder***. Now, he who is perplexed and wonders believes himself to be ignorant…they took to philosophy to escape ignorance … " (Aristotle 982b)

Thales' (c. 624–546 BCE) views about water as the root element of all things on Earth was supported by long-held mythic accounts (both Egyptian, Babylonian, and Biblical creation stories that begin with water). Despite this, he was immediately superseded by the multiple novel theories of Pythagoras, c. 570 BCE; Anaximander, c. 547 BCE; Anaximenes, c. 530 BCE; Heraclitus, c. 505 BCE; Xenophanes, c. 500 BCE; all within 80 years. For example, instead of anthropomorphic creator gods, the pre-Socratic philosopher Anaxagoras, c. 499–428 BCE, thought *nous* "mind" controlled the universe.

Clearly, the rise of naturalistic inquiries into what was going on in the world had been stimulated by the economic changes in trade, mercantilism, manufacture. They were creating an urban, socially fluid environment, leading Athens to the social reforms of Drakon, Solon Pisistratus and Cleisthenes from c. 500 BCE. It was still a small urban world and the old 'agricultural estates' and the feudal aristocrats who owned the land and the power were being outflanked by these new economic and social patterns. And the gods, seemingly did not mind.

In Athens the philosophical schools were slow to form. The Sophists, teachers mostly of social wisdom, because there must have been much agitation over the class structure and the character of democracy versus oligarchy, made their welcome appearances in Athens. Especially so was the visiting political relativist Protagoras, c.490-420, under the controversial sponsorship of Pericles.

The ultimate philosophical and intellectual revolution was precipitated by the fall of Athens as a power and the subsequent execution of Socrates, a sophistic personality that put pins into the verbal claims for truth of those claiming wisdom. This sense of the conventionality of language and its contextual misuses was the pivot of a newer dialectical sophistication.

Plato's early dialogues in his Academy pursued Socrates' moral inquiries into convention. Later, having absorbed the general Hellenic search for a naturalistic understanding of physical and human reality, Plato's delving became deeper and metaphysical, conjectural. His student, Aristotle, coming from the unsophisticated north, with a family medical background, was more interested in the naturalistic investigations swirling around the Greek cities, from Italy and Sicily, to the Miletian East: they who were ever more backed up against the westerly advances of the Persians.

With the death of Plato in 347 BCE, Aristotle was now receiving assistance from his former student, Alexander. Aristotle lectured students in his newly established Lyceum. He did not write his own notes, but his students wrote down his encyclopedic investigations of our natural world, again both in the physical and human sense. The death of Alexander in 323 BCE, and a consequent revolt in Athens against Macedonian control, catapulted Aristotle out of Athens, so that as they quoted him, "Athens should not sin twice against philosophy." He died a year later in 322 BCE, while living with a daughter on a wealthy island off the coast of Greece. He had died too young—aged only sixty-two. Plato had lived until age eighty.

Aristotle's students at the Lyceum subsequently organized his lectures into a magnum philosophical opus, which reflected the genius of this thinker, envisioning the incorporation of a naturalistic philosophy in the guidance of humanity into a more rational future. Not only had Aristotle created a great philosophical synthesis of knowledge, but he also stimulated a newer intellectual study of nature as a discipline within itself—science.

Greek science

The timeline is interesting. On or about 330 BCE, Alexander, in his conquest, founded a new city on the Egyptian Mediterranean. It was named after him, and upon his death was taken over in the division of the Persian Empire by one of his generals, Ptolemy, a Greek. The town rapidly grew into a great export/import city, attracting Greeks, Jews and the ambitious of all backgrounds, as its wealth exploded. Now, with Athens being a relatively backwater town, Alexandria would rival only Rome as a great magnet for leadership.

This came in a descendant of the first of Ptolemy's line, Ptolemy II Philadelphus. He, like many other leaders of the great civilizations, saw their leading forms of knowledge as a key to their glorification. He founded a Library and a Museum. There was also a royal endowment for scholars to come and do research, even in the creative arts, no holds barred.

The Alexandrian Library quickly acquired a large number of papyrus scrolls, due largely to the Ptolemaic kings' aggressive and well-funded policies for procuring texts. It is unknown precisely how many such scrolls were housed at any given time, but estimates range from 40,000 to 400,000 at the Library of Alexandria, c. 250 BCE. Alexander himself, perhaps undergirded by his association with Aristotle, aimed at extending this abstract Greek civilization into his world at large.

Medical school researcher, Hippocrates of Kos, lived from 460–370 BCE. A successor of his school was *Herophilus* of Chalcedon, who conducted his research in Alexandria, in about the beginning of the third century, 325–280 BCE. Without question, he was the greatest anatomist of antiquity, and, according to the Greek, Galen, c. 220 CE (Roman imperial period), Herophilus was the first to practice human dissection.

Among his most important achievements were a detailed description of the brain, with an attempt to distinguish between the functions of its various parts; the discovery of the significance of the pulse and its use in diagnosing illness; and the discovery that the arteries contain blood alone, not a mixture of blood and air as Aristotle had taught, and that their function is to carry blood from the heart to all parts of the body. The value of this last discovery in laying the basis for a knowledge of the circulation of the blood can hardly be overestimated.

(Left) Figure 287: Greek medical practice. Physician treating a patient (Attic red-figure aryballos, 480–470 BC)

Euclid

Euclid (c. 323–285 BCE) of Alexandria collected important mathematical and geometric knowledge in one book: *The Elements.* This is a textbook rather than a reference book, so it does not cover everything that was known. In it, he gave definitions, postulates, and axioms, and called axioms "common notions." *Euclid* presented geometry as an axiomatic system: Every statement was either an axiom, a postulate, or was proven by clear logical steps from axioms and postulates. He further recorded some of his own original discoveries, such as the first-known proof that there are infinite prime numbers.

Aristarchus

The most famous of the earlier Alexandrian astronomers of this time was *Aristarchus* (310–230 BCE), who is sometimes called the "Hellenistic Copernicus." As a result of his discovery that the apparent immobility of the "fixed" stars is due to their vast distance from the earth, he was the first to have any adequate conception of the enormous size of the universe. His chief title to fame comes from his deduction that the earth and the other planets revolve around the sun.

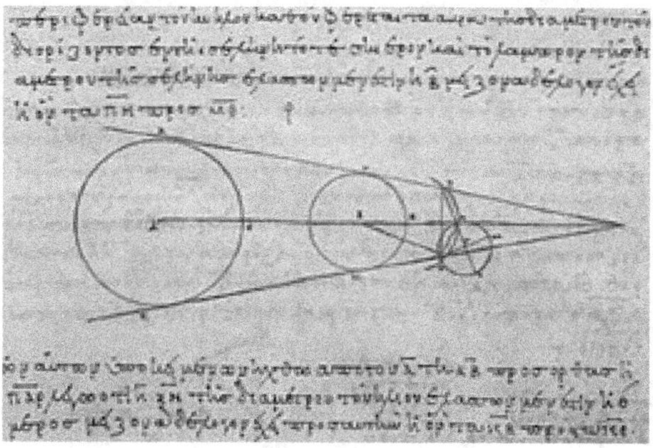

(Left) Figure 288: *Aristarchus's* 3rd-century BC calculations on the relative sizes of (from left) the Sun, Earth and Moon, from a 10th-century CE Greek copy

Archimedes

Archimedes, c. 212 BCE at age c. 75, in Syracuse, anticipated modern calculus and analysis by applying concepts of infinitesimals and the method of exhaustion to derive and rigorously prove a range of geometrical theorems, including the area of a circle, the surface area and volume of a sphere, and the area under a parabola. He also studied the specific gravity of floating objects, principles of the lever, the pulley and the screw.

Other mathematical achievements include deriving an accurate approximation of pi, defining and investigating the spiral bearing his name, and creating a system using exponentiation for expressing very large numbers. He was also one of the first to apply mathematics to physical phenomena, founding hydrostatics and statics, including an explanation of the principle of the lever. He is credited with designing innovative machines, such as his screw pump and propeller for ships, compound pulleys, and defensive war machines, including the burning lens, to protect his native Syracuse from invasion.

Archimedes died during the Siege of Syracuse, when he was killed by a Roman soldier despite orders that he should not be harmed.

Eratosthenes

Hellenistic geography owed most of its development to *Eratosthenes* (c. 276–194 BC), the astronomer, poet, philologist, and librarian of Alexandria. By means of sun dials placed some hundreds of miles apart, he calculated the circumference of the earth with an error of less than 200 miles. He produced the most accurate maps that had yet been devised, with the surface of the earth divided into degrees of latitude and longitude. Furthermore, he propounded the theory that all of the oceans are really one, and he was the first to suggest the possibility of reaching India by sailing west.

(Left) Figure 289: *Eratosthenes teaching in Alexandria* by Bernardo Strozzi (1635)

Hipparchus

Hipparchus, who did his most valuable work in Alexandria, died c. 120 BCE. His chief contributions were the invention of the astrolabe, the preparation of the best chart of the heavens known to antiquity, the approximately correct calculation of the diameter of the moon and its distance from the earth, and the discovery of the precession of the equinoxes. He laid the foundations of both plane and spherical trigonometry.

Heron

The record of inventions credited to Hero or *Heron* of Alexandria, a Greek who lived in the last century BCE, almost surpasses belief. The list includes a fire engine, a siphon, a force pump, a hydraulic organ, a slot machine, a catapult operated by compressed air, a thermoscope, and even a steam engine. How many of these inventions were really his own is impossible to say, but there appears to be no question that such contrivances were actually in existence during his time or soon thereafter. Nevertheless, the total progress in applied science was comparatively slight, probably for the reason that human labor continued to be so abundant and cheap that it was not worthwhile to substitute the labor of machines.

Ptolemy

Ptolemy of Alexandria (a Greek with no relation to the dynasty) was the last of the Hellenistic astronomers, 100–170 CE (Roman Empire). Although Ptolemy made few original discoveries, he systematized the work of others. His principal writing, the *Almagest*, was based upon the geocentric theory. It was originally entitled the *Mathematical Treatise* (Μαθηματική Σύνταξις, *Mathēmatikē Syntaxis*) and then became known as the *Great Treatise* (Ἡ Μεγάλη Σύνταξις, *Hē Megálē Syntaxis*). The second achievement is *Geography*, which is a thorough discussion of the geographic knowledge of the Greco-Roman world.

Hellenes: the modern civilization of *Homo sapiens sapiens*

The Sumerians created the institutions of modern urban, literate life. They were practical in the sense of searching for the natural causes of things. They allowed for the full flowering of the *Hss* brain in terms of culture. But for the understanding of the larger relationships of their physical (material), as well as social world, they had to, as with all other early civilizations, search in the pantheon of the gods for the *reasons why*.

After some three thousand years of literate urban experience, in war and in peace, a people arose out of very difficult and non-opportune economic possibilities for urban power, but with a secular look at the possible causes of things. Yes, the gods were there in full panoply, but the priests and priestesses were in short power supply, existing mostly to celebrate or mourn ancient historical moments of the people. This ethnicity was late into its cultural self-consciousness, coming down towards the Mediterranean Sea in waves after 1800 BCE. The soil was mostly rocks—a great stimulus for practical thinking.

There was something about the ethnic character of the Greeks, that was so different geographically, linguistically and emotionally from their ethnic Indo-European neighbors to the west, the Latins. All of cultural life was turned into an intellectual contest to understand material reality, and then use it for survival. It took time for them to come to terms with what they were perceiving intellectually—first as poetry (Homer, Sappho, Hesiod, Pindar) and then as flood of wonder about the physical and human world that they occupied. Always, war and the personal arête of killing in social victory was the accompanying drumbeat of life's rhythm.

We have to ask ourselves, after seeing how they finally blended into obscurity in the later Greco-Roman world, about their historical meaning. The seeds of their intellect and philosophical depth was a formulable element in the successive religious interregnum of the medieval. The Greek creation of ideas and methods never lost its power. Even if we don't hear much talk about them today, explicitly, we still feel their presence in the very nature of our civilizational discourse. The Sumerians gave us the scope and energy of civilizational life. The Greeks, even in their tangible disappearance from the centers of civilizations to come, gave us the intellectual possibilities of a naturalistic vision. They were modern beyond 2021.

Chapter 15

Universality: Rome

Roman character

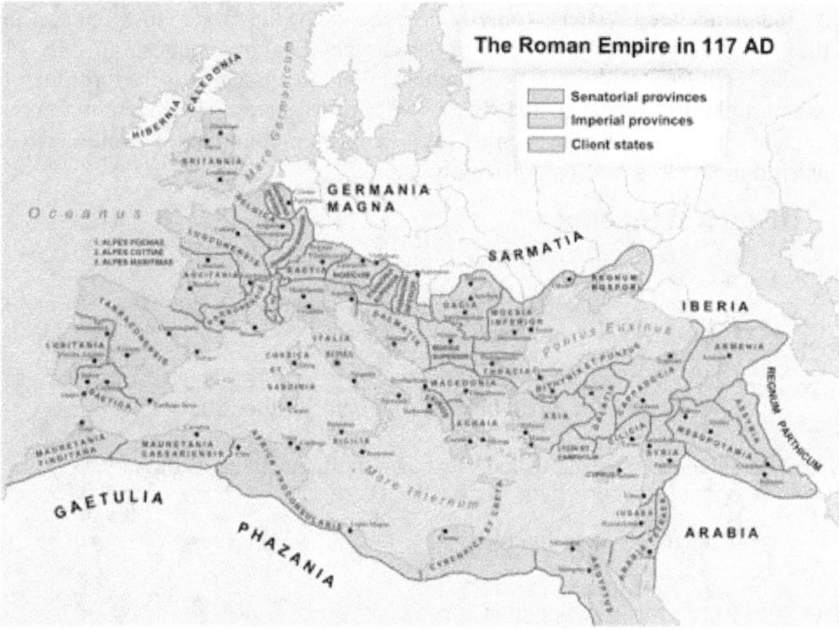

Figure 290: Events in c. 168 BCE:

At the time when Antiochus approached Ptolemy and meant to occupy Pelusium, Caius Popilius Laenas, the Roman commander, on Antiochus greeting him from a distance and then holding out his hand, handed to the king, as he had it by him, the copy of the senatus-consultum, and told him to read it first, not thinking it proper, as it seems to me, to make the conventional sign of friendship before he knew if the intentions of him who was greeting him were friendly or hostile. But when the king, after reading it, said he would like to communicate with his friends about this intelligence, Popilius acted in a manner which was thought to be offensive and exceedingly arrogant. He was carrying a stick cut from a vine, and with this he drew a circle round Antiochus and told him he must remain inside this circle until he gave his decision about the contents of the letter. The king was astonished at this authoritative proceeding, but, after a few moments' hesitation, said he would do all that the Romans demanded. Upon this Popilius and his suite all grasped him by the hand and greeted him warmly.

(Polybius c. 140 BCE)

Dominance

Between 341 and 338 BCE, the Romans faced a rebellion by their neighboring Latin allies. (The Battle of Chaeronea was then being fought in 338 BCE, province of Boeotia in Greece; result: Philip of Macedon destroyed Greek city independence.) After Rome emerged victorious over the Latins, the settlement they imposed underpinned subsequent Roman conquests of Italy and overseas territories. The Latins, and other Italian allies, hence were forbidden to conduct diplomacy or enter into treaties with other states. They were not taxed, except in having to provide men to fight in Roman-commanded armies, which bolstered the latter's ranks significantly.

Historical timeline

Early Events	
1100 BCE	Etruscans reached north-central Italy
800–500 BCE	Greeks established colonies throughout southern Italy
800 BCE	Phoenicians established Carthage on the north coast of Africa
About 753 BCE	Village of Rome founded
600 BCE	Rome was a province of Etruria
509 BCE	Romans revolted against the Etruscan kings and created the system of government of the Senate and the Assembly
THE CONQUEST OF ITALY	
494 BCE	first disputes between patricians (wealthy landowners who controlled the Senate) and plebeians (ordinary citizens)
450 BCE	"Law of the 12 Tables" provides written Roman law (Periclean law in Athens}
390 BCE	Gaulic (Celtic) invasion sacked Rome
282–272 BCE	War with Macedonian Pyrrhus in Italy
265 BCE	Rome completed the occupation of the Italian peninsula

THE CONQUEST OF THE MEDITERRANEAN	
264–241 BCE	First war with Carthage (First Punic War)
238 BCE	Conquest of Sardinia
229–228 BCE	First Illyrian War (Balkans)
219 BCE	Second Illyrian War
218–201 BCE	Second Punic War (Hannibal crossed the Alps)
215–205 BCE	First Macedonian War (Now in Greece)
200–197 BCE	Second Macedonian War (End of Greek independence)
200–191 BCE	Gaul invasion of northern Italy
192–189 BCE	Syrian War
171–168 BCE	Third Macedonian War
149–148 BCE	Fourth (and final) Macedonian War (Revolt, Greece/Macedonia now a Roman province)
149–146 BCE	Third Punic War and final defeat of Carthage

(Jones)

Note: Nine years after the conviction and execution of Socrates, 399 BCE, by the Athenian populists, Rome was sacked by the Gauls, who lived south of the Alps.

(Left) Figure 291: The Praetorian Guard, c. 1st century CE, legionnaires and elite soldiers. Marble relief, Louvre, Paris. Figures wearing headresses similar to that of the sea peoples who attacked Egypt, c. 1200 BCE, the Philistines

Figure 292: Egyptian illustration of the repulsing of the sea peoples, they, especially shown lower left with unique headress

Contributions of Rome

Ancient Roman civilization has given the modern world some of its unique contemporary institutions and beliefs. Their impact can be seen in our modern languages, religion, society, technology, law, politics, government, warfare, art, literature, architecture and engineering. Rome learned to professionalize and expand its military system, creating a form of government called *res publica*—the inspiration for modern republics such as the United States and France. It achieved impressive technological and architectural feats, such as the construction of an extensive system of aqueducts and roads, as well as the construction of large monuments, palaces, and public facilities.

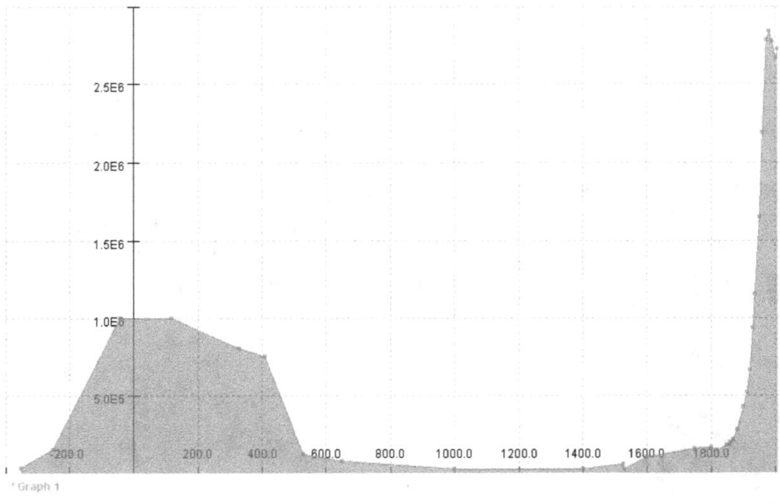

Figure 293: Graph of the population of city of Rome through history by David Galbraith. Note that at c. 200 BCE, city of Rome's population was c. 350K

Art of war

When Pyrrhus the Macedonian invaded Apulia in Italy (279 BCE), the two armies met in the Battle of Asculum, southeast of Rome, where Pyrrhus won a costly victory. The consul Publius Decius Mus was the Roman commander, and while his able force was ultimately defeated, they managed to almost break the back of Pyrrhus' Epirot (southwestern Greece) army, which guaranteed the security of the city of Rome itself. In the end, the Romans had lost 6,000 men and Pyrrhus 3,500, including many officers. Pyrrhus later famously commented on his victory at Asculum, stating, "If we are victorious in one more battle with the Romans, we shall be utterly ruined." It is from reports of this semi-legendary event that the term "Pyrrhic victory" originates.

(Left) Figure 294: Pyrrhus and his elephants at Asculum 279 BCE, southeast of Rome, near the Adriatic Sea

As German scholar Hans Delbruck notes in *History of the Art of War*: " ... the superiority of the Roman art of warfare was based on the army organization, a system that permitted very large masses of men to be concentrated at a given point, to move in an orderly fashion, to be fed, to be kept together. The Gauls could do none of these things."

Historian Sir Edward Creasy, in his *The Fifteen Decisive Battles of the World,* had this to say about the Iberian conflicts: "The war against the Spaniards, who, of all the nations subdued by the Romans, defended their liberty with the greatest obstinacy ... the Romans in both provinces were so often beaten, that nothing was more dreaded by the soldiers at home than to be sent there ... " Iberian cavalry were at the center of Hannibal's early success—c. 215 BCE. Final subjugation of Iberia (Spain and Portugal) took place in the days of Augustus, 15 CE.

Speaking of the decline of the heavy infantry, the Roman historian Vegetius (c. 400–450 CE) lauded the old fighting units, and lamented how the heavy armor of the early days had been discarded by the weaker, less disciplined, barbarized forces: "Those who find the old arms so burdensome, must either receive wounds upon their naked bodies and die, or what is worse still, run the risk of being made prisoners, or of betraying the country by their flight. Thus, to avoid fatigue, they allow themselves to be butchered shamefully, like cattle."

Historian Arther Ferrill notes that even towards the end, some of the old infantry formations were still in use. Such grouping was increasingly ineffective, however, without the severe close-order discipline, drill and organization of old times. At the Battle of Chalons (c. 451 CE), Attila the Hun rallied his troops by mocking the once-vaunted Roman infantry, alleging that they merely huddled under a screen of protective shields in close formation. He ordered his troops to ignore them and to attack the allies of Rome—the powerful Alans and Visigoths instead.

It was a sad commentary on the force that had once dominated Europe, the Mediterranean and much of the Middle East. It is true that at Châlons, the Roman infantry contributed to the victory by seizing part of the battlefield's high ground. Nevertheless, its day had already passed in favor of the mass levies of the barbarian federates.

There are concrete reasons for the power and efficiency of late Republic and Imperial Roman military, both in their tactical techniques and their strategic planning and operations. The Romans were able to copy and adapt the weapons and methods of their opponents more effectively. Some weapons, such as the *gladius* (short sword), were adopted outright by the legionaries. Publius asserts that the *pilum* (thrown javelin) was of Samnite origin, and the shield was based on Greek design. In other cases, well-trained units of enemy forces were invited to serve in the Roman army as auxiliaries after peace was made. In the naval sphere, the Romans followed some of the same methods they used with the infantry, dropping their ineffective designs and copying, adapting and improving on Punic warships, and introducing heavier marine contingents (infantry fighters) on to their ships.

Figure 295: A legionnaire stabbing with his gladius.

Roman officers were given much room to improvise when needed by unexpected tactics developed by opponents. Compared to the tightly packed spearmen of the Greek phalanx, the Roman heavy infantry, through their training and discipline, and operating in conjunction with the cavalry, could quickly adopt a number of methods and formations, depending on the situation. From the so-called *Testudo* formation during siege warfare, to a hollow square against cavalry attack, to mixed units of heavy foot, horse and light infantry against guerrillas in Spain, to the classic "triple line" or checkerboard patterns, versatility meant something. Against more sophisticated opponents, the Romans also showed great flexibility at times, such as the brilliant adjustments Scipio made against Hannibal at Zama (202 BCE, Tunisia). These included leaving huge gaps in the ranks to trap the charging elephants, and then the recall, reposition and consolidation of a single battle line that advanced to the final death struggle against the Carthaginian veterans of Italy.

In their Republican maturity, Roman discipline, organization and logistical systemization sustained combat effectiveness over long campaigns. Notably, the Roman system of *castra*, or fortified camps, allowed the army to stay in the field on favorable ground and be rested and resupplied for battle. Well-organized Roman logistics also sustained combat power, from routine resupply and storage, to the construction of military roads, to state-run arsenals and weapons factories, to well-managed naval convoys that helped stave off defeat by Carthage.

A general's death in battle did not cause the legions to crumble. Others stepped to the fore and carried on. In their defeat by Hannibal at the River Trebia (218 BCE, Italy), 10,000 Romans cut their way through the debacle to safety, maintaining unit cohesion when all around was rout: a testimony to their tactical organization, discipline, and confidence in their officers.

The Romans were more persistent and more willing to absorb and replace losses over time than their opponents. Unlike other civilizations, the Romans kept going relentlessly until, typically, their enemies had been completely crushed or neutralized. The army acted to implement policy, and were not allowed to stop unless they received a command from the emperor or a decree from the senate.

Demography, psychics and war

Tribal independence and lack of structure, particularly in Hispania, gave Roman tenacity and material weight to eventually wear down most opposition. The tribes of Europe did not have a state or economic structure able to support lengthy campaigns, and therefore could often (but not always) be made to change their minds about opposing Roman hegemony. The defeat in the Teutoburg Forest (9 CE, Goth *Arminius* destroyed several Roman legions, this in the reign of Augustus) might seem like an exception, but even here, the Romans were back on the warpath five years later, with major forces against their Germanic opponents. This commitment to persist in search of ultimate victory did not always work, especially towards the end of the Empire. However, their vision of long term campaigning most often was its own reward.

Where the Romans faced another large state structure, such as the Parthian Empire, they found the military road rocky indeed and were sometimes forced to an impasse. Nevertheless, the distinct pattern of Roman tenacity holds. Rome suffered its greatest defeats against sophisticated Carthage, notably at Cannae (216 BCE, southeastern Italy), and was forced to avoid battle for a lengthy period. We must remember that the Carthaginian leadership was Hellenic in its brilliance. Yet, in time, Rome rebuilt its forces on land and at sea, and persisted in the struggle, astonishing the Punics, who expected it to sue for peace. Against the Parthians (eastern Turkey to Eastern Iran), crushing defeats did not stop the Romans from inflicting serious defeats on the Parthians themselves, for they invaded Parthian territory several times afterwards. Even though Parthia proper was never totally conquered, Rome ultimately secured a rough hegemony in the area, and managed to successfully destroy Parthian forces in Mesopotamia on numerous occasions (Parthian Wars, 53 BCE–217 CE).

Military leadership was often politically achieved, and success was mixed. Over time, it did secure Roman military success. Leadership debacles are common in Roman military history, from the routs against Hannibal, to the demise of the unlucky Crassus against the Parthians (53 BCE in today's Turkey). The Roman polity's structuring however, produced a steady supply of men willing and able to lead troops in battle—men that were held accountable for defeat or malfeasance. It was not unusual for a losing general to be prosecuted by political enemies in Rome, with some having their property confiscated and barely escaping death. The senatorial oligarchy, for all its political maneuvering, interference and other faults, provided the functions of oversight and audit over military matters, that over the course of time, shaped final results. The record is a mixed one, but whether under a boisterous Republic or under imperial autocrats, Rome produced enough competent leaders to secure its military dominance for over a millennium. Some of the best leaders came from both eras, including Marius, Sulla, Scipio, Agricola, Drusus,Caesar, Trajan and others.

Note should be taken here of the large number of junior officers the Romans typically used to assure coordination and guidance. The initiative of such men played a key part in Roman success. Effective leadership was also bound up with the famous Roman *centurians*, the backbone of the legionary organization. While not all such men could be considered models of perfection, they commanded substantial respect.

The women of Roman Italy produced the continuing manpower supply These massed divisions enabled the legions to stay in the field and continue fighting after defeats and to launch new campaigns. Against Hannibal for example, Rome suffered huge losses, but still vastly outnumbered Hannibal's forces. This was due not only to defensive operations under Fabius, but also to the aggressive deployment of new armies under Scipio to take the battle to the Carthaginians in Africa. Other enemies of Rome came up against this massive manpower reserve and faltered over time: from small tribes, city-states or kingdoms fighting to maintain their independence, to major empires that confronted the Romans. The huge pool of fighting men gave the Romans much more room for error or setbacks, compared to their opponents.

Civic culture

Caeser's legionaires were an exemplification of the power of citizenship, and its civic culture. First, it stimulated a long tradition in the Roman military with consistent motivation and cohesion. Such culture included but was not limited to:

a) the valuing of Roman citizenship
b) the broad-based muster of free males into mass infantry units (as opposed to widespread use of foreign contingents, slaves or mercenaries)
c) loyalty to those fighting units (the Legion) which remained characteristically Roman in outlook and discipline.

Citizenship conveyed certain valuable rights in Roman society, and was another element that helped to promote the standardization and integration of the infantry. The citizen under arms—the legion soldier—was primed to reflect and practice the Roman ideal of *virtus, pietas, fides*: self-discipline, respect and faithfulness to engagements. Many historians saw the implementation of such ideals as mixed, but it was “a trilogy, driving every aspect of military, domestic, economic and social life.” As such, it was a strong force for cohesion among Rome's infantrymen.

Political universalism

Roman universalism was unique among ancient peoples, in that Rome willingly incorporated newly conquered people into their own society, freely giving citizenship to outsiders in order to Romanize them and make them willing participants (instead of unwilling subjects or enemies) in the Roman imperial system. Romans preferred government on the cheap and as such chose to administer new lands and peoples indirectly, through indigenous collaborators, who were awarded Roman citizenship or other benefits. This tactic was considered *divide and rule* because they literally divided up conquered peoples into their component units (usually tribes and city-states), made separate alliances and treaties with each, and induced each, through a complex system of rewards, to keep an eye on the others and provide for the common defense.

The geographical position of Rome was such that from its earliest centuries, it was suronded by many powerful, independent city-states and tribes that were intent on Rome’s destruction. In the rich alluvial plain of Latium alone, Rome lived among at least twelve independent, Latin-speaking nations. To Rome’s north was the ancient, wealthy, and highly civilized Etruscan confederation. To the south, around the Bay of Naples and beyond, were the large Greek cities of Cumae, Neapolis (modern Naples), and Posidonia. Thanks to its position on both the last available crossing of the Tiber River near the sea and the great salt trade route, Rome became a natural contact point for all of these cultures. Because of its location, and the rich trade in salt and other commodities, Rome was coveted by all its neighbors. And Rome reciprocated.

Once Rome moved beyond Italy, the Romans became much more guarded in their grants of citizenship. The system of divide and rule still applied, but now the Romans added a new dimension: the perks for the conquered began with possible tax exemption, rather than citizenship. Once Rome moved beyond Italy, it no longer rewarded former enemies quite so generously. Because of the costs incurred with overseas expansion, Rome could not afford to allow its overseas subjects to go untaxed.

During the imperial period, Rome did not offer citizenship to the newly conquered, as it had to their closer Latin brothers, but rather alliance and confederation. These allies of what the Romans would come to call *Latin-rights* status would contribute troops to the common defense, in return for which Rome would grant some of the perks of citizenship: (1) the right to do business at Rome; (2) the right to appeal the actions of Roman officials in Roman courts; and (3) the right to marry Roman citizens, the children of which unions would then be legal Roman citizens.

Strategic taxation

The Roman senate and the oligarchic leadership wanted both power and peace. They wanted to preserve that mutual amalgamation that had controlled Italy so well. The solution was selective taxation. Rome would reward especially loyal or strategic allies with tax-exempt status. This tax exemption, or partial tax forgiveness, was always held out as a reward for special clients or whole communities. Over time, many individual clients, cities, and tribes passed through tax exemption and "Latin rights" to full citizenship. By 212 CE, all areas under Roman control were given full Roman citizenship by Emperor Caracalla (188–217 CE), he of North African and West Asiatic ethnicities.

Because of the complex military and political relationships, and as a result of this system of selective taxation, the Roman tax code was bewilderingly complex, and the Roman government, always hesitant to increase the bureaucracy, required a cost-effective, non-governmental way to collect revenue.

As Cicero's speeches against the Sicilian governor, Gaius Verres (ca. 115–43 BCE), and private letters to his friends make clear, the tax farmers often lent money to individuals at exorbitant rates (48 percent per annum and higher), so that the borrowers could pay taxes to those same tax-farmers. Because the Roman government wanted revenues without bureaucracy, and local Roman governors like Verres wanted money in the form of bribes and company shares, the tax-farming companies were allowed to grow into pseudo governmental entities that for a short time during the Late Republic, acted as if they were the Roman state. From 133 to 44 BCE, the independent, private tax companies collected taxes, lent money, fielded troops, and controlled the economies of Roman possessions outside of Italy.

(Left) Figure 296: Pompey and Mt. Vesuivus

Legacy of the Roman Empire

In so many ways, the Roman mind was completely practical in searching for solutions to this vast military/political enterprise. The inventions and innovations which were generated by the Roman Empire profoundly altered the lives of the ancient people and continue to be used in cultures around the world today. Advancements in the construction of roads and buildings, indoor plumbing, aqueducts, and even fast-drying cement were either invented or improved upon by the Romans. The calendar used in the West derives from the one created by Julius Caesar, and the names of the days of the week (in the romance languages) and months of the year also come from Rome.

Apartment complexes (known as "insula"), public toilets, locks and keys, newspapers, and even socks all were developed by the Romans, as were shoes, a postal system (modeled after the Persians'), cosmetics, the magnifying glass, and the concept of satire in literature. During the time of the empire, significant developments were also advanced in the fields of medicine, law, religion, government, and warfare. The Romans were adept at borrowing from, and improving upon, those inventions or concepts they found among the indigenous populace of the regions they conquered. It is therefore difficult to say what is an "original" Roman invention and what is an innovation on a pre-existing concept, technique, or tool. It can safely be said, however, that the Roman Empire left an enduring legacy which continues to affect the way in which people live even today. (Mark 2008)

In Chapter 14, we noted the mathematical and engineering innovations of the Greek, *Heron*, 10–70 CE, who was active in Alexandria, during the early transition to Roman control. As stated there, *Heron* published a well-recognized description of a steam-powered device, called an *aeolipile* (sometimes called a "Hero engine"). Among his most famous inventions was a wind wheel, constituting the earliest instance of wind harnessing on land. Sadly, as was characteristic of Rome, its energies were hardly speculative and theoretical, and thus there was little practical movement to innovate on the basis of such theories. Giant corporations are slow in radically changing material or intellectual directions. So too, Rome.

Then, we have another Greek, *Galen*, c. 129–200/216 AD), in Pergamon (then a Greek city in now Western Turkey). He was a Greek physician/surgeon and philosopher in the Roman Empire. Arguably the most accomplished of all medical researchers of antiquity, Galen influenced the development of various scientific disciplines, including anatomy, physiology, pathology, pharmacology, and neurology as well as philosophy and logic.

Roman exemplars

Figure 297: Aqueduct of Segovia, Spain

Figure 298: The Colosseum in Rome (Emperor Vespasian, 80 CE)

Figure 299: Arles France, Aqueduct

Figure 300: Pantheon, Rome, 113–125 CE, Completed by Emperor Hadrian

Figure 301: The Pantheon dome. The coffered dome has a central ulus as the main source of natural light

(Left) Figure 302: Pantheon

Roman concrete

Also called *opus caementicium*, it was a material used in construction during the late Roman Republic until the fading of the Roman Empire. Roman concrete was based on a hydraulic-settling cement. Recently, it has been found that it materially differs in several ways from modern concrete, which is based on Portland cement. Roman concrete is durable due to its incorporation of volcanic ash, which prevents cracks from spreading. By the middle of the 1st century (50 CE), the material was being used universally, often brick-faced, although variations in aggregate allowed different arrangements of materials. Further innovative developments in the material, called the Concrete Revolution, contributed to structurally complicated forms, such as the Pantheon dome—the world's largest and oldest unreinforced concrete dome. Roman concrete was normally faced with stone or brick, and interiors might be further decorated by stucco, fresco paintings, or thin slabs of fancy-colored marbles. Made up of aggregate and cement, like modern concrete, it differed in that the aggregate pieces were typically far larger than in modern concrete, often amounting to rubble, and as a result, it was laid rather than poured. Some Roman concretes were able to be set underwater, which was useful for bridges and other waterside construction.

There is controversy as to when Roman concrete was first developed, but it was becoming widespread and customary in use from about 150 BC; some scholars believe it was developed a century before that.

(Left) Figure 303: Alcantara Bridge, Spain

Education

Early Roman education was rooted in the family. By the late Rpublic, however, the political needs of a swelling demography argued for governmental intrusion and tuition-based schools dominated the late Republic and the Empire. The Roman education system was based on the Greek System, and many of the private tutors in the Roman system were Greek slaves or freedmen. The educational methodology and curriculum used in Rome was copied in its provinces, and provided a basis for education systems throughout later Western civilization.

Organized education was relatively rare early on, and there are few primary sources or accounts of the Roman educational process until 100–150 CE. Due to the extensive power wielded by the *paterfamilias* over Roman families, the level and quality of education provided to Roman children varied drastically from family to family; nevertheless, Roman popular morality came eventually to expect fathers to have their children educated to some extent, and a complete, advanced education was expected of any Roman who wished to enter politics.

Figure 304: Relief found in Neumagen near Trier (Roman Germany), a teacher with three *discipuli* (180–185 CE)

At the height of the Roman Republic and later in the Roman Empire, the Roman system of education gradually found its final form. Formal schools were established, which served paying students; very little that could be described as free public education existed. Both boys and girls were educated, though not necessarily together.

(Left) Figure 305: Roman fresco of a blond maiden reading a text, 9 CE, Pompei, Italy

Prior to c. 250 BCE, the Roman system of education was closely bound to the Roman social institution of *patria potestas*, in which the father acted as head of the household (paterfamilias), and had, according to law, absolute right of control over his children. It was the father's duty to educate his children, and should he be unable to fulfill this duty, the task was assumed by other family members. It was not until 272 BCE, with the capture of Greek Tarentum during the annexation of Greek Sicily in 241 BCE, and the period following the First Punic War, that Romans were exposed to a strong influence of Greek thought and lifestyle, and found leisure to study the arts.

In this time period, a Greek captive from Tarentum, named Livius Andronicus, was sold as a slave and employed as a tutor for his master's children. After obtaining his freedom, he continued to live in Rome and became the first schoolmaster (private tutor) to follow Greek methods of education and would translate Homer's Odyssey into Latin verse in Saturnian meter.

As Rome grew in size and in power, following the Punic Wars, the importance of the family as the central unit within Roman society began to deteriorate, and with this decline, the old Roman system of education carried out by the paterfamilias deteriorated as well. The new educational system began to center more on the one encountered by the Romans with the prominent Greek and Hellenistic centers of learning, such as in Alexandria. In time, there would come to exist an elite literary educational system.

Greek cultural dominance

The Romans were definitely not anti-intellectual. They focused much of their brain power on law, politics, and war (not necessarily in that order).

Roman philosophy was mostly based on Greek philosophy. There were Stoics, Sophists, Cynics, Skeptics, Epicureans and Neo-platonists. Stoicism was probably the most prevalent type of philosophy practiced by Romans. Starting in the middle of the 2nd century BCE (150 BCE), private Greek Culture was increasingly in ascendancy, in spite of tirades against the "softening" effects of Hellenized culture from

(Left) Figure 306: A bust of young Marcus Aurelius (Capitoline Museum). Anthony Birley, his modern biographer, writes of thebust: "This is certainly a grave young man."

the conservative moralists. By the time of Augustus (c. 10 CE), cultured Greek household slaves taught the Roman young (sometimes even the girls). Chefs, decorators, secretaries, doctors, and hairdressers all came from the Greek East. Greek sculptures adorned Hellenistic landscape, gardening on the Palatine or in the villas, or were imitated in Roman sculpture yards by Greek slaves. The Roman cuisine is preserved in the cookbook, *Apicius* (c. 50 CE), is essentially Greek. Roman writers disdained Latin for a cultured Greek style. Only in law and governance was the Italic nature of Rome's accretive culture supreme.

Meditations: Emperor Marcus Aurelius, Stoic in philosophy, c.170 CE

Never let the future disturb you. You will meet it, if you have to, with the same weapons of reason which today arm you against the present.

To understand the true quality of people, you must look into their minds, and examine their pursuits and aversions.

Everything we hear is an opinion, not a fact. Everything we see is a perspective, not the truth.

Very little is needed to make a happy life; it is all within yourself, in your way of thinking.

You have power over your mind—not outside events. Realize this, and you will find strength.

The universe is change; our life is what our thoughts make it.

Death smiles at us all, but all a man can do is smile back.

When you arise in the morning, think of what a precious privilege it is to be alive—to breathe, to think, to enjoy, to love.

He who lives in harmony with himself lives in harmony with the universe.

If it is not right do not do it; if it is not true do not say it.

The object of life is not to be on the side of the majority, but to escape finding oneself in the ranks of the insane.

(Marcus Aurelius c. 170 CE)

Decline and fall

Gibbon, 1776–1789, gave a classic formulation of reasons why the Fall happened. He began an ongoing controversy by attributing a significant role to Christianity in the Western Roman Empire's fall. This view is no longer accepted by modern Roman historians. However, he did give great weight to other causes of internal decline as well, and to the attacks from outside the Empire.

> The story of its ruin is simple and obvious; and, instead of inquiring why the Roman Empire was destroyed, we should rather be surprised that it had subsisted so long. The victorious legions, who, in distant wars, acquired the vices of strangers and mercenaries, first oppressed the freedoms of the Republic, and afterwards violated the majesty of the purple. The emperors, anxious for their personal safety and the public peace, were reduced to the base expedience of corrupting the military discipline which rendered them alike formidable to their sovereign and to the enemy; the vigour of the military government was relaxed, and

> finally dissolved by the partial religious institutions of Constantine; and the Roman world was overwhelmed by a deluge of Barbarians.
>
> **(Gibbon 1737–1794, Chapter 8)**

From at least the time of Henri Pirenne,1862–1935, scholars have described a continuity of Roman culture and political legitimacy long after in 476 CE, Odoacer, the Gothic, takes over of what is left of Western Roman Empire. Pirenne postponed the demise of classical civilization to the 8th century. He challenged the notion that Germanic barbarians had caused the Western Roman Empire to end, and he refused to equate the end of the Western Roman Empire with the end of the office of emperor in Italy. He pointed out the essential continuity of the economy of the Roman Mediterranean, even after the barbarian invasions, and suggested that only the Muslim conquests represented a decisive break with antiquity.

The more recent formulation of a historical period, characterized as "Late Antiquity," emphasizes the transformations of ancient to medieval worlds within a cultural continuity. In recent decades, archaeologically-based argument even extends the continuity in material culture and in patterns of settlement as late as the eleventh century. Observing the political reality of lost control, but also the cultural and archaeological continuities, the process has been described as a complex cultural transformation, rather than a fall.

Going back a few decades, in 410, Alaric took Rome by starvation, sacked it for three days (there was relatively little destruction, and in some Christian holy places, Alaric's men even refrained from wanton wrecking and rape), and invited its remaining barbarian slaves to join him, which many did. The city of Rome was the seat of the richest senatorial noble families and the centre of their cultural patronage; to pagans, it was the sacred origin of the empire, and to Christians, the seat of the heir of Saint Peter, Pope Innocent I, the most authoritative bishop of the West. Rome had not fallen to an enemy since the Battle of Allia (Gauls 390 BCE), over eight centuries before. Refugees spread the news of this late fall and their stories throughout the Empire; and the meaning of the fall was debated with religious fervor.

Both Christians and pagans wrote embittered tracts, blaming paganism or Christianity respectively for the loss of Rome's supernatural protection, and blaming Stilicho's (military defender of Rome) earthly failures in either case. Some Christian responses anticipated the imminence of Judgement Day. Bishop Augustine in his book, *City of God*, ultimately rejected the pagan and Christian idea that religion should have worldly benefits; he developed the doctrine that the *City of God* in heaven, undamaged by mundane disasters, was the true objective of Christians. More practically, then Emperor Honorius was briefly persuaded to set aside the laws forbidding pagans to be military officers, so that one Generidus could re-establish Roman control in Dalmatia (Yugoslavian coast). Generidus did this with unusual effectiveness; his techniques were remarkable for this period, in that they included training his troops, disciplining them, and giving them appropriate supplies even if he had to use his own money. However, the penal laws were reinstated no later than 25 August 410, and the overall trend of repression of paganism continued.

Figure 307: Inscription honouring Honorius, as *florentissimo invictissimoque*, the most excellent and invincible, c. 417 CE–, Forum Romanum. (Right)

In that era, historian Procopius, c. 550 CE, wrote that the Emperor Honorius in Ravenna received the message from one of the eunuchs, evidently a keeper of the poultry, that Roma had perished. Honorius cried out and said, "And yet it has just eaten from my hands!" For he had a very large cockerel called Roma; the eunuch, comprehending his words, said that it was the city of Roma which had perished at the hands of Alaric, and the emperor with a sigh of relief answered quickly: "But I thought that my fowl Roma had perished." So great, they say, "was the folly with which this emperor was possessed." (Procopius, 2.25–26)

Going back to another, much earlier disaster, the *Battle of the Allia* was fought between the Senones (one of the south of the Alps Gallic tribes that had invaded northern Italy) and the then Roman Republic It was fought at the confluence of the rivers Tiber and Allia, eleven Roman miles (16 km) north of Rome. The Romans were routed and subsequently, the Senones sacked Rome. The common date given for the battle is July 18, 390 BCE.

Those who had been officers of state decided to meet their fate wearing their ceremonial dresses and "the insignia of their former rank and honour and distinctions." They sat on their ivory chairs in front of their houses. The next day, the Senones entered the city. They passed through the open Colline gate and went to the Roman Forum. They left a small body to guard there against any attack from the Capitoline and went through the streets for plunder. They did not meet anybody. People moved to other houses. The Gauls returned to the area of the Forum. Livy memorably described Gauls' encounter with the elderly patricians: "The houses of the plebeians were barricaded, the halls of the patricians stood open, but they felt greater hesitation about entering the open houses than those which were closed. They gazed with feelings of real veneration upon the men who were seated in the porticoes of their mansions, not only because of the superhuman magnificence of their apparel and their whole bearing and demeanour, but also because of the majestic expression of their countenances, wearing the very aspect of gods. So they stood, gazing at them as if they were statues, till, as it is asserted, one of the patricians, M. Papirius, roused the passion of a Gaul, who began to stroke his beard—which in those days was universally worn long—by smiting him on the head with his ivory staff. He was the first to be killed, the others were butchered in their chairs. After this slaughter of the magnates, no living being was thenceforth spared; the houses were rifled, and then set on fire."

"The Senones [Gauls] conducted the siege 'with great slackness' and concentrated on preventing the Romans from slipping through their lines. The patrician clan of the Fabii held an annual sacrifice on the Quirinal Hill. Gaius Fabius Dorsuo came down the Capitoline carrying the sacred vessels, passed through the enemy pickets and went to the Quirinal. He duly performed the sacred rites and returned the Capitoline." Livy comments, "Either the Gauls were stupefied at his extraordinary boldness, or else they were restrained by religious feelings, for as a nation they are by no means inattentive to the claims of religion."

At 390 BCE, the population of Rome proper was in the 100–150,000 range, but enough wealth circulated for the aristocracy to have purchased ivory-handled walking sticks.

Pax Romana, 800 years of inviolable untouchability, Rome. Compare Athens: Peace from the ancient Mycenaean period. Then, 480 BCE, Xerxes burns down the wooden Acropolis and city proper. In 404 BCE, the Spartans tear down the protective walls of Athens but spare the Acropolis and the city proper. After 338 BCE a vassal city of Macedon.

Figure 308: Roman mosaic of female athletes playing ball at the Villa Romana del Casale of Piazza Amerina, Roman Sicily, of 4th century CE. Think of the Minoans, also the Greeks, who still lived in Roman-occupied Sicily.

(Above) Figure 309: Portrait Bust of a Man, Republican Rome, 60 BCE

(Above) Figure 310: Roman portraiture is characterized by its "warts and all" realism

(Left) Figure 311: Veristic portrait bust of an old man, head covered *(capite velato)*: either a priest or pater familia (marble, mid-1st century BCE)

All three above portrait busts were probably carved by itinerant Greek artists

(Left) Figure 312: Detail of the gold glass medallion in (Brescia Museo di Santa Giulia) most likely from Alexandria, 3rd century CE. Probably owned by a Roman family living in Alexandria Egypt.

Figure 313: Pompeian painter with painted statue and framed painting.

Figure 314: For Livia, Wife of Emperor Augustus 1st Century CE, Fresco

Culture/Entertainment

Although much of ancient Roman life revolved around ***negotium*** (work and business), there was also time available for ***otium*** (leisure). Ranging from swimming to playing board games to attending theatre performances, or even athletics, the forms of entertainment enjoyed by Romans in ancient times were not much different from those that exist today.

One of the most popular recreational areas in Rome was the Campus Martius, a vast floodplain that acted as a playground for the youth of Rome. Here, they engaged in a variety of sports and activities, including jumping, wrestling, boxing, racing and throwing. Roman girls and women generally did not participate in these activities.

There were several other activities to keep the people of Rome content and to prevent uprisings. These consisted of chariot races at the *Circus Maximus*, musical and theatrical performances, public executions, beast hunts, and gladiatorial combat. Rome's most famous amphitheatre, the Colosseum, could accommodate up to 50,000 spectators. There are also accounts of the arena being flooded to hold naval battles (called *Naumachia*) for the public to watch. These forms of entertainment, all of which were massive public spectacles, were a distinct and important part of ancient Roman culture.

Figure 315

Roman entertainment is a byword for the decadence of the late empire, leading to its downfall, when it spent more time on amusement than reforming the military or rooting out corruption. Few others did mass entertainment better then the Romans. Their coliseums still inspire modern-day sporting arenas. Other forms of Roman entertainment could be found in the amphitheater, the hippodrome or the theatre.

The Colosseum in Rome was the largest amphitheatre in the Empire. It was here that people gathered to see the fights between gladiators, slaves, prisoners and wild animals like lions. Modern stadia and our own entertainment culture that achieves similar popular success, will tell us much about the future of our contemporary civilization.

The Emperors encouraged people to go to see the fights, as it stopped them from being bored and criticizing their ruler. The fights were very violent and ended when the loser died. Sometimes, when the arena was flooded, there would be ingenious fights with boats. The cells where the animals and prisoners were kept was underneath the floor of the main arena. The Colosseum even had a lift to bring them up to the arena. **(Fife 2012)**

Roman entertainment: the hippodrome

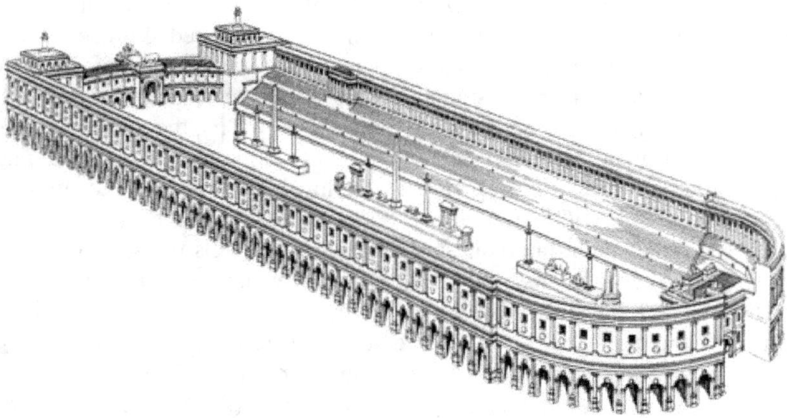

Figure 316

This was where the Romans went to see the chariot racing. The Circus Maximus was the largest hippodrome in Rome and could hold up to 250,000 people. Chariots were pulled by **two to four** horses, and were driven seven times around the ring at extremely fast speeds. Sometimes, accidents happened, and drivers were often trampled to death. There were four teams—red, white, blue and green—and fans of each team would wear their team's colours.

After the gothic overcoming of imperial Rome and the advance of Papal Christianity, Chariot racing became a major sports event in Byzantium/Constantinople, the seat of Eastern Christianity.

(Left) Figure 317: Mosaic depicting a theatrical troupe preparing for a performance

Literature

During the late Republic and the reign of the early emperors of Rome, there was a golden age of historical literature. Works such as the *Histories of Tacitus* the *Gallic Wars by Julius Caesar* and *History of Rome* by Livy have been passed down through generations. Unfortunately, in the case of Livy, much of the script has been lost and we are left with a few specific portions: the founding of the city, the war with Gaul and Hannibal, and its aftermath.

In the ancient world, poetry usually played a far more important part of daily life than it does today. In general, educated Greeks and Romans thought of poetry as playing a fundamental part of life. Initially in Rome, poetry was not considered a suitable occupation for important citizens, but the attitude changed in the second and first centuries BC. In Rome, poetry considerably preceded prose writing in date. As Aristotle had earlier pointed out, poetry was the first sort of literature to arouse people's interest in questions of style. The importance of poetry in the Roman Empire was so strong that Quintilian (35–110 CE), the greatest authority on education, wanted secondary schools to focus on the reading and teaching of poetry, leaving prose writings to what would now be referred to as the university stage. Virgil (70–21 BCE) represents the pinnacle of Roman epic poetry. His *Aeneid* was produced at the request of Maecenas and tells the story of flight of Aeneas from Troy and his settlement of the city that would become Rome. Lucretius (99–55 BCE), in his *On the Nature of Things* (explaining Epicurean atomistic moral philosophy), attempted to explicate science in an epic poem. Some of his science seems remarkably modern, but other ideas, especially his theory of light, are no longer accepted. Later, Ovid (28–60 CE) produced his *Metamorphoses*, written in dactylic hexameter verse, the meter of the epic, attempting a complete mythology from the creation of the earth to his own time. He unifies his subject matter through the theme of metamorphosis. It was noted in classical times that Ovid's work lacked the gravitas possessed by traditional epic poetry.

Catullus (84–54 BCE) and the associated group of neoteric poets produced poetry following the Alexandrian model, which experimented with poetic forms challenging tradition. Catullus was also the first Roman poet to produce love poetry, seemingly autobiographical, which depicts an affair with a woman called Lesbia. Under the reign of the Emperor Augustus, the works of Horace (65–25 BCE) were lauded, he who continued the tradition of shorter poems, with his Odes and Epodes. Martial (40–102 CE), writing under the Emperor Domitian, was a famed author of epigrams—poems which were often abusive and censured public figures.

The Roman discovery of *Hss*

Roman World Populations: 14 CE, c. 45.5 million; 164 CE, c. 61.5 million; c.200 CE (the highest, equaling China's population), c. 60 million persons.

The differences between the Greeks and the Romans before they were part of a worldwide civilizational amalgam, were at times attributed to their ancient language separation. *Centum* = Latin, Central European group; *Satem* = Greek, part of an eastern constellation. The differences in the Roman/Latin personality and the Hellenic vision has also been attributed to their individualistic reshaping of the original universal Cro-Magnon prototypes. Both ethnicities are of Indo-European derivation. Thus, they represented a chronologically later coming to be of a great linguistic family, as compared to the Sumerians, Mesopotomian Semites, and Egyptian Hamites, as well as the *Han* linguistics of the Chinese, and the late Hindu *satem* branches of the Indo-Europeans.

Others, possibly more correctly, have looked to the early social shapings that faced each ethnicity: climate, ecology, challenging neighbors and neighborhoods. For one, the Athenians in Attica had many hundreds of years of peace before facing the Indo-European Persians in 490 BCE. Then, of course, came the Peloponnesian civil wars of Greek versus Greek. Add the warlike rustic northern Greeks, the Macedonians, all before the final Roman triumph.

Rome had richer agricultural resources that could ultimately support a large population on the Italic peninsula. Greece had a good level of rocks to be surmounted with grapes, olive oil and genius (the potters wheel). The Italic female was early trained to provide bundles of boys, who, trained into warriors, learnt through losing.

In the Roman north were the Etruscans, possibly migrants from Anatolia (language as yet unidentified, possibly Indo-European *satem*), and even further north of Rome were bands of tough Gauls, linguistically close to Latin, ever eager to head down from the southern Alps for some Roman loot, 390 BCE. South of Rome from Naples were a series of Greek colonies, apparently settled after the Etruscans arrived in Italy. These city states, as well as the Etruscans, did much to civilize the Latins. Rome had much competition. But it also had the Tiber River.

A thousand years after the final Gothic conquests of political Rome, 476 CE, Italy was producing another great civilizational contribution—the Renaissance. The Italians, now an amalgam of indigenous Romans and German migrants, all speaking Italian, produced one of the great esthetic epochs: poetry, painting, sculpture, music, philosophy, and the beginnings of science. Was it the absorbed genetics of the two ethnicities, else, time, new stimulations and learnings, a wonderful climate, the Mediterranean Sea, or even the Papacy, that undergirded this world view?

The Romans attacked the issue of urban living through the development of one of the greatest engineering and technological systems ever seen: the road system, aqueducts, the arch/dome, pantheon, colosseum, concrete, and even urban sanitation/sewage systems. There was also the law, universal and applicable in all the Roman political settings. Then, the Julian calendar, the two official languages, Latin and Greek, plus the allowance of the ethnic traditions and languages of the conquered. Its educational system was implemented worldwide, to bring the outlying domains and local principalities together in one political and secular belief system. Religious diversity was tolerated, as long as it did not breed revolution and secession, e.g. the Jews.

(Left) Figure 318: 1st-century CE Roman bridge in Vaison la Romaine, France

The transformation of Rome into a Christian society, both in the east and west, was a slow process of internal, cultural disintegration. The early republican, semi-democratic traditions that bled into imperial rule allowed for no peaceful consensus transitions, only the murder of the temporary rulers; such transitions of leadership were always accompanied by new oligarchic underpinnings of power. Christianity represented a revolution of ideological commitment from the classes below. This Roman contribution to history revealed so clearly that the disintegration of belief always leads to the dissolution of political and military power.

The mystery of *Hss*

It is a puzzle, but it is reality. Even people living close to each and over a long period of time might be hostile to the culture of their neighbors, to experientially dislike the personalities of the individuals that they meet across the fence. And sure enough, their languages, here Greek and Roman, were different, and this applies even to much closer members of the same ethnicity—the twang of the Texan, the drawl of the Carolinean, the Brooklynese of a president.

The *Hss* mystery is of ethnicity and culture—the enormous, untouched possibilities of variance in every area of cultural life, coming out of humans who are equally adept at differential equations or composing a violin sonata. That is why there will never be a homogeneous world-wide civilizational culture of smart, creative people. From experience, we know that our brain is not designed that way. Thus, we don't have to be afraid of Harvard choosing high-powered think alikes. By the time they are seniors, the place will be as differentiated in terms of talents as between China and Japan. The most important thing is that these adjoining peoples should be equally able to design a bridge, do high powered math equations, and play fearsome chess. Yes, even without cheating/stealing, they should be capable of building a high-powered atom bomb, else discover a cure for cancer, even manage a grocery store. Both the Romans and the Greeks had these *Hss* intellectual capabilities, but in different amygdale proportions and thus with different cultural emphases.

Chapter 16

Christianity

The petrification of Rome.

Roman Emperor Theodosius II legislated again, in 438, against Paganism and forbade Pagan sacrifice. Once more, Theodosius threatened those who failed to comply with death and the confiscation of their property. In this law, as the Emperor explicitly admitted, Pagan sacrifices were still being openly celebrated:

> Hence our clemency perceives the need of keeping watch over the pagans and their heathen enormities, since by natural depravity and stubborn lawlessness, they forsake the path of true religion. They disdain in any way to perform the nefarious rites of sacrifice and the false errors of their baleful superstition by some means or other in the hidden solitudes, unless their crimes are made public by the profession of their crimes to insult divine majesty and to show scorn to our age. Not the thousand terrors of laws already promulgated. nor the penalty of exile pronounced upon them deter these men, whereby, if they cannot reform, at least they might learn to abstain from the mass of their crimes and the multitude of their sacrifices. But their insane audacity transgresses continually; our patience is exhausted by their wicked behavior so that if we desired to forget them, we could not disregard them.
>
> (von Harnack)

Timeline: decline of Paganism

CE 313: By the Edict of Milan, Constantine and Licinius establish toleration of Christianity.

329–335 CE: Constantine orders destruction of Hellenic temples at Didyma, Mt. Athos, Aigeai, Baalbek, etc. According to pagan rhetor, Libanius in his address to Theodosius, Pro Templis, 6 and 37, Constantine I " … did not bring any change to the traditional [pagan] worship ... one could notice that [during the reign of Constantine] all rituals were performed properly" and "he [Constantine] did not prohibit [pagan] sacrifices." According to Zosimus in *History*, Constantine erected two pagan temples in his capital, Constantinople.

361–363 CE: Emperor Julian, the last pagan Emperor, proclaims restoration of Hellenic worship.

389–391 CE: Emperor Theodosian decrees, Christianity becomes official state religion of the Roman Empire.

440-450 CE: All Hellenic monuments, altars and Temples of Athens, Olympia, and other Greek cities are destroyed. As German historian Gregorovius, in his *History of Athens* (1889) states, the pagan monuments of Athens and Greece were the best-preserved among other monuments in the late Roman Empire.

410–529 CE: Despite the Platonic Academy being destroyed in the first century BCE by the Roman, Sulla in 86 BCE, it was not until the early 5th century (c. 410 CE) that a revived academy (which had no connection with the original Academy) was established by some leading Neoplatonists. In 529, the emperor Justinian ended the funding of the revived Neoplatonic Academy because of active paganism of its professors. Other schools continued in Constantinople, Antioch and Alexandria, which were the centers of Justinian's empire. Modern historians such as Blumental and A. Cameron claim that the Neoplatonic Academy continued to exist for almost 50–60 years after 529 CE.

651 CE: Harran (Present day Turkey) is conquered by the Arabs. Sabians, descendants of Hellenistic tradition by virtue of their Hermetism are judged a people of a book, the Hermetic Corpus and as such are permitted to survive. *The Nabatean Agriculture* (Nabateans Christianized Arab inhabitants of today's Jordan) was said by Maimonides, c. 12th century CE, to have made an accurate record of the beliefs of the Sabians (unaffiliated religious group in Harran with neo-Hellenic philosophical beliefs) in the Harranian area.

804 CE: Hellenes of Laconia (ancient Sparta), Greece, resist the attempt of Tarasius, Patriarch of Constantinople, to convert them to Christianity.

The political traditions of the Roman imperium were only slightly modified by Christianity. The role of the late emperors could be defined not as governing in terms of day to day decisions, but rather setting goals for their subordinates that needed to be achieved. A legitimate Roman emperor was not a secular ruler in our modern understandings of the term, but one chosen directly by the supreme creator, the divinity of the entire cosmos, to maintain the key pillars of rational Roman civilization—education, city life, the rule of written law, and now the welfare of the Christian Church—by which humankind was destined to be brought into line with the divine plan. In return for which, the divinity would guarantee *that* Divine victory: the role of Christianity in subsequent Roman military conquests.

Hellenistic influences on christianity.

The well-known Talmud scholar, Daniel Boyarin, has argued that Paul's theology of the spirit is more deeply rooted in Hellenistic Judaism than generally believed. In *A Radical Jew: Paul and the Politics of Identity* (1994), Boyarin argues that the Apostle Paul combined the life of Jesus with Greek philosophy to reinterpret the Hebrew Bible in terms of the Platonic opposition between the ideal (which is real) and the material (which is false). Judaism is a material religion, in which membership is based not on belief, but rather descent from Abraham, physically marked by circumcision, and focusing on how to live this life properly. Paul saw in the symbol of a resurrected Jesus the possibility of a spiritual, rather than corporeal, messiah. He used this notion of messiah to argue for a religion through which all people—not just descendants of Abraham—could worship the God of Abraham. Unlike Judaism, which holds that it is the proper religion only of the Jews, Pauline Christianity claimed to be the proper religion for all people.

In his use of the the Platonic distinction between the material and the ideal, Paul showed how the spirit of Christ could provide *all* people a way to worship the God who had previously been worshipped only by Jews and Jewish proselytes, although Jews claimed that he was the one and only God of all. Boyarin roots Paul's work in Hellenistic Judaism and insists that Paul was thoroughly Jewish, but argues that Pauline theology made his version of Christianity appealing to Gentiles. Boyarin also sees this Pauline reworking of both Jesus' teachings and Pharisaic Judaism as essential to the emergence of Christianity as a distinct religion, because it justified a Judaism without Jewish law.

Council of Jerusalem, great as was the success of Barnabas and Paul in the heathen world, the Judcaic-oriented authorities in Jerusalem insisted upon circumcision as the condition of admission of members into the church, until, on the initiative of Peter, and of James, the head of the Jerusalem church, it was agreed that acceptance of the Noachian Laws—namely, regarding avoidance of idolatry, fornication, and the eating of flesh cut from a living animal—should be demanded of the heathen desirous on entering the Church. Paul successfully argued that Gentile converts did not need to follow Jewish customs, get circumcised, follow Jewish dietary restrictions, or otherwise observe *Mosaic law*.

Christian fathers

At this time, the Christian intellectual world was divided:

Greek-speaking fathers of the Church: Justin Martyr, Athanasius of Alexandria, Basil of Caesarea, Gregory of Nazianzus, John Chrysostom, Cyril of Alexandria, and Maximus the Confessor.

Latin Fathers: Tertullian, Cyprian, Jerome, Ambrose of Milan, Augustine of Hippo, and Gregory the Great.

Christian Philosophers

Justin Martyr: Christian apologist and philosopher whose work often focused on the doctrine of the Logos and argued that many Stoic and Platonic philosophical ideas were similar to ideas in the Old Testament

Tertullian: Before he converted to Christianity, Tertullian was a Pagan philosopher; after that change of direction, he remained a prolific writer in the second century CE, and is commonly called the "Father of the Western Church." He was the first church father to use the term *Trinitas* in reference to the Godhead and developed the doctrine of traducianism—the idea that the soul was inherited from the parents, the idea that God had corporeal (although not fleshly) existence, and the doctrine of the authority of the gospels. He fought voraciously against Marcionism (a Roman heresy that rejected the Hebrew Bible), and considered Greek philosophy to be incompatible with Christian wisdom. Toward the end of his life, he joined the heterodox sect of Montanism, and thus has not been canonized by the Catholic Church.

Irenaeus of Lyons: He vociferously argued in his writings for the unity of God, and against Gnosticism (a dualistic belief in the evil of matter and the good of the spirit). He argued that original sin is latent in humanity, and that it was by Jesus' incarnation as a man that he "undid" the original sin of Adam, thus sanctifying life for all mankind. Irenaeus maintained the view that Christ is the Teacher of the human race, through whom wisdom would be made accessible to all.

Clement of Alexandria (150–215 CE): Theologian and apologist who wrote on Greek philosophy, using ideas from pagan literature, Stoic and Platonic philosophy, and Gnosticism to argue for Christianity.

Origen: Origen of Alexandria, c. 225 CE, student of Clement, was influential in integrating elements of Platonism into Christianity. He incorporated Platonic idealism into his conceptions of the Logos and the two churches—one ideal and one real. He also held a strongly Platonic view of God, describing him as the perfect, incorporeal ideal. He was later declared a heretic for subscribing to the "too-Platonistic" doctrine of the pre-existence of the soul. Despite this, Origen is a Church Father and is widely regarded as one of the most important Christian theologians of all time.

Augustine, Bishop of Hippo (now Tunisia): (c. 430 CE) developed classical Christian philosophy, influenced the whole of Western thought, largely by synthesizing Hebrew and Greek thought. He drew particularly from Plato, the Neoplatonism of Plotinus, and Stoicism, which he altered and refined in light of divine revelation of Christian teaching and the Scriptures. Augustine wrote extensively on many religious and philosophical topics; his use of an allegorical method of reading the Bible further developed such doctrines: of hell as endless punishment, original sin as inherited guilt, divine grace as the necessary remedy for original sin, baptismal regeneration and consequently infant baptism, inner experience and the concept of "self," the moral necessity of human free will, and individual election to salvation by eternal predestination. Augustine has been a major influence in the development of Western theology, and in particular, his works, *City of God* and *Confessions*, laid the foundations for Western Philosophy, influencing many philosophers and making him one of the most influential figures in the history of philosophy.

Athanasius of Alexandria: father of trinitarian orthodoxy, involved in the formation of the Nicene Creed (d. 298 CE), and vehemently opposed Arius, the bishop of Alexandria and his large following across Europe and Western Asia, who held that Christ was a created being.

Medieval unification

Figure 319: Map of mediaeval universities established by Catholic students, faculty, monarchs, or priests

In achieving a medieval consensus, first came the affirmation of the existence of a personal-infinite God. This God, acknowledged as the Creator of the world, was both transcendent and immanent. He took a continuing interest in *His* creation. He ruled over and intervened in the course of history, either by miracle or by Providence.

Medievals saw the universe as an ordered place. This order was thought to be transcendently imposed from the mind of the Creator. The universe was also governed by moral categories, because God's order included *His* moral law. Since God is just, violation of *His* law was thought to bring retribution.

The elite and the masses viewed human beings as sinful, finite, and contingent. They understood that they needed divine help and forgiveness. Their recognition of human limitation and flaw, however, was balanced by a vision of human dignity based upon the imago dei.

Westerners during the Middle Ages believed that Jesus Christ was the God-man. They confessed Him to be the unique revealer of the Father: "God of God, Light of Light." They worshiped Him as the unique redeemer of humankind, whose death on the cross and resurrection from the dead make salvation possible. Central to the medieval consensus, was the necessity of faith and religion. To the medieval mind, faith was the primary avenue for the acquisition of knowledge. Medievals thought that all of reality was so penetrated by mystery that mere observation of facts would never reveal truth. The Scriptures and the Church held a privileged position as repositories of truth. (Bauder 2010)

Institutional learning

Literacy and socio-political organization suffered, following the fall of Roman Empire in 476 CE. Monks sought refuge at the far fringes of the known world: like Cornwall, Ireland, or the Hebrides. Disciplined Christian scholarship carried on in isolated outposts, like Skellig Michael in Ireland, where literate monks became some of the last preservers in Western Europe of the poetic and philosophical works of Western antiquity. By around 800AD, these outliers were producing illuminated manuscripts, such as the Book of Kells, by which old learning was re-communicated to Western Europe. The Hiberno-Scottish mission, led by Irish and Scottish monks like St Columb, spread Christianity back into Western Europe during the middle centuries, establishing monasteries through Anglo-Saxon England and the Frankish Empire during the Middle Ages.

Following the collapse of the Western Roman Empire, these small monastic communities were practically the only outposts of literacy in Western Europe. In time, stimulated by the Frank, Chalemagne, c. 787 CE, the Cathedral schools later developed into Europe's earliest universities. The Church has established thousands of primary, secondary and tertiary institutions throughout the world in the centuries since. The Church and clergymen have also sought at different times to censor texts and scholars.

By the late 11th century, beginning with the efforts of Pope Gregory VII, the Church successfully established itself as "an autonomous legal and political ... [entity] within Western Christendom." For the next several hundred years, the Church held great influence over Western society; church laws were the single "universal law ... common to jurisdictions and peoples throughout Europe," giving the Church "preeminent authority." With its own court system, the Church retained jurisdiction over many aspects of ordinary life, including education, inheritance, oral promises, oaths, moral crimes, and marriage. As one of the more powerful institutions of the Middle Ages, Church attitudes were reflected in many secular laws of the time.

The Catholic Church was very powerful, essentially internationalist and democratic in it structures, and run by monastic organizations, generally following Benedictine rule. Men of a scholarly bent usually took Holy Orders and frequently joined religious institutes. Those with intellectual, administrative or diplomatic skill could advance beyond the usual restraints of society—leading churchmen from faraway lands were accepted in local bishoprics, linking European thought across wide distances. Complexes, like the Abbey of Cluny, became vibrant centers with dependencies spread throughout Europe. Ordinary people also trecked vast distances on pilgrimages to express their piety and to pray at the site of holy relics.

The first significant renewal of learning in the West, as noted above, came when Charlemagne, advised by Candidus, Peter of Pisa and Alcuin of York, attracted the scholars of England and Ireland, and by imperial decree in 787 CE established schools in every abbey in his empire. This development continued to the period from the middle of the eleventh century to the middle of the fourteenth century and is known as the "High medieval" or "scholastic" period. Of early significance was Johannes Scotus Eriugena (c. 815–877), successor of Alcuin of York as head of Charlemagne's Palace School. He was an Irish theologian and Neoplatonic philosopher.

Surely it is clear that the intellectual power of Greek learning had now gone beyond pagan Rome and had become a domineering element in the successor theological world of universal Christianity. *The Hss program of civilizational discovery was thus continued*

Scholasticism and Higher Learning

It is generally agreed to begin with Saint Anselm of Canterbury (1033–1109) an Italian philosopher, theologian, and church official who is famous as the originator of the ontological argument for the existence of God ("that than which nothing greater can be thought") He stimulated schools, from which the name Scholasticism is derived. They became centers of medieval learning.

From this continuing stimulus, the universities developed in the large cities of Europe and rival clerical orders within the Church began to battle for political and intellectual control over these centers of educational life. As most universities were subject to the approval of the church or even papacy, the rediscovery of the teachings of Aristotle became a point of contention. The University of Paris' existence was approved of by the Pope in 1215. Oxford: patterned after Paris around the same time. They both taught a version of the ancient liberal arts, medicine, law, and theology.

The two main orders founded in this period were the Franciscans and the Dominicans. The Franciscans were founded by Francis of Assisi in 1209. Their leader in the middle of the century was Bonaventure, a traditionalist who defended the theology of Augustine and the philosophy of Plato, incorporating only a little of Aristotle, with more neoplatonist elements. Following in the steps of Anselm, Bonaventure supposed that reason can discover truth, only when philosophy is illuminated by religious faith. Other important Franciscan writers in the 13th and 14th centuries CE were Duns Scotus, Peter Auriol, and William of Ockham.

(Left) Figure 320: Thomas Aquinas 1225–1274

Parallel with the Franciscans was the Dominican order, founded by St Dominic in 1215, who placed more emphasis on the use of reason and made extensive use of the new Aristotelian sources derived from the East, and Moorish Spain. The great representatives of Dominican thinking in this period were Albertus Magnus and (especially) Thomas Aquinas (1225–1274), whose brilliant synthesis of Greek rationalism and Christian doctrine eventually came to define Catholic philosophy. Coming from a southern Italian aristocratic family (lord of Aquino), Thomas tried to blend theology and philosophy, reason and faith.

St. Thomas believed that reason could not contradict faith, because the same God had given both faith and reason to humanity. In fact, rather than contradicting it, he said that faith expands upon reason.

In the University of Paris, Aquinas placed more emphasis on reason and argumentation, and was one of the first to use new translations of Aristotle's metaphysical and epistemological writings. This was a significant departure from the Neoplatonic and Augustinian thinking that had dominated much of early Scholasticism. Aquinas showed how it was possible to incorporate much of the philosophy of Aristotle without falling into the "errors" of the commentator Averroes, an important Islamic Aristotelian from Spain.

(Left) Figure 321: Plato, Seneca, and Aristotle from *Devotional and Philosophical Writings*, c. 1330

Medieval architecture

Figure 322: Early medieval secular architecture in pre-Romanesque Spain: the palace of Santa María del Naranco, c. 850 CE (Right)

The Romanesque, which dominated architecture during the 11th and 12th centuries, was the first pan-European style since Roman Imperial Architecture, and examples are found in every part of the continent. The term was not contemporary with the art it describes, but rather, is an invention of modern scholarship based on its similarity to Roman architecture in forms and materials. Romanesque is characterized by a use of round or slightly pointed arches, barrel vaults, and cruciform piers supporting vaults. Romanesque buildings are widely known throughout Europe.

Figure 323: The Tower of London (1078). William the Conqueror built the central White Tower as his stronghold and residence

Figure 324: Speyer Cathedral, begun by Conrad II, Holy Roman Emperor in 1030, as an expression of imperial power and architectural innovation

Figure 325: *Castello di Venere*, Erice (12th–13th century), is one of many built by the Normans in Sicily, Italy.

Figure 326: Many towns, such as San Gimignano, (Siena-Tuscany) were enclosed with walls, 1200–1300 CE, causing crowding and the building of tower houses in Romanesque and Gothic styles.

Gothic architecture can be recognized in a number of 11th and 12th century building projects, particularly in the Île de France (Paris) area. There, character is first acknowledged in the distinctively Gothic style at the 12th century abbey church of Saint-Denis in Saint-Denis, near Paris. Verticality is emphasized in Gothic architecture, which features almost skeletal stone structures with great expanses of glass, pared-down wall surfaces, supported by external flying buttresses, pointed arches to give vertical shape, ribbed stone vaults, clustered columns, pinnacles and sharply pointed spires. Windows contain stained glass, showing stories from the Bible and from lives of saints. Such advances in design allowed for the construction of ever taller cathedrals. It became something of an inter-regional contest to build a church as tall as possible. Variations included these early, Brick Gothic styles:

Figure 327: The rebuilt ambulatory at the Abbey of Saint-Denis (1140–1145)

Figure 328: Exterior of Saint-Denis Basilica, with its three portals and remaining tower (1140–45)

Figure 329: The west front at Noyon Cathedral, begun 1145, showing influence of Saint-Denis

Figure 330: The apse of Noyon Cathedral (1155)

Figure 331: Laon Cathedral, begun 1155

Figure 332: Facade of Notre Dame de Paris, facade begun about 1200

High Gothic Architecture 1200–1270

This culminating architectural tradition fulfills that pulse in *Hss* humans, going as far back as Stonehenge and then to the Parthenon, the Pantheon and then to this medieval explosion of that drive to control, to master, even to understand,under very different intellectual guidance. Always, such construction gives evidence of the *Hss*' exploratory intellectual vision of human possibility.

Figure 333: Overview of Chartres Cathedral (1194–1220)

Figure 334: Choir of Chartres Cathedral with the new, stronger four-part rib vault

Figure 335: Bourges Cathedral with flying buttresses (1195–1230)

Figure 336: Facade of Amiens Cathedral (1220–1266)

Figure 337: Reims Cathedral from the northwest (1211–1345)

Figure 338: Amiens Cathedral choir and altar

Falling Buttresses: Beauvais Cathedral and the Limits of Gothic Architecture

By its very nature, building involves limits. The great cathedrals of Medieval Europe were the apogee of technology during their time—the peak accomplishment of the contemporary architects, masons, and artisans. As the centuries passed, the mighty churches became larger and more ambitious—the buildings soared ever higher on incorporated increasingly lofty flying buttresses, until finally, the builders reached the limits of stone and iron and mortar.

(Left) Figure 339: Cathédrale Saint-Pierre de Beauvais

This failure happened at the Cathédrale Saint-Pierre de Beauvais (hereafter Beauvais Cathedral), an incomplete cathedral which many people regard as the most daring achievement of Gothic architecture. Work was commenced on the cathedral in 1225. From the very beginning, the building was designed to be the tallest and most splendid church in the world. This magnificence was probably partially intended as an act of defiance by France's northern barons, who were allied with the episcopacy in a struggle with the French throne—a struggle for power which culminated when the northern lords revolted against Louis VIII and attempted to kidnap his son Louis IX (who escaped the plot to ascend first to the throne and later to sainthood, and after whom Saint Louis, Missouri is named). Unfortunately, the grandiose architectural plans were hampered by funding problems and by structural flaws.

Figure 340: Cathédrale Saint-Pierre de Beauvais from the west (Right)

Although Beauvais Cathedral was intended to be taller than other cathedrals, the original planners also designed the flying buttresses to be thinner. This was both to allow them to soar higher (since they weighed less) and to allow maximum light into the stained glass windows of the building. Unfortunately, the design did not work out. In 1284, part of the choir vault collapsed, along with multiple flying buttresses. Modern structural engineers (21st century) believe that the failure was the result of resonant vibrations—an unhappy mixture of spindly buttresses which were just the wrong length to cope with the region's high winds. The cathedral was built and rebuilt off and on throughout the following centuries, with mixed results. In 1573, the structure suffered another major setback, when the 153-meter-tall central tower collapsed.

(Right) Figure 341: The Plan of Cathédrale Saint-Pierre de Beauvais (the black is extant, the gray is missing)

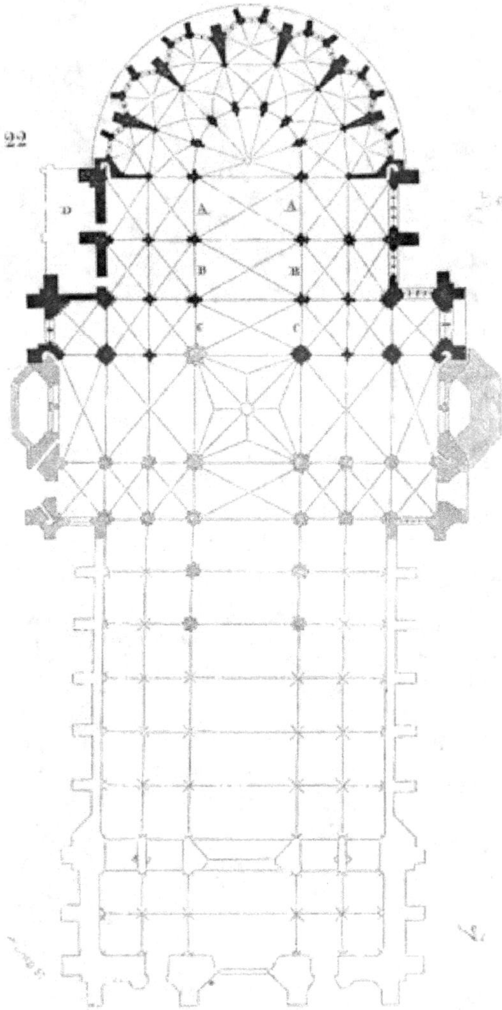

Today, only the transept and choir remain, but they are indeed magnificent. Unfortunately, the structure is still in peril. In the 1960s, the cathedral's caretakers removed iron bars which were laterally connecting the buttresses, in hopes of making the cathedral look even more graceful. Unsurprisingly, this action caused the transept to separate from the choir. Steel rods were quickly added, but, being more rigid than iron, they seem to have increased the rate of fissure. A wide number of sundry modern braces were added throughout the eighties and nineties, and in 2001, a team of architects from Columbia University scanned the entire edifice. They hope to use their comprehensive imaging resources to design less unwieldy solutions to the cathedral's many problems, but, at present, the world's most ambitious gothic edifice remains a masterpiece of beauty and a failure of function.

Figure 342: Inside Beauvais Cathedral

Figure 343: A fantasy of the early 16th Century, inspired by Beauvais Cathedral, Northern France, pen and brown ink and watercolour, on paper (Albert Edward Richardson 1934)

The Italian Renaissance

Christian art seems never to reveal its nod to classicism. But where early Christians utilized Roman religious symbols, Renaissance Christians embraced both the art and philosophy of the pagan ancients. As the grey of the middle ages was ending, their discovery of Greco-Roman artwork complemented and mirrored a shift in ideology in Europe: "The development of the Julian High Renaissance was supported by the pope's interest in the grandeur of ancient Rome … and by the sensational discovery of the sculptural group of 'Laocoön and His Sons.'" (Hartt 494) In form, Renaissance artists began to replicate the artists of antiquity in their idealization of the human form. And although the Catholic Church continued to rule over the continent—especially in the South—the proliferation of the intellectualism associated with Ancient Greece began to flourish in the Mediterranean in conjunction with Catholicism.

These two ideologies—which were at odds during the disciplines of the Middle Ages (classical philosophy and religion)—take on a new partnership in the Renaissance period: "Artists and philosophers, each in their own way, combined Christian belief and ancient philosophy into a balanced, rational, humanistic system." (Stokstad 315). These interests even shaped the Church's theological stance on ancient doctrine: "This [fusion of art and religion] was made possible because of a rare confluence of interests between the artistic and theological cultures of the papal court, especially concerning human nature and the human body." (Campbell 349). And thus, art and religion began to modernize together, under the veil of Christian Humanism.

Artists like Michelangelo and Raphael took Christian art to a radical new place with works such as *The School of Athens* and *The Creation of Man*. Gone were the days where religious art was strictly functional, although that role still held its place: "Michelangelo aimed to demonstrate art's capacity to represent and even reveal Christian principles" (Campbell 348). In the Renaissance, painters not only asserted their technical dominance, but also pushed forward the Renaissance ideology (both in art and religion), which placed artists in the realm of philosophers. In the Renaissance, for the first time in the history of Christian art, there was a decidedly

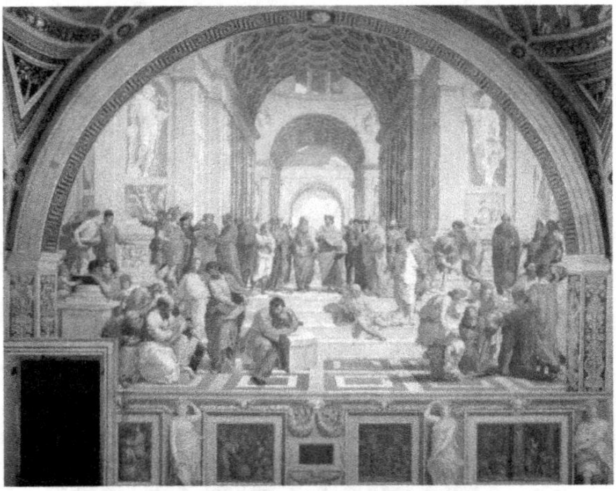

Figure 344: The School of Athens: Raphael, Stanza della Segnatura,

pro-artistic tradition developing. That is, the artists of the Renaissance are notably different from the anonymous artists of earlier eras, in that they asserted their own philosophies in tandem with their religious imagery. [Faith Imagined—Internet]

Music

Johannes Ockeghem

Figure 345: Johannes Ockeghem—Franco-Flemish

Born	c. 1410/1425---1497 Saint-Ghislain, Netherlands (now Belgium)
Died	6 February 1497 (aged 71-87) Tours, France
Notable work	*Prolation Mass* *Mass in any mode* *Requiem*

From the time of the early melismatic, plain song of Gregorian Chant during the 9th-10th centuries CE, composers ventured to free up the melodic and contrapuntal character of both religious and secular songs and masses such that they stimulated and extended the intellectual and emotional dimensions of their musical lives, within the church and outside, in the fields and towns. The stimulation was of the ancient classical modes, but new notation and rhythmic devices allowed for repetition and performance by ever diverse bodies of singers, including, with time, instrumentalists as soloists and in mixed groups of singers.

Ockegham's compositions were a mix of enormous complexity of voices, rhythms, usually sung by choirs of young boys, all taught to be able to reproduce this difficult complexion of emotional aural beauty and intellectual challenge. Ockegham lived at the end of the 15th century, temporally close to a new revolution in music—the Baroque style. His writings and those of his fellow polyphonic composers represent a monument of the powerful emotional and intellectual challenge that late in time-Renaissance, Catholic (Papal) Christianity presents to history.

Figure 346: An illuminated opening from the Chigi codex featuring the Kyrie of Ockeghem's *Missa Ecce ancilla Domini*

Biography of the greatest Christian Renaissance composer

Franco-Flemish composer, Ockeghem, was born in the village of St. Ghislain (near Mons, Belgium) in the diocese of Cambrai, but his date of birth is unknown. Long ago, Fétis thought he was born around 1430, assuming that he was a choirboy in Antwerp in 1443; Plamenac suggested 1425, then later agreed with Riemann and Van den Borren, who picked 1420; Leeman Perkins wrote in 1980 that he must have been born closer to 1410, since he was thought to have lived about 90 or 100 years.

It is unknown where he got his training as a musician. Lack of any documentary evidence is only compensated for by two compositions, in which Ockeghem refers to the composer Gilles de Binche (Binchois), who was for more than 30 years a distinguished member of the Burgundian court chapel of Duke Philip the Good. Shortly after Binchois' death in 1460, Ockeghem composed *Mort tu as navré de ton dart,* a lament whose text suggests personal acquaintance with Binchois. Presumably written earlier, is Ockeghem's polyphonic mass setting based on the Tenor of one of Binchois' most famous chansons: *De plus en plus se renouvelle.* Probably composed for a Marian feast, it is the Sanctus section of Ockeghem's mass which seems to reveal the composer's intention. The first twenty-seven notes of Binchois' Tenor are presented in a compound triple time (tempus perfectum cum prolatione maiori—like 9/8), but sung in a slow tempo against the other voices. In Binchois' chanson, these twenty-seven notes form a lower counterpoint in 6/8 (tempus imperfectum cum prolatione maiori), against the melody to which the first three lines of the original rondeau text are sung:

De plus en plus se renouvelle	(Ever more and more,
ma douce dame gente et belle	my sweet lady, kind and lovely,
la voulenté de vous veir.	the desire to see you renews itself in me.)

The first appearance of Johannes Ockeghem in the surviving records is the year (1443–1444) that he spent as a *vicaire-chanteur* (an adult singer in the cathedral choir) at Notre Dame in Antwerp. Two years later, he was singing in the chapel of Charles I, Duke of Bourbon, whose wife was the sister of Philip the Good, Duke of Burgundy. It is possible that he had gone there directly from Antwerp. He remained for at least two years, until he moved to his next known position, the chapel of Charles VII, King of France, where he is named in the list of singers in 1451. Among the chapelains (singers in the chapel) who are not priests, he was already listed premier chapelain by 1453.

From that date onwards, there exists documentation of various rewards made by the king for offerings of what may have been Ockeghem's own compositions: *ung livre de chant* (1454) or ung *chanson bien richement enluminée* (1459). However, he must have rendered much greater service to be nominated in 1459 as *trésorier* of the Abbey of St. Martin in Tours, of which the French kings were hereditary abbots. This function made Ockeghem first in rank among the administrators of the vast royal property in and around that city, and secured him a considerable annual income. But already in April 1461, he was released from the obligation to reside in Tours; apparently his function as *trésorier* gave him sufficient status to become a reliable official at the French court, which was always en route between the various royal estates.

Ockeghem remained at the royal court after Charles VII died, serving Louis XI after 1461 and Charles VIII after 1483. Under Louis XI, Ockeghem became *Maistre de chapelle de chant du roy* (1464), a title renewed by Charles VIII in 1483. In that capacity, he developed into a highly prolific composer of polyphonic mass cycles, of which thirteen have been transmitted in a single manuscript, copied around 1500, and now preserved in the Vatican Library (Chigi 234). During the early 1460s, the young Flemish composer Antoine Busnoys apparently refined his musical abilities in Tours with Ockeghem; his famous motet *In hydraulis*, hails Ockeghem as the successor of Pythagoras and "a true image of Orpheus." In 1462 and 1464, Ockeghem visited the Chapter of Notre Dame in Cambrai, and stayed in the house of the great composer, Guillaume Dufay. In spite of his residence in Tours, Ockeghem was a canon of Notre Dame in Paris from 1463 to 1470. He must have been ordained a priest in Cambrai some time before 1472, during a visit to Dufay. He is still mentioned in court documents in 1488, when he was present for the foot-washing ceremony of Holy Thursday.

In documented travels to Spain in 1470 and to the Low Countries, visiting Damme and Brugge (August 1484), Ockeghem may well have assumed a double role: as both the highly esteemed composer, traveling with his fellow chapel singers, and the experienced, trustworthy official, representing the interests of the French royal court.

Because of his great age, he probably spent his last years in retirement. He died on February 6, 1497.

Ockeghem's output was comparatively modest in size: ten complete Masses, some partial Masses and single movements, and also the earliest extant polyphonic Requiem, a handful of motets and twenty-two chansons. His best music can be found in the Masses, many of which are preserved in the "Chigi Codex," a lavishly ornate manuscript in the Vatican Library. In some, he followed the cantus firmus technique of Dufay's late Masses, but others are written, sometimes ingeniously, without any pre-existent musical material; the Missa Prolationum, for instance, consists of a series of mensuration canons. But such structural ingenuity, far from being drily academic, is beautifully and artfully concealed in an almost timeless flow of counterpoint that is the hallmark of Ockeghem's style. The comparative paucity of cadences creates a floating effect, but the ends of sections are marked by intricate rhythmic climaxes. He wrote for a choir whose four parts were equal in importance and often rather low in range, but hardly developed the idea of imitation, soon to be so important. His chansons are more conventional in style, in three parts with the focus on a melodious top line.

Where Dufay's music impresses us by its grace, soaring majesty, and formal clarity, Ockeghem presents an entirely different musical personality: moody, flamboyant, enigmatic. In common with so many composers of the era, Ockeghem was trained as a singer. His deep bass voice was greatly admired and may explain his fondness for exploring previously unnavigated subterranean registers. A characteristic aspect of Ockeghem's music is its integration of old medieval techniques of hidden structures into the new contrapuntal Renaissance style. One notable example is the use of canon in his *Missa Prolationum* (Mass of the Time Signatures). Each movement of this four-part piece contains two different canons sung simultaneously, with each part moving at a different speed. No less astonishing are his *Missa Cuiusvis toni* (Mass in any mode) which may be sung in any one of four modes, and the complexity works. Ockeghem also created tet *Deo gratias,* a canon for four nine-part choruses (thirty-six total parts). Yet, despite the enormous technical feat involved in composing such works, Ockeghem created music of contemplative vastness and inward rapture. [HOASM_Ockeghem—Internet]

(Above) Figure 347: Bronze, life size Christ and St. Thomas (or Doubting of Thomas) Andrea del Verrocchio 1435-1488. Sculptured 1467-83

Significance of the Christian civilizational vision

First in the human mind are the inchoate questions of the what and the why of this world outside of us—the powers above, and the inadequacies of humans below. Then came the discovery of literacy and its power to make ideas concrete and the awareness and implication of grave and tangible human fears and hopes. Only with consciousness could we begin to build a world of defense, also a world of concrete physical and emotional meanings. This world of meanings as we write them down and also build edifices, exemplifies our concurrence with these mysterious natural and now human powers. But they didn't have to be filled with a pantheon of gods. Philosophical speculation, and eventually scientific method will try to unravel these mysterious symbolic realities that we are only products.

So we build and create under a given intellectual canopy. We have the Ziggurats, Sphinx, Parthenon, Pantheon. Today, many of those gods who stimulated these brilliant and tangible edifices have disappeared into the mists of the past. The civilizations which they explained and guided remain as testimony to the power of those citizens, as they stretched their thinking beyond the rituals, the priests, mullahs, rabbis, ministers, to build a civilization that incorporated these very diverse symbolic realms of meaning—those that the theogeny would tolerate.

Holy meanings are out there—physically, monumentally—but also, most importantly, are being written down. At some point in time, all literate civilizations get to the point of self reflection. Perhaps they realize the relativism and temporal fragility of their theological or philosophical premises. Thinkers, if not priests, feel the need to tell the world and history about their religious, intellectual as well as physically tangible achievements. It happened in Assyria, Nineveh, as a great library was created. This institution was developed at the cusp of a crash into dust of an important political, national entity, Assyria. Because of an insight and the power of one ruler, history understands much better the premises of life, achievements, as well as failures from the dim past of Sumer to that bloody moment in Assyria, c. 630 BCE.

It was the same with Athens and the Greeks a few hundred years down the line. The great second shock to once invincible Athens and its arrogant political/cultural arête was the execution of Socrates, the epitome of working class equality and educational achievement. This execution was perpetrated by his own *hoi polloi* working class, democratically elected equals. Their gods still represented tradition, achievement, heritage, and ethnic significance. But they did not represent the questions and issues of life in that present. Thus, Socrates' questioning spawned the Academy and the Lyceum, and many other schools of investigation and learning into the puzzlements of the world around. And when the Hellenistic world of Greece/Macedonia had expanded to its peak of power stasis, another Greek, Ptolemy, created a great museum as a depository and laboratory of ideas and experiments of the philosophical/scientific world. Here too, a repository of the new had then been intellectually created alongside the ancient cultural pantheon.

The libraries in Nineveh and Alexandria have passed into oblivion, but their historical and intellectual contributions to our own civilization are still lurking, there for us to find. So too with Christianity. The tangible physical presence of priests, ministers, nuns, bishops, and the now emptying churches are still here. There is a Pope in Rome, an Archbishop in London, and wealthy Protestant ministers in Houston. Ultimately, the power of Christianity will be seen as once maintaining and mining the intellectual framework of the past, preserving the ideals of *Hss* humanity in the medieval universities, the philosophical theology that connected this past with the then present. Most importantly, early on in the areas of architecture, the visual arts, even jewelry and the beautification of the physical dimensions of the person, the great musical advances, it peaked in a final burst of creativity before ceding the political/social world to secular domains. Christianity allowed for the specialness of the Renaissance, and then unleashed the creative powers of the north in the Reformation.

When we compare the condition of the free mind in the period between 400 and 1600 CE (yes, even considering the heresy burnings), as compared with recent 20th and 21st century ideologies which have brutally repressed the free-thinking domains of its citizens, and then committed incomparable genocidal barbarities on intellectuals and plebes alike, the era of Christian dominance evokes the pastoral.

Figure 348: Tablet containing part of the *Epic of Gilgamesh* (Tablet 11 depicting the Deluge), now part of the holdings of the British Museum. From library at Nineveh, founded by Ashurbanipal, of Assyria, 668–627 BCE

Figure 349: ***The Great Library of Alexandria,*** **Nineteenth-century artistic rendering of the Library of Alexandria by the German artist, O. Von Corven**

Figure 350: A miniature depicting the Schola Medica Salernitana from a copy of Avicenna's (1025 CE) ***Canons.*** **Avecenna was a Persian Muslim medical researcher at Salerno University, in Italy, founded c. 850 CE**

Chapter 17

The Jews: An Outlier *Hss*

Why the Jews?

The Jews never had the opportunity to form their own long-lived civilization. Always a member, they were most often a dissident member of the greater civilizations of their eras. This dissidence from major civilizational orientations—religious, philosophical, moral—allowed them to survive as an alternative *Hss* model, mostly without national political independence. Here, described as an exemplification of the critical importance of an *Hss* commitment framed around the Roman description of the Jews as a "nation of philosophers" and the subsequent emblematic, "people of the book."

This of course, was before the *Holocaust* of the Jews by the peoples of Europe. Beyond the tangible barbarism horror of the 20th century, the lesson we should all learn, is that of the fragility of the *Hss* intellect and its historical civilizational achievements.

Origin of the Jews

Probability has it that three groups contributed to the forging of an ethnic identity that later became the tribes and nations of the Jews.

Figure 351: House Altar depicting Akhenaten, Nefertiti and three of their daughters, limestone, New Kingdom, Amarna period, 18th dynasty, c. 1350 BCE

The first consists of the corvee workers in the Delta cities of the Nile, led by a possible religious follower of Pharaoh Akhenaten. Thus Moses, (c. 1275–1250 BCE) committed to Egyptian and near desert tribal circumcision, led the rabble out of Egypt, they still in possession of the gold loot stolen from their Egyptian employers. Moses in his flight of refuge to the desert, where he met his future wife also absorbed a new variant monotheism—that of a warrior God, *Yahweh*. After forty years of struggle with a very unmanageable crowd, where he impotently introduced his Ten Commandments, he did achieve visual contact with the land of milk and honey—Caanan—only to have *Yahweh* disallow his further entrance into this new and rich "holy land."

Figure 352: The 12 Tribes of Israel, c. 1100 BCE

The second group consists of the Shasu nomads of the desert lands, so described in Egyptian inscriptions. Considered less barbaric by the Egyptians, they occupied the Sinai, northern Arabia, present day Jordan, in the Negev areas. The blond-haired David, to be a king of a momentarily united people, owes his probable origin to one of these desert tribes, the inspiration for the "J." version of the Biblical Pentateuch.

The third group consists of the "Apiru" (Hebrews) of Egyptian documentation, going further back in history before Moses, when the Indo-European (Hittite, Hurrian) towns of Canaan (Jebusites, Jerusalem) appealed to their Egyptian protectors to save them from these nomadic barbarians. They were of mixed ethnicities, and followers of the supreme Semitic god, "El." These were the Hebrew tribes, perhaps the ten that formed the future nation of Israel.

Their Israelite descendants probably wrote a more intellectual version, the "E" elements of the Pentateuch, the "J" and "E" versions were subsequently amalgamated and revised many times into the time of the Babylonian exile (586–538 BCE). Samara, the final capital of Israel (Jerusalem was in Judea), was captured by the Assyrians in 722 BCE.

Perhaps for one hundred years, the rule of Saul, David, and Solomon united Israel/Judea. The breakup into two ethnicities, northern Israelites ("E" of the Pentateuch), and the southern Judeans ("J" of the Pentateuch). They went their cousinly way from c. 900 to 722 BCE for Israel (overrun by Assyria), and to c. 586 BCE for the defeat of Judea by the Chaldean Babylonians. Chaldean retribution for treachery against the peace treaty by radical Jews led to the destruction of the 1st Temple, its gold and silver confiscated, the elites exiled into Babylon, and others (Jeremiah) fleeing to Egypt.

Four centuries of divided monarchical rule, each nation experiencing a bonanza of prophetic wanderings and writings, saw a natural falling away from the monotheistic teachings of the holy writings. Yet, among the intellectuals, the priesthood of Solomon's Temple, prophets in Israel who fought against its materialism, the original vision of one holy principle demanding a purer morality, thus upholding the Commandments, burned through the nights.

Figure 353: Artistic representation of Solomon's Temple in Jerusalem

Two kingdoms fail—historical moments

The northern kingdom of Israel, with its capital Samaria and with greater wealth than Judah, fell to the Assyrians in c. 722. As per Assyrian policy, large populations were exiled, these the northerly ten tribes; strangers, other ethnic peoples were imported into the territory.

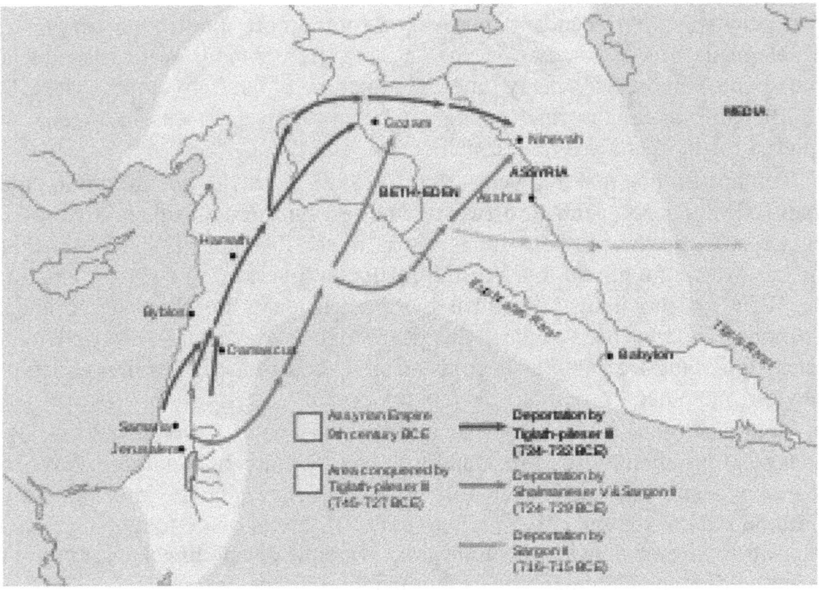

Figure 354: Deportation of the Northern Kingdom of Israel by the Assyrian Empire.

Figure 355: Judean prisoners being deported into Assyrian exile after the capture of Lachish. Note: Women and children ride into exile, Assyrian soldiers, Judean men, walk. The Assyrian "Siege of *Lachish*" and conquest of the cityof *Lachish* in 701 BC, by the Assyrians over the Judeans; wall relief by Sennacherib in Nineveh. The rest of Judea, Jerusalem, its capital, remained out of Assyrian hands

After engulfing Israel:

In 722 BCE the armies of a new Assyrian king, Sennacherib descended on Judah, sweeping down along the Phoenician coast to deny all separatist national claims, after destroying well-fortified Lachish in the Judean Sheppalah. These armies arrived at the walls of Jerusalem. And in a colloquy worthy of the Athenian historian, Thucydides, he,c 420 BCE, Rabshakeh, the Assyrian emissary, the Tartan (Viceroy—second in command) along with both the Rabshakeh (Chief Butler—ran the royal court) and Rebsaris (Chief Eunuch—led the armies) address the representatives of the Judean King Hezekiah.

Rabshakeh: "See, you are relying now on Egypt, that broken reed of a staff, which will pierce the hand of anyone who leans on it."

The Jews replied: "Then Eliakim, son of Hilkiah, (in charge of the palace) and Shebnah (the secretary) and Joah" (the recorder), said to Rabshakeh, "Please speak to your servants in the Aramaic language, for we understand it; do not speak to us in the language of Judah within the hearing of the people who are on the wall." But the Rabshakeh said to them, "Has my master sent me to speak these words to you, and not to the people sitting on the wall, who are doomed with you to eat their own dung and to drink their own urine?"

Then the Rabshakeh stood and called out in the language of Judah, "Hear the words of the great king, the King of Assyria!" Thus says the king: "Do not let Hezekiah deceive you, for he will not be able to deliver you out of my hand. Do not let Hezekiah make you rely on the LORD by saying, The LORD will surely deliver us ... Come now, make a wager with my master, the king of Assyria: I will give you two thousand horses, if you are able on your own part to set riders on them. How then can you repulse a single captain among the least of my master's servants, when you rely on Egypt for chariots and horsemen?" (2 Kings 18)

However, the Assyrians withdrew, either the angel of the LORD struck them down, else Hezekiah bribed them with more silver and gold than he had earlier given. Else, they tired of an unsuccessful siege, the Siloam tunnel bringing water to besieged Jerusalem. In the end the salvation may have been due to Sennacherib's indolent generalship.

The resident prophet Isaiah has the LORD say to the Assyrians: "Because you have raged against me and your arrogance has come to my ears, I will put my hook in your nose and my bit in your mouth; I will turn you back on the *by* which you came." (Kings 19:28). The reward for a holy king, Hezekiah.

Mystery. Who was this Rabshakeh, Tartan/Viceroy and Hebrew-literate representative of the Assyrian king? Could he have been one of those Samaritan deported Israelites enlisted like their charioteers into Assyrian hegemony? Was the relief of Jerusalem due to the secret intervention of a Yahwist lieutenant of an incompetent king?

(Itzkoff 2004)

A historical moment of victory, yet the plurality of gods further infects the Jerusalemites, so too their desire for goodies of that world. The later prophet, Jeremiah (c. 626 BCE), speaks about the many gods of Judah, “which are as numerous as the lands.” “Their worship of the northern storm god Hadad (Baal)” permeated the streets of the holy city.

The plurality of deities was such that in an inscription by the Assyrian king Sargon II, who completed the conquest of the kingdom of Israel, it is mentioned that after capturing the capital Samaria, his troops brought back “the [statues of] gods in which [the Israelites] had put their trust.”

As the new Yahwh cult evolved and spread, he was worshipped in temples across the land. Early 8th-century inscriptions found at Kuntillet Ajrud probably refer to different gods and cultic centers by invoking “Yhwh of Samaria and his Asherah” and “Yhwh of Teman and his Asherah.” Only later, under the Judean reign of King Josiah at the end of the 7th century BCE, would the Yhwh cult centralize worship as YHWH at the Temple in Jerusalem.

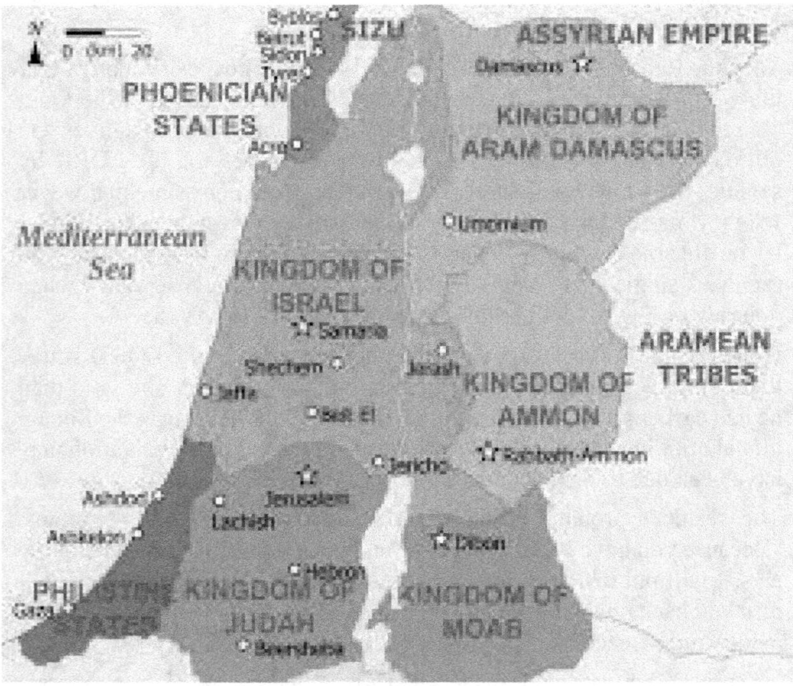

Figure 356: Near East Political Map, c. 650 BCE

(Left) Figure 357: Phoenician figure, probably of the Canaanite goddess Astarte (also known as Ashera), 7th century BCE. (García 2016)

Jeremiah's warning prophesy: "Take for yourself a scroll and write upon it all the words that I have spoken to you concerning Israel and concerning Judah ... Perhaps the house of Judah will hear all the evil that I plan to do to them, in order that they should repent, each man of his evil way, and I will forgive their iniquity and their sin." (c. 600 BCE)

Assyria was defeated and extinguished militarily by a Neo-Babylonian/Chaldean resurgence, c. 610 BCE.

End of Judah, 1st temple

After Nebuchadnezzar (Babylonia/Chaldeans) defeated the Judean defenders of Jerusalem, c. 597 BCE he appointed the very respected Jew, Gedaliah, as governor of the remnant of Judah, the Yehud Province, with Chaldean guards stationed at Mizpah. The Bible reports that, on hearing this news, Jews who had fled to Moab, Ammon, Edom, and in other countries returned to Judah.

A band of Judeans led by Ishmael, son of Nethaniah, and the ten men who were with him, treacherously murdered Gedaliah. Then, most of the Jews who had joined Gedalia and many Babylonians whom Nebuchadnezzar had left with Gedaliah (Jeremiah 41:2–3) ran for the desert. This population that had remained and those who had returned then fled to Egypt for safety, here included Jeremiah. In Egypt, they settled in Migdol (it is uncertain where the Bible is referring to here, probably somewhere in the Nile Delta), Tahpanhes, Memphis (called Noph), and Pathros (the vicinity of Thebes).

Nebuchadnessar, in revenge for these traitorous actions, appropriated the wealth of gold and silver within the Temple, burned it down, and exiled to Babylonia (mostly along the Euphrates River) the remaining elite who had not fled. These deportations are dated to 597 BCE for the first, with others dated at 587/586 BCE, and 582/581 BCE respectively. Nebuchadnessar did not touch the farmers and workers who produced the grapes and wheat, the productive life of Judea, so that the province could continue to thrive. In Babylon, the elite exiles quickly became prosperous owners of material and human wealth.

Transitions: Persia

Indo-European Persians and Medes (today—Kurds) defeated the Babylonian/Chaldeans (539 BCE) and occupied Jerusalem. Ezra, in the Book of Ezra describes how he latterly, c. 458 BCE, after the Persian King Artaxerxes gave him permission, travelled to help rebuild the Jerusalem Temple, along with a new group of Judean exiles living in Babylon (Ezra 8.2–14), where he is said to have enforced observance of the Torah. He was described as exhorting the Israelite people to be sure to follow the Torah Law, so as not to intermarry with people of different religions, and he enforced sets of commandments, as described in the Pentateuch of the Jews. His marriage prohibitions were based on religiosity not ethnicity.

(Left) Figure 358: Ezra Reads the Law to the People, one of Gustave Doré's illustrations for La Grande Bible de Tours

Some years later, the Persian king, Artaxerxes, sent Nehemiah (a Jewish noble in his personal service to the Persian king), c. 445 BCE, to Jerusalem as governor with the task of rebuilding the city walls. Once this task was completed, Nehemiah had Ezra read the Law of Moses (the Torah) to the assembled Israelites in Hebrew with simultaneous translation into Aramaic for the multitude who, by then, did not know Hebrew. At this time in history, the people and priests entered into a covenant to keep the law and separate themselves from all other peoples.

Figure 359: Gustave Doré, Nehemiah Views the Ruins of Jerusalem's (Right)

Even as the Jews were fleeing Jerusalem in one way or another in 586 BCE, Solomon's Temple being burned to the ground in back payment by the Chaldean Babylonians for Judean perfidy, Solon of Athens was attempting to break the ancient feudal traditions of Athens to meet a new economic, military and social world being created around the Mediterranean.

The Jewish *Diaspora* had begun. Babylon, along the Euphrates, and elsewhere was yet rich with economic possibilities. The Persian conquest of the Middle East some fifty years earlier opened up vast new possibilities for Jews in the newly tolerant Persian domains. Judea was only a province of this empire, situated alongside Samaria, once the nation of Israel, they with a whole new Assyrian population now adopting the religion of the land, Yahwism.

Transitions: Greeks

Switch from Persia 538 BCE, liberating Jerusalem to Ezra and Nehemiah, 458–445 BCE, and we have the Persian empire now recoiling from their defeat by the Greeks, and the beginning of an era by "the new kid in town, the Greeks in the Eastern Mediterranean," with their coins, manufactures, language and intellectual persuasions. Alexander arrived in c. 330 BCE. With his sudden death in 323 BCE, his generals, Ptolemy in Egypt and **Seleucus** in Judea, continued the Greek imperium. Jerusalem lay mostly dormant, but a new city off the coast of Egypt was built and named after Alexander. Into this city Jews, Greeks, and Egyptians poured.

Hebrew texts commenced to be translated into Greek in Alexandria in about 280 BCE and continued until about 130 BCE. These early Greek translations of the Torah and other sacred documents, supposedly commissioned by Ptolemy II Philadelphus, were called the *Septuagint* (Latin: "Seventy") from the supposed number of translators involved (hence its abbreviation "LXX"). This Septuagint remains the basis of the Old Testament as recognized by the Eastern Orthodox Church. As Jews in the Diaspora moved from the Judean homeland, they increasingly used the *Septuagint.*

The Hebrew Bible varies in many places from the Masoretic (Rabbinical) Text and includes numerous books no longer considered canonical in some traditions: 1 and 2 Esdras, Judith, Tobit, 3 and 4 Maccabees, the Book of Wisdom, Sirach, and Baruch. Early modern Biblical criticism typically explained these variations as intentional or ignorant corruptions by the Alexandrian scholars, but most recent scholarship holds that it is simply based on early source texts differing from those later used by the Masoretes in their work.

The *Septuagint* was first used by Hellenized Jews whose knowledge of Greek was better than Hebrew. Even Philo, the orthodox Jewish philosopher of Alexandria, who negotiated with the Romans, c. 33 BCE, was barely knowledgeable of Hebrew. The texts came to be used predominantly by gentile converts to Christianity and by the early Church as its scripture, Greek being the *lingua franca* of the early Church.

(Left) Figure 360: Woodcut from *Die* Schedelsche *Weltchronik* (Nuremberg, 1493)

Transitions: rabbinical

After the Roman destruction of the 2nd Temple in 70 CE, the ground was taken out from the power and influence of the traditionalist Sadducees, who had major control of the Temple. The Pharisees, themselves somewhat represented in Temple adjudications, also scattered. Rabbi Yohanen ben Zakkai escaped from Jerusalem in a coffin to Yabneh, supposedly sending a message to General Vespasian, soon to be Emperor, that he would cooperate with the Roman authorities. The principle was based on Isaiah, 10:34, "And he shall cut down the thickets of the forest with iron, and Lebanon shall fall by a mighty one." (Vespasion/ Lebanon Cedars of the Temple) There in coastal Yabneh, the ancient Philistine town where Samson and David had cavorted, the pharisaic tradition of home and synagogue study and prayer rooms, was amplified into a school of Torah study and reinterpretation of the ancient scrolls. But ben Zakkai also instituted a curriculum of Greek wisdom studies.

From here and around the various diaspora centers of *rabbinical-* (teacher in Aramaic) studies, the influence of Jabneh was great, as a protection of the faith:

1. The development of the technique of Midrash, as a way to deeply analyze and subject Scripture to careful exegesis.
2. Following that, a study of Halakhah, the legal tradition and rulings by sages given over the generations.
3. Haggadah: philosophy, theology, lore, ethical guidance, even fables that add dimension to the understanding of Scripture.

A halakhic example in the Mishnah is the decision that civil cases are to be tried by three judges, because in Scripture, the case of a thief was followed three times by the word *Elohim* (the Lord), when discussing the juridical procedures to be followed.

The Jews did not have a 'closed canon' of Scripture during the time of Christ, before 100 CE, or even after Jabneh. Even during the time of Christ there were competing opinions on what books actually belonged in the Jewish Bible. There were various collections in existence. Sadducees and Samaritans accepted only the Pentateuch, the first five books, whereas the Pharisees accepted a fuller canon including Psalms and the Prophets. The later Masoretic text did not contain the deuteron-canonicals, whereas the widely used Greek *Septuagint* did.

"This uncertainty continued well into the second century. The discussion over the books of the canon of the Old Testament continued among the Jews long after Jabneh (after 90 CE the intellectual capital of the Pharisees), which demonstrates that the canon was still under discussion in the third century—well beyond the apostolic period. The challenges to canonicity at Jabneh involved only Ecclesiastes and the Song of Solomon, but the debate over the canon continued past Jabneh,. Even the Hebrew canon accepted by Protestants today was disputed by the Jews for two hundred years *after* Christ." (Ray 2004)

Transitions: Rome

Julius Caesar reformulated a policy of allowing Jews to follow their traditional religious practices. This policy was followed, and extended, by Augustus, first emperor of Rome, who reigned 27 BCE–14 CE. This gave Judaism the status of a *religio licita* (permitted religion) throughout the Empire.

Jews retained their traditions and customs, including kosher dietary customs. During the late Hellenistic and Roman era, pork, a meat forbidden by Jewish law, became a common and frequent cuisine for Romans. Since Jews had the status of a *religio licita*, allowing them to continue their practices, they refused to eat pork. Scholars have argued that by refusing pork, Jews symbolically separated themselves from Romans and created a cultural identity for themselves.

Important Roman writers of the 1st–2nd century CE, Ovid, Juvenal and Tacitus, in their dislike of the Jews, migrants, slaves, prisoners, now mostly citizens, had to admit the devotion to one god and his moral demands was striking.

According to Theodor Mommsen, in the first century CE, there were no fewer than 1,000,000 Jews in Egypt, in a total of 8,000,000 inhabitants; of these, 200,000 lived in Alexandria, whose total population was 500,000. Adolf Harnack (1902) reckons that there were 1,000,000 Jews in Syria (which included Lebanon) and the areas east of the Euphrates at the time of Nero in 60s CE, and 700,000 in Judea, and he allows for an additional 1,500,000 in other places, thus estimating that there were, in the first century, 4,200,000 Jews in the world. Many scholars see this estimate as too high.

The themes of Aristotelian philosophy still resounded in the Roman/Greek intellectual world, 1st century CE. The theme of the "unmoved mover," Aristotle's logical solution for the "beginning of things," so similar to the Deist 18th-century concept of God as the primeval "watch winder" and then nature takes over, resounded amongst the Greek intellectuals' seeing the Jews as a "nation of philosophers," given their monotheistic insistence. Thus, there was busy social intercourse at the various synagogues both in Rome and in this Greco-Roman civilizational era, regardless as to what was happening in Jerusalem itself with the Roman destruction of the Sadducean 2nd Temple, 70 CE. (Itzkoff 2004, pp. 144–145)

Of course, when the proselytism of Paul and the new Jewish "heresy," Christianity, began to compete as a mystery religion, a substitution for the now decadent Roman polytheistic religious tradition (the Roman soldier by the 1st Century CE had taken up and spread an eastern mystery religion, Mithra), given the power and corruption of its pagan elite, Judaism with its demands of circumcision, dietary and other strict rituals could not compete for the pagan masses who saw in Christianity personal salvation, fascinating incipient trinitarianism, and also strict moral prohibitions.

The Judaic revolution—Mishnah/Talmud

(Left) Figure 361: Synagogue Beth Alpha, c. 520 CE. floor mosaic, "Binding of Isaac"—6th century-Byzantine Palestine, note blond Jews inscriptions written in Hebrew, Greek and Aramaic

The breakup of priestly centrality by the Chaldeans (586 BCE) was speeded up by the lack of political ties between the new Temple and the governing authorities—Babylonians, Persians, Greeks, and Romans. Yet, the intellectual demands of the new world in the making required broader literacy and societal skills. This was met by a new group of Jewish leaders. Naturally, the new alien rulers wanted to use the priestly class as a vehicle of socio/political control, and thus a party of *Sadddducees* was formed. All the while, out in the field, the *Soferim* (scribes) *Pharisees* (stand apart) formed their own guild of scholar/leaders in competition, interpreting the vast incursion of writings and interpretations that the Hellenes stimulated into what the *Pharisees* called the "oral law," that which was written or relevant long after the Pentateuch, and now critical in coping with the new world. Of course, there were other groups: *Essenes*, and the *Qumran*.

The scribal tradition of Torah learning and questioning that democratized learning away from the Temple, began with the returnees from Babylon, then under Persian hegemony, Ezra/Nehemiah. It lasted approximately 400 years and transitioned into the pharisaic movement, deeply influenced by the literacy and literary traditions of the Greeks. The rabbis took up this motif with a vengeance, first in pharisaic competition with the ascetic communities, *Essenes* and Christians, as well as the establishment Sadducees, but especially after the destruction of the 2nd Temple in 70 CE.

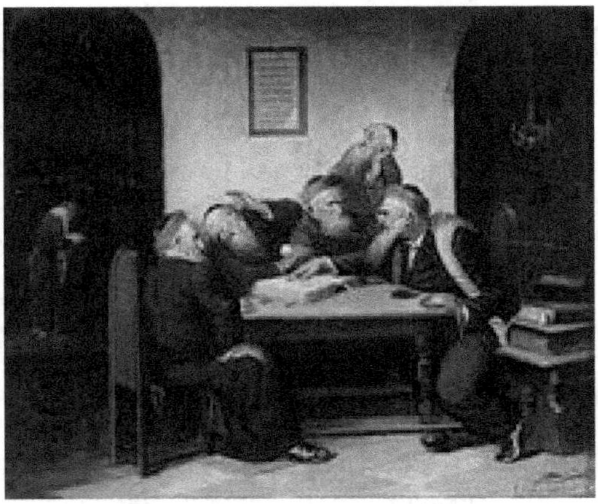

Figure 362: A Controversy Whatsoever on Talmud (Schleicher c. 1870)

The rabbis and the followers of *Jesus*, the Messiah, were all that now remained of the Judaic religious infrastructure. A wholly new pathway emerged. The Torah study groups in the communities, the *Proseuche* (prayer rooms) and *Synagogue* (*meeting* room) that had spread from the Greek-speaking Alexandrian community to other Diaspora communities in Parthia, the Greek and Roman world and then back into Judea, now became the institutional pivot for the renewal of Israel.

The year 200 CE was a propitious one for the Jews. True, Christianity was spreading amongst the masses throughout the Empire. Great thinkers such as Clement and Origen of Alexandria were expanding Christianity's philosophical focus to compete with both Judaism and the pagan Hellenistic tradition. Judaism had had its secular knocks. After the destruction of the 2nd Temple by the Romans, the Jews had had a series of uprisings against Roman power in Alexandria, Babylonia, Cyreneica. These were all unsuccessful, especially against the Roman Emperor Trajan. Again, to obtain their liberation from the heavy rule of Rome, the unsuccessful, bloody Bar Kokhba revolt was initiated against Emperor Hadrian, a devout Hellenist.

Bar Kokba was supported by the religious establishment, even from Jabneh, the supposed home of the pacifists. The greatest of these rabbis, Akiba, even tried to place Bar Kokba, the leader of the revolt, as the messiah once predicted by Isaiah. Thr victories by Bar Kokba were ephemeral. Far worse then that of Masada of the first Judeo-Roman war, 70 CE, the massacres of Jews by Hadrian in his victory, 136 CE, took the life of Akiba, he who once declaimed in a rabbinical debate over: "Which is greater, learning or deeds … ' Akiba's 'victory',…learning is greater because learning leads to great deeds" [BT Qid, 40b]

The Severan Imperial dynasty in Rome, 195–235 CE, originated in Syria, where there was much opportunity to encounter the Judaic tradition. They assisted the Jewish worshippers of the empire in many ways, even causing the final Severan, Alexander, to be called the "archisynagogos." It was in this time period that we encounter Judah Ha Nasi (the Prince), c. 217 CE, in the Galilee, now the center of much writing and debate over the "oral law."

There were many "mishnahs" "study by repetition" in this time period, 90–200CE, compendia of the laws and wisdom of the oral tradition, the *Pharisees*. But Judah was unique—he had great wealth, for a Jew. He had under his control large Galilee agricultural plantations, many slaves, and a body guard of foreign origin. Yet, his ambition was to find the wisest rabbinical masters in the area, even from the Euphrates in Babylon, to organize the "oral law" and the writings of the post-Temple tradition down as structure: "a constitution of belief and behavior that could regulate and secure the future of 'Israel.'"

The six orders of the Mishnah of Judah deal soley with issues of Jewish religious life together:

1. *Zeraim* ("Seeds"), dealing with prayer and blessings, tithes and agricultural laws (11 tractates)
2. *Moed* ("Festival"), pertaining to the laws of the Sabbath and the Festivals (12 tractates)

3. *Nashim* ("Women"), concerning marriage and divorce, some forms of oaths and the laws of the nazirite (7 tractates)
4. *Nezikin* ("Damages"), dealing with civil and criminal law, the functioning of the courts and oaths (10 tractates)
5. *Kodashim* ("Holy things"), regarding sacrificial rites, the Temple, and the dietary laws (11 tractates) and
6. *Tohorot* ("Purities"), pertaining to the laws of purity and impurity, including the impurity of the dead, the laws of food purity and bodily purity (12 tractates).

A quote from Judah's legal thinking:

The saving of a man's life takes priority over a woman's ... The covering of a woman's nakedness takes priority over a man's. A woman's ransom has priority over a man's. A man in danger of being forcibly sodomized has priority over a woman in danger of rape. A priest takes precedent over a Levite, a Levite over an Israelite, an Israelite over a *mamzer*, a *mamzer* over a *natin* [from nathin, a Gibeonite descendant], a *natin* over a proselyte, a proselyte over a freed slave.

Under what circumstances?

When all of them are equivalent [in other regards].

But if the *mamzer* was the disciple of a sage, and a high priest was an ignoramus, the *mamzer* who is a disciple of a sage takes precedence over a high priest who is an ignoramus.

(*Mishna: Horayot*, 3, 5.)

Mamzer is a child born of a forbidden sexual union. Although commonly translated as "bastard," the term is defined in rabbinic literature as a child born of incest or adultery. Not being personally responsible for their condition, a mamzer is usually unlucky.

The Tosefta, another and slightly later compendium admitted to orthodoxy adds:

A sage takes precedence over a king

For if a king dies any Israelite is suitable to mount the throne.

{Tosefta *Horayot*, 2, 8.}

Judah Ha Nasi:

He who translates a verse with exact literalness

is a falsifier, and he who makes additions is a blasphemer.

Figure 363: An early printing of the Talmud (Ta'anit 9b); with commentary by Rashi, Shlomo ben Yitzhak. He was a medieval French scholar and author of a comprehensive commentary on the Talmud

The fact that the Mishna was largely ahistorical, Halakhah (law), without source commentary, seemed to necessitate amplification. It needed explaining, from both the Biblical and Post-Biblical writings. The Gemara, rabbinical commentaries of the Talmuds (study/complete), both Palestinian and Babylonian, now written in Hebrew and Aramaic, provided principles, sources, relationships, the Mishnah had merely implied, "according to everyone." Still, the Mishnah was the accepted summa of over three hundred years (150 BCE–200 CE) of commentary on the meaning for "today" of the written and unwritten Torah. (Itzkoff, p. 156)

In the Mishnah, there remained unresolved disputes inherited from the earlier Tannaim (post-Temple scholars). There was only sporadic use of biblical proof texts, **and** occasional narratives. All of these needs to amplify began to reflect in Jewish communal life the value of a study as a religious ritual in its own right. (Itzkoff, op site: fn. 71, p. 157) The Mishnah's statements seem to divorce scripture from the law, even when repeating the essence of the stories, events, parables that appeared in scripture. Both Talmudes (study/complete), Palestinian and Babylonian, supplied proof text from scriptures to the laws of Mishnah. "How do we know these things?" They then cite text proofs. [op. cit.: fn. 72, p. 157.]

The "Jerusalem" Talmud was completed c. 350 CE in Aramaic. The Babylonian Talmud, written over many generations, **was** completed c. 500CE, also written in Aramaic.

It is commonly accepted that of the two Talmudic compilations—one in the aforementioned towns of the Galilee, the "Jerusalem" Talmud, in Tiberias in the Galilee; the other in the Babylonian/Euphrates River towns of Nehardea, Pumbeditha, and especially Sura—the Babylonian Talmud was more complete.

The Jerusalem Talmud has fewer legal (halakhic) analyses, has more words of blessing and consolation. Interestingly, the Talmud of the Galilee has more Greek terms than the Babylonian Talmud, which occasionally uses Greek words as puns. It is puzzling that the "Jerusalem" omits any discussion of ritual Temple practices.

Jerusalem Talmud: "What was reduced to writing does not give us a work carried out after a preconceived plan, but rather represents a series of jottings answering to the needs of the various individual writers, and largely intended to strengthen the memory." (op. cit. fn. 81, p. 157)

It is fair to argue that life under Rome, especially during the persecutions of Constantine, Constantius, and Theodosius II, from 325 CE, throughout the fourth century when Christianity became the only official religion of the empire, weighed heavily on the life and work of the Galilean Rabbis.

Rabbi Abba bar Aggvu was the ordained Babylonian student of Judah Ha-Nasi, who, after his teacher's death, returned to Babylon. Under Parthian/Sassanian rule (217–227 CE), policies continued that, in the beginning, were far more tolerant of Jews than would have been those of the Romans and the Christians to the south. Rabbi Abba was the first of the deeply intellectual Babylonian Jewish leadership in the 70-mile enclave along the Euphrates—Pumbeditha and Nehardea in the north, Sura in the south—an area that became almost a nation of Jews in itself, and by the local Jews, called "Israel."

Rabbi Abba's famous epigram to his Babylonian flock: "Do not marry two women; if you do marry two, then take a third."

An important dimension of the growth of Talmudic thought was the institution of the Kallah, a twice-a-year Babylonian assemblage, from the mid-third century CE, under Rab Samuel and Abba, in the Academies (Yeshivoth), starting with Nehardea, then Sura and Pumbeditha, finally in Baghdad under the Muslims. Here were gathered vast numbers of students of all ages and attainments, to discuss questions and decisions on issues of Torah or Mishnaic-Talmudic interpretation. "Thus the Babylonian academies combined the function of specialist-law schools, universities, popular parliaments. They were a unique product of rabbinism; and the authors of the system were also the compilers of its literary expression, the Talmud." [op. cit. fn. 83, p. 160.]

The value of the study of God's law was as important in Talmudic religion as the content of that study; the hair-splitting dialectics of medieval rabbis reflects their interest in internally consistent knowledge, in making distinctions between areas of knowledge, and in the integration of teachings from widely disparate fields into a single web of unifying principles. It gave Rabbinic studiousness a scholastic tinge that continued to sharpen as later centuries wore on. Text commentaries over the centuries were jammed into each page of the Talmud. Rabbinic intellectualism turned into disciplined argument: the interplay between proof and refutation.

PAGE FROM TRACTATE KIDDUSHIN OF THE BABYLONIAN TALMUD, SABBIONETTA, 1559.
(From the Sulzberger collection in the Jewish Theological Seminary of America, New York.)

Figure 364

Cultural endogamy

"When a man is brought in for judgment in the world to come, he is asked, 'Did you deal in good faith? Did you set aside time for the study of Torah? Did you engage in procreation? Did you look forward to salvation? Did you engage in the dialectics of wisdom? Did you look deeply into matters?'" (op. cit. fn. 1, p. 167.)

In fulfilling this intellectual and moral devotion, each male would thereby be transformed into a rabbi hence into a saint. When all Jews had become rabbis, they then would no longer lie within the power of history. The Messiah would come. (op. cit. fn. 7, p. 170.)

Talmud study became the oral Torah given by God to Moses. This study became universal for Jewish boys and men. No Jew was considered incapable of adding new and original ideas to this learning. (op. cit. fn.8, p. 170.)

Rabbis came anew, from all segments of the community, often as wanderers from distant academies. Their power was vested in their learning, their wisdom and personal probity. " ... only those properly and sufficiently learned in Torah should be allowed to assume leadership over the community, since only such leaders could be trusted to guide the people in a divinely ordained direction." This divine direction was formed from Talmudic learning, including all parts of the Holy Bible, Mishnah, Gamara, plus other writings. (op. cit. fn. 9, p. 171.)

The rabbinic sages sought the company of their communicants, **the** other Jews. They lived in villages, not in monasteries, owned property, **and** became part of village communal life in administration or business. Entrance to rabbinic schools was open to all, depending on ability and inclination. They actively proselytized for new candidates for their schools. The rituals of the table, saying blessings, **and** prayers could be learned from the rabbi, **and were not** secrets from the ordinary Jew. The ideal was for all male Jews to become rabbis. The rabbis wanted to transform the entire Jewish community into an academy (*yeshivah*), where the whole Torah was studied and kept. (op. cit.fn. 10, p. 171.)

Thus, the Jewish community took on the character of an evolutionary outlander in defense, nurturing the ancient mammalian attributes of high intelligence, obscurity, alertness **and** adaptability. Literacy was its great reservoir of perseverance in the cultivation of the moral and social discipline necessary to surviving **and** existing on the edge of the larger political, military, **and** even religious powers that now dominated.

No longer could Judaism easily proselytize for new recruits, as in the post-Babylonian exilic and Hellenistic era. The inclination was now negative. A deeply centripetal ethnicity arose, whose social profile was increasingly anathema to the gentile world. No longer would they struggle with sectarianism. With few worries over schismatic Jews (Samaritans, Karaites), the great concern was with the powerful pagan outsider.

Lod (Lydda), c. 300 CE: Rabbi Hanina bar Hama to Rab. Hoshaya while touring the synagogues of Lod: "How much [money[have my ancestors invested here [in these buildings]?" R. Hoshaya: "How many souls have you lost here? There is no one studying Torah." (wrong prioritization of values). (op. cit. fn. 11, p. 171.)

"[Regarding] a small town in Israel [Galilee], they]the townspeople], built for themselves a synagogue and academy and hired a sage and instructors for their children. When a nearby town saw] this], it]also] built a synagogue and academy, and likewise hired teachers for their children." (Seder Eliyahu Rabbah 11) [op. cit. fn. 12, p. 171.]

In the post-Temple period, delegations of rabbis were sent throughout Palestine to establish for the towns teachers of Scriptures and the Oral Law. In one place, they found neither. They summoned the elders of the town: "Bring us the guardians of the city." The sentries were brought. "These are not the guardians of the city, these are its destroyers … " "And who are its guardians?" the people asked. The sages replied: "The teachers of Scriptures and teachers of the Oral Law." (op. cit. fn. 13 p. 171.)

"The popularization of education fostered a high regard for learning and the learned man. The illiterate became an object of disrepute and were called *am ha-aretz* (people of the land), a term used to designate a grossly ignorant person." (op. cit fn. 14, p. 172.)

The economically well-endowed father would search out a young rabbinical student of great intellectual promise to be the future husband of his daughter. Academic intelligence would be joined to entrepreneurial discipline and then to the Biblical injunction to "procreate." The community would be enriched with functioning and contributory members, the family success insured by a large number of competent youngsters to ennoble the life of Yeshivah and the Synagogue.

To the bridegroom, from the Rabbis: "One should always sell everything one possesses in order to marry the daughter of a scholar; if he does not find a scholar's daughter, he should marry the daughter of the great men of the generation (the communal leaders); if he does not find a daughter of great men, he should marry the daughter of archisynogogi (lay head of the synagogue); if he does not find a daughter of archisynagogi, he should marry the daughter of charity supervisors; he should marry the daughter of elementary teachers, but he should never marry the daughter of the people of the land (*am ha-aretz*/illiterates)." (op. cit. fn. 15, p. 173.)

St. Jerome, c. 406 CE, commented as follows on the condition of the scholarly in an increasingly pressured Jewish community in Palestine, its Talmud incomplete, its people fleeing to less threatening climes: " … among the Hebrews, those who meditate day and night on the law of the Lord and have no portion in this earth except God alone, are maintained by the services of the synagogues and the rest of the world. This is done out of a sense of justice, lest some people live in well-being and others in care, but that the superfluity of the former alleviate the misery of the latter." (op. cit. fn. 18, p. 174.)

Christians, Muslims, Jews

The roads from Jerusalem led in all geographic directions for the Jew. With only their holy literature to give them identity, they followed these roads northeast to Babylonia, west to Spain and northwest into Europe proper. Now, only their intellect and communal discipline protected them.

Christianity: During the Carolingian period, c. 800 CE, Jews had a vital function as importers of goods from the East, and their laws and customs were generally tolerated, although they were not allowed to proselytize to Christians. These peaceful relations would end with the beginning of the First Crusade, **when** thousands of Jews in communities all along the Rhine were attacked and killed under the presumption that if they were going to attack enemies of the Christ in

Jerusalem, they should attack Christ's enemies around them in Germany, ushering in a lasting anti-Semitism that included pogroms, blood libel, and being blamed as the cause of catastrophes like the Black Death and the Mongol invasion of Europe. Despite this bigotry, German Jews continued to practice, refine, and evolve their religious and social customs, including the development of the Yiddish language and an identity as Ashkenazi Jews.

Jews were highly engaged in the export trade, particularly under Charlemagne, traveling to Palestine. When the Normans disembarked on the coast of Narbonnese Gaul, they were taken for Jewish merchants. One authority said the Jewish traders boasted about buying whatever they pleased from bishops and abbots. Isaac the Jew, who was sent by Charlemagne in 797 with two ambassadors to Harun al-Rashid, the fifth Abbasid Caliph, was probably one of these merchants. He was said to have asked the Baghdad caliph for a rabbi to instruct the Jews **on** whom had been allowed to settle at Narbonne.

(Above) Figure 365: Towns under attack, Mainz, Speyer Worms, Cologne. Massacre of the Jews of Metz during the First Crusade, May 1096, painting by Auguste Migette (1802–1884).

Art Resource, New York/Scala

(Left) Figure 366: Averroes--- 1126 (Cordoba, Spain)-d.1198 (Marrakesh, Morocco): "Reflect, you have a vision" (Quran, 59.2) and "they give thought to the creation of heaven and earth" (Quran, 3:191),

Figure 367: Depiction of Maimonides teaching students about the "measure of man" in an illuminated manuscript. Maimonides: Paraphrase on the Messiah: "He shall not tire nor be weary, till he establishes judgment in the earth: and the isles shall hope for his law." (Isa. 42:4)

Figure 368: Thomas Aquinas (c. 1225–1274, near Naples) Detail from *Valle Romita Polyptych* by Gentile da Fabriano (c. 1400). Christ said to Thomas, "You have written well of me, Thomas. What reward would you have for your labor?" Thomas responded, "Nothing but you, Lord." (c. 1273)

In the early Middle Ages, Christian Europe was eagerly attempting to build a political and economic infrastructure that could replace the disintegrated Roman institutions. The Christian church was endeavoring to reproduce the Diocesan political structure of the Romans. The threat of Muslim expansionism even beyond Spain, to France, Sicily, and southern Italy, mobilized northern Europe. Charlemagne, founder of the Holy Roman Emperor, and his son Louis the Pius gave private charters to Jewish merchants, c. 800–825 CE. They both invited the Jews to settle in the empire, to promote trade and crafts. In Mainz, a large community of Jewish families had settled from Italy and France.

Gershom ben Yehudah (c. 960–1028), the most important rabbinic leader of this community, was interpreting the Talmud and Bible on his own, **with** no gaonic correspondence from Babylon or Jerusalem. In an "Enactment" to Ashkenazic Jewry, "It is a commandment to listen to the words of the sages," {B.T., *Yev*. 20a} and seeing the need to live within a majority Christian moral environment, he ruled against polygyny and the forced divorce of a woman, the former permitted by the Hebrew Bible and the Talmud. (op. cit.fn. 48, p. 180.) Jews in Muslim nations did not obey Gershom's dictum.

Three generations later, as these communities prospered, the mobs accompanying the First Crusade, c. 1095, in Worms, Speyer, Cologne, **and** Mainz, all Jewish communities on the Rhine River trading routes were butchered. Jewish life from then on was centered to the southwest, in Champagne, Troyes, on the Seine River.

There, "Rashi," Shelomoh ben Yitshaq, 1042–1105, never a rabbi with a congregation, but a wine dealer, led a revival of Talmudic analysis. His commentary on the Bible and the Babylonian Talmud became a basis for study and analysis in the yeshivas of Western Europe, until the dissolution of the ghetto. (op. cit. fn. 49, p. 180.)

Ashkenazi (European), Jewry cut off from the liberal intellectual winds of Spain, southern France, and Italy, impoverished and isolated on the land or in towns, subject to constant pogroms, and always in flight, became invisible to the outside world as the Jews strove to survive as a community by passionately cultivating its populist Talmudic literacy and scholasticism.

Islam

During the Muslim era, from c. 650 CE, there was much Babylonian autonomy for Jews. These were wealthy communities in touch commercially, and in the crafts and trade, with the world. The redaction of Talmudic learning, the questions and *responsas* that came into these yeshivahs, with donations, for their rabbis to study and solve issues about halakhic (religious law) as well as haggadic (social behavioral) issues, their replies which were read in the synagogues of Jews throughout the world, indicated the preeminence, the intellectual and moral dominance of the Babylonian academies for world Jewry.

It is important to emphasize the relative freedom given the Jewish communities in the early centuries of Muslim rule. At the beginning of the eleventh century at the former Pumbeditha Yeshivah, now in Baghdad, then under the governance of Sherira Gaon (c. 1038), a German Jewish traveler witnessed the Kallah discussions: the twice-yearly convention of Talmudic scholars and students that had become a tradition since the middle of the third century CE, under the Sassanians. He reported that two thousand participants were in attendance at these conclaves under the direction of the gaon and his *meturgemans* (broadcasters). (op. cit. fn. 26, p. 176.)

For nine centuries—from Philo (c. 54 CE) to Saadia ben Joseph (882–942 CE)—Jewish philosophical activity had ceased. Alexandrian Jewry had lost its creative impulse, and the Academies of Palestine and Babylonia felt little inclination toward philosophy; only faint traces of Platonic, Neo-Pythagorian, and Stoic thought, probably through the intermediary of Philo, can be found in Talmudic and general Midrashic literature. [op. cit. fn. 28, p. 176.]

Saadia was a unique light in an era gradually tending toward prosperity and stagnancy. He was born in Fayum in Egypt, but studied in Babylonia. A brilliant controversialist, he was appointed Gaon of Sura Yeshiva at age 46, c. 928 CE.

The intellectual influences that moved his mind came from the Babylonian Islamic philosophers of Baghdad and Basra, the *Mu'tazilah*, a doctrine involving *Kalam*, 'speech about theological matters, especially the boundaries and powers of reason to deal with religious knowledge. Here Saadia, in his *Book of Beliefs and Convictions, (Creed and Faith)*, endeavored to argue that reason and revelation lead to the same religious truths, here echoing Philo, almost a thousand years earlier, and himself anticipating the culminating medieval philosophical *summa* of Jewish philosophy—the Spaniard, Maimonides (**c.** 1135 Cordoba Spain) **c.** 1204 Fustat/old Cairo, Egypt).

The Jewish community of Spain was organized about two centuries after the Arab conquest of 715 CE. In c. 950 CE, Hasdai ibn Shaprut (roughly contemporary with Saadia in Sura/Baghdad), its leader, established Talmudic law, probably Babylonian in origin, within a Spanish-Jewish population now estimated to have grown to **c.** 250,000. This was a long-established pre-Christian population, added to by in-migration from various parts of a then more intolerant world, this especially after the Arabs ousted the hostile post-Visigothic rulers. (op. cit. fn. 40, p. 179.) Singer sees Shaprut as initiating the intellectual leadership of Spain, over a declining Babylonia. (op.cit. fn. 41, p. 179.) Solomon ibn Gabirol was the first Jewish philosopher in Spain (c. 1020–1070 CE) ("Avecibron" in the Christian world). Gabirol wrote in Hebrew, and in these generations was as well-known and influential throughout the Mediterranean as was Saadia. (op. cit.fn. 42, p. 179.)

Shortly after the Muslim conquest of Spain, the Ommayad Caliphate broke up into a number of competing "caliphates," some under Slavic and Germanic military auspices. The theme was tolerance and acceptance of the skills that the Jews brought with them. Spain prospered economically, and soon a cultural renaissance developed, with a literature of poetry, philosophy, **and** scientific studies, in which Jewish names predominate.

The important Spanish Islamic Aristotelian judge, physician, and philosopher Averroes (1120–1198) of Cordoba attempted to carve out a place for secular reason and philosophy within the Koranic domain. Ibn Khaldun, Tunis-born, a

fourteenth century CE Muslim thinker of Berber heritage, stated that: "most of the learned men in Islam were non-Arabians." (op.cit. fn. 43, p. 179.) The opponents of Averroes regarded him as a Judaizer (op. cit. fn. 44, p. 179.), exiling Averroes to the largely Jewish town of Lucena, near Cordoba, 1196.

Moses ben Maimon, a younger contemporary of Averroes from Cordoba, at that time with a very large and prosperous Jewish community (Maimonides, 1135–1204), fled that city with his family, c. 1148, when the fundamentalist Almohades sect from North Africa overran southern Spain and inaugurated a horrendous assault on the Jews. Many other Jews fled north into the receptive Christian provinces, **who were** eager to obtain the skills the Jews would provide for them, and thus **obtain** the means to recapture Spain for Christendom.

Maimonides settled in Egypt, after much wandering, **and** wr**ote** philosophy that attempted to reconcile the corpus of sacred Jewish texts with the new secular philosophical revival of Greek learning that was swirling through the civilized world at this time. Eventually, he became court physician to Saladin, the Kurdish ruler of these domains. Maimonides was to become the greatest of the medieval Jewish scholars, and one of the great achieving minds of history.

Even as Maimonides wrote his *Guide for the Perplexed*—in Arabic, but with a Hebrew script—and prepared to pass his position of leadership of the Fustat (Old Cairo) Jewish community to his brilliant and worthy young son, Abraham, he would write to the rabbis of Lunel in southern France, led by his friends from Cordoba, the Bene Tibbon family (Samuel and Judah), c. 1200 CE, that he was looking forward to European leadership in the evolution of Jewish learning. (op. cit. fn. 45, p. 179) Maimonides, on his son Abraham, the latter to be the author of *Comprehensive Guide for the Servants of God*: " … for he is humble and meek more than anyone else; add to this his fine demeanor, his sharp intellect, and his beautiful character; with God's help he will be renowned as one of the great." (op. cit. fn. 46, p. 180)

He was correct, but the sun was setting for Jewry in Africa and Europe. Palestine itself was then long in decline. Another generation or two of the *Maimom* intellect and then, cultural stagnancy. As Goitein documents in detail, the entire Islamic world, from North Africa to Babylonia and Persia, were caught up in polygyny, concubinage and the slave trade from Africa below the Sahara, as well as the mercantile degradation of women. (Goitein 1900-1985)

The Jewish widows of Fustat (Old Cairo) were deeply philanthropic toward their beloved slave girls, such as Dahab (gold) and Sitt al-Sumr (Queen of the brown-skinned), whom they often manumitted and promoted to become members of the Jewish community, when they then married into, their children becoming orthodox Jews. (op. cit. fn. 47, p. 180)

An era ends

The Muslim east and south withered in a wave of slave importations and conversions. Catholic Spain and Portugal threw out their productive and successful Jewish minority. Out of sight of the world, Ashkenazi Judaism soon took on a new cast. For almost a millennium, it had nurtured its intellectual and civilizational achievements in the dilapidated yeshivahs and synagogues of the shtetls, its stigmatized faithful moving from town to town, country to country, wherever they might find a year, a decade or two, of tolerance; here to maintain their discipline and await their due reckoning for obedience to and in the study of their Law. "A Jew, however poor, if he has ten sons, will put them all to letters, not for gain as the Christians do, but for the understanding of God's law—and not only his sons but his daughters too." (School of Abelard, c.1150 CE). [op. cit. fn. 65, p. 191.]

For a thousand years, to this point in history, c. 1100–1500 CE, the Jews had wandered, strangers even in their own Holy Land. But they carried with them the scrolls and documents of their moral, civic, and family life, the Torah, the Hebrew Bible, the Talmud, and the writings of the great scholars of the Diaspora. This was their education, and thus the cultural fabric out of which they wove a pattern of survival strategies in foreign domains. All the while, they stubbornly defended their superiority as the "Chosen People," dedicated to serve the Lord. The dialogue that they engaged in with their God, through their scriptures and writings, became the Jewish secret weapon for survival, the skill of abstract intellectual *Hss* thinking, always disseminated to the entire community. No longer was there room for an ignorant *am ha-aretz* in this terribly hostile, often violent, genocidal world.

Chapter 18

The *Hss* Search for Universal Law

Perspective

The Greeks: The great influence of Greek rationalism, exploration of human cognitive creative behavior. Political ineptness, loss of democracy led to the militarism of Macedonia. But still, there was the spread of Hellenism, and the long-enduring ideal of *paideia*, the *arete* of education for the body, mind and soul.

Rome: Roman military dominance built on Italic fecundity and the teachings of warfare. Rome was untouched for 800 years, 390 BCE–410 CE. There was political unity under Roman cultural military/political hegemony. With the decay of Roman religious paganism, imperial military autocratic corruption, the inner cultural life of Rome, both West and East, was fulfilled in a new discovery of human need—Christianity, a deeper world of belief and moral life.

Gothic Tribes

Figure 369: Detail of the votive crown of Reccesuinth from the Treasure of Guarrazar, hanging in Madrid; the hanging letters spell [R]ECCESVINTHVS REX OFFERET (meaning [King R. offers this] Visigothic crown to the Catholic Church), Nicene Christianity—7th century CE

(Left) Figure 370: The above Votive crown of the Visigoth King Reccesuinth, made of gold and precious stones in the 2nd half of the 7th century

Chritianity filled up the veins of Roman political universalism. The Gothic tribes provided much military assistance for the Papacy after the barbarian conversions and the ensuing rise of national states.

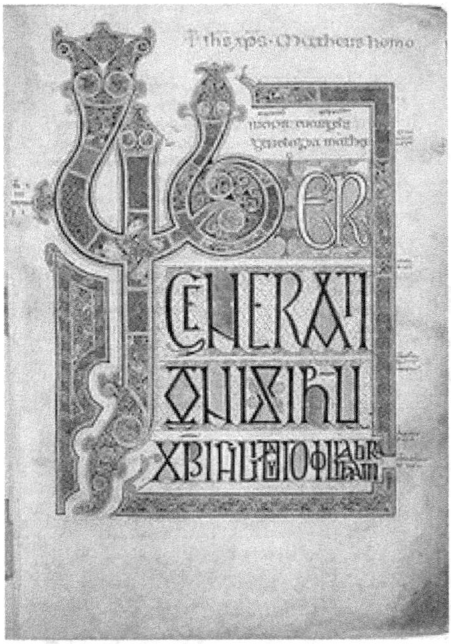

(Above) Figure 371: Book of Lindisfarne, Northumbria, c. 715 CE; decoration invades the text of the beginning of Matthew: "Liber generationis Jesu Christi filii David, filii Abraham."

For most medieval scholars who believed that God created the universe according to geometric and harmonic principles, scientific thought (particularly geometry and astronomy) were linked directly to the divine. To seek these principles, therefore, would be to seek God.

First renaissance—800 CE+

The coming of Charlemagne created a dual centering of belief/faith and military stability, conversion of barbarians, repulsion of Muslims first from Gaul and then continuing in Spain. The natural secular search for survival, economic, nationality/ethnicity began to create a new secular dynamic, leading to c. 1100 CE's universities, trade, craft, cathedrals/architecture, written music, sculpture, fine jewelry adornment, feudal aristocratic secular visions.

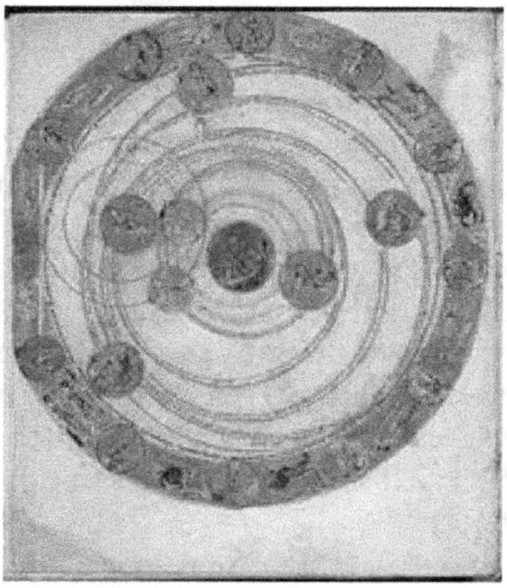

(Left) Figure 372: Ninth century diagram of the observed and computed positions of the seven planets on 18 March 816 CE

The new *Gothic* cultural period, after c. 1100 CE, saw major technological advances, including the adoption of gunpowder, the invention of vertical windmills, ethnic village spectacles, mechanical clocks, and greatly improved water mills. Building techniques were also refined (Gothic architecture, medieval castles), and agriculture in general (three-field crop rotation).

The development of water mills from their ancient origins was impressive, and extended from agriculture to saw mills, both for timber and stone. By the time of the *Domesday Book*, 1086 (under the rule of William the Conqueror), most large villages had turnable mills: around 6,500 in England alone. Water-power was also widely used in mining for raising ore from shafts, crushing ore, and even powering bellows.

European technical advancements from the 12th to 14th centuries were either built on long-established techniques in medieval Europe, originating from Roman and Byzantine antecedents, or adapted from cross-cultural exchanges through trading networks with the Islamic world, China, and even India. Often, the revolutionary aspect lay not in the act of invention itself, but in its technological refinement and application to political and economic power.

Though gunpowder, along with other weapons, had been started by Chinese, it was the Europeans who developed and perfected its military potential, precipitating European expansion and eventual imperialism in the Modern Era. Saddles on horses, improved designs for the solid-treed saddle as well as the arrival of the stirrup, horseshoe and horse collar were significant advances in medieval society.

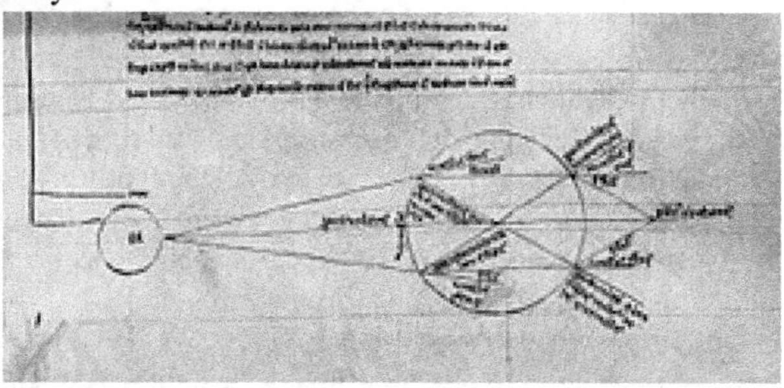

Figure 373: Optical diagram showing light being refracted by a spherical glass container full of water (from Roger Bacon, *De multiplicatione specierum*), c.1280–90 CE

Why the stirrup?

The introduction of the stirrup not only made the mounted warrior supreme in medieval warfare, but may also have initiated complex and far-reaching social and cultural changes in Europe. Some scholars credit the birth of feudalism and its subsequent spread into Northern Italy, Spain, Germany and into the Slavic territories to this use of the stirrup. It is argued that the rising feudal class structure of the European Middle Ages derived ultimately from the use of stirrups: "Few inventions have been so simple as the stirrup, but few have had so catalytic an influence on history. The requirements of the new mode of warfare which it made possible found expression in a new form of western European society dominated by an aristocracy of warriors endowed with land so that they might fight in a new and highly specialized way." (Anonymous scholar)

Other scholars dispute this assertion, suggesting that stirrups may provide little advantage in shock warfare, but are useful primarily in allowing a rider to lean farther to the left and right on the saddle while fighting, and simply reduce the risk of falling off. Therefore, it is argued, they are not the reason for the switch from infantry to cavalry in medieval armies, nor the reason for the emergence of feudalism.

For the comfort of the horse, all stirrups require that the saddle itself be properly designed. The solid tree of the saddle distributes the weight of the rider over a greater surface area of the horse's back, reducing pressure on any one area. If a saddle is made without a solid tree, without careful engineering, the rider's weight in the stirrups and leathers can create pressure points on the horse's back and lead to soreness. This is especially noticeable with inexpensive bareback pads that add stirrups by means of a strap across the horse's back with a stirrup at each end.

This analysis begins by tracing the research of the 19th century German historian Heinrich Brunner (1840–1915), who claimed that the switch to mounted warfare occurred *after* the Battle of Tours with a Saracen army in 732. Brunner pointed out that Pepin the Short began demanding horses as tribute from the Saxons in 758, citing this as evidence of an increasingly cavalry-dependent army.

Brunner also claimed that the Muslim incursion into Europe prompted Charles Martel to confiscate church lands to support cavalry. Linguistic changes are also evidence of a drastic change in weapons to support claims that this change to mounted shock combat occurred in the early 8th century. The *francisca* (Frankish throwing ax) was replaced by long swords and lances—weapons designed to be used from horseback. The lance is thought to be the strongest evidence that the Franks had adopted the stirrup by this time.

"The feudal class of the European Middle Ages existed to be armed horsemen, cavaliers fighting in a particular manner which was made possible by the stirrup." It is believed that the stirrup made the knight. The Frank short throwing spear and the battle ax, thrown or in hand, were crucial at Tours, but the cavalry attacks of the Saracen Muslims all throughout Gaul made it clear that the well-designed cavalry attack with spurs defined the future defeat of the Saxons by Pepin c. 758 CE, grandson of Martel and son of Charlemagne.

Figure 374: Carruca (Heavy Plough) northern Europe, 6–9th cenuries

Also significant in this respect were advances in maritime technology. Advances in shipbuilding included the multi-masted ships with lateen sails, the sternpost-mounted rudder and the skeleton-first hull construction. Along with new navigational techniques, such as the dry compass, the Jacob's staff and the astrolabe, allowed economic and military control of the seas adjacent to Europe and enabled the global navigational achievements of the dawning Age of Exploration.

Islam battles christian Europe

Constantinople: In spring 718 CE, two Arab fleets sent as reinforcements were destroyed by the Byzantines after the Arab's Christian crews defected, and an additional army sent overland through Asia Minor was ambushed and defeated. Coupled with attacks by the Bulgars on their rear, the Arabs were forced to lift the siege of Constantinople on 15 August 718 CE.—Greek fire..

Kingdom of Asturias—Visigothic Spain, with a large population speaking Latin, c. 720 CE, defeated a Muslim army. Politically, Asturias was relatively undeveloped.

The Battle of Toulouse (721) resulted in the victory of an Aquitanian Christian army led by Duke Odo of Aquitaine, over an Umayyad Muslim army besieging the city of Toulouse, and led by the governor of Spanish Muslim occupier of Al-Andalus, this under Al-Samh ibn Malik al-Khawlani.

The Battle of Bordeaux was fought in 732 between an Umayyad army led by Abdul Rahman Al Ghafiqi, then governor of Al-Andalus, and Aquitanian forces led by Duke Odo. Odo was defeated by Al Ghafigif. This was the farthest north of Islam in Gaul.

Battle of Tours (October 10, 732), also called the "Battle of Poitiers" and, by Arab sources, the "Battle of the Highway of the Martyrs" (Arabic: معركة بلاط الشهداء, *Ma'arakat Balāṭ ash-Shuhadā'*) marked the victory of the Frankish and Burgundian forces under Charles Martel over the invasion forces of the Umayyad Caliphate, led by the same Abdul Rahman Al Ghafiqi, Governor-General of al-Andalus, shortly after the latter's victory at Bordeaux.

Muslim cavalry, mostly loot raiders, almost broke through Martel's (Greek-like) phalanx, but he was saved by his guard. Almost half the Islamic cavalry force left the battle when they thought that Frankish irregulars were at their rear defense camp to release slaves and take back loot. Al Ghafigi was killed in the retreat from the bloody, wooded Tours hillside. The long term consequence blunted Muslims, and allowed dominant Franks under Charlemagne to take over northern Italy from Gothic Lombards, c. 774 CE.

The Saracen's attack north coincided with the fall of Narbonne on the Mediterranean. This last Muslim enclave in Gaul was reconquered by the Franks, 759 CE. The challenge of transforming a once-Christian Hispania drained much Islamic energy, and thus the north of Spain was also soon to fall into Frankish hands, in the case of Catalonia-Barcelona, c. 785–801 CE.

Intellectual murmurs

Charlemagne, as noted above, at around 800 CE, assisted by the English monk Alcuin of York, undertook a program of cultural revitalization and educational reform. The chief scientific aspect of Charlemagne's educational reform concerned the study and teaching of astronomy, both as a practical art that clerics required to compute the date of Easter, and as a theoretical discipline. From the year 787 on, decrees were issued, recommending the restoration of old schools and the founding of new ones throughout the empire. Institutionally, these new schools were either under the responsibility of a monastery, a cathedral or a noble court.

The scientific work of the period after 800 CE was not so much concerned with original investigation as it was with the active study and restoration of ancient Greek and Roman scientific texts. This investigation paved the way for the later effort of Western scholars to recover and translate ancient Greek texts in philosophy and the sciences.

After the collapse of the Western Roman Empire due to the political and economic chaos caused by the migrations from the north and east, an intellectual decline took place in the western part of Europe from the CE 400s. In contrast, the Byzantine Empire resisted the attacks from the barbarians, and preserved and improved upon the learning.

John Philoponus, a Byzantine scholar in the CE 500s, was the first scholar ever to question Aristotle's teaching of physics and to note its flaws. John Philoponus' criticism of Aristotelian principles of physics served as an inspiration to medieval scholars as well as to Galileo Galilei, who, ten centuries later, during the ultimate scientific revolution, extensively cited Philoponus in his works while making the case as to why Aristotelian physics was flawed.

In the Byzantine Empire, many Greek classical texts were preserved. Many Syriac translations were done by groups such as the Nestorians and Monophysites—eastern Byzantine and Sassanian Christians. They played a role when they translated Greek classical texts into Arabic under the Caliphate, during which many types of classical learning were preserved, and, in some cases, improved upon.

Universities

The *Trivium/Quadrivium*, **the** rediscovery of Greek intellectualism about a secular world, stimulated technological advances **and** the crusades. The production of wealth undergirded the power of the Vatican Popes, **and** thus an inner dynamic, which, by the 15th century, had broken loose from the question of reason versus faith. Reason was necessary, and to be explored as part of human nature.

Inevitably, the human curiosity about nature and its workings bred an intellectual dynamic which went beyond the philosophical debates over the meaning of the rediscovered Greek manuscripts, translated into Latin, Hebrew, and Arabic from the 11th century CE. Scientific thought was on the horizon.

Thomas Bradwardine, and his partners, the Oxford Calculators of Merton College, c. 1340 CE, distinguished kinematics from dynamics, emphasizing kinematics, and were investigating instantaneous velocity. They formulated the mean speed theorem: a body moving with constant velocity travels distance and time equal to an accelerated body whose velocity is half the final speed of the accelerated body. They also demonstrated the above theorem—the essence of "The Law of Falling Bodies"—long before Galileo, who has gotten the credit for this.

In his turn, Nicole Oresme, c. 1370, Paris, showed that the reasons proposed by the physics of Aristotle against the movement of the Earth were not valid and adduced the argument of simplicity for the theory that the Earth moves, and *not* the heavens. Despite this argument in favor of the Earth's motion, Oresme fell back politically, on the commonly held opinion that "everyone maintains, and I think myself, that the heavens do move and not the earth."

The University of Paris cleric Jean Buridan (c. 1295–1358), described as "perhaps the most brilliant arts master of the Middle Ages," contrasted the philosopher's search for "appropriate natural causes" with the common folk's erroneous habit of attributing unusual astronomical phenomena to the supernatural.

Modern historians of science believe that scientific assumption of methodological naturalism can now be also traced back to the work of these medieval Christian thinkers. This is further supported by Jacques de Vitry's early 1218 mention of the compass being used at sea, which indicates a broader knowledge and understanding of invented instruments such as the compass, and its naturalistic uses in medieval northern Europe.

Renaissance acquisition of ancient knowledge

Peter Dear has argued for a two-phase model of early modern science: a *Scientific Renaissance* of the 15th and 16th centuries, focused on the restoration of the natural knowledge of the ancients; and a *Scientific Revolution* of the 17th century, when scientists shifted from recovery to innovation. (Dear 2008)

The rediscovery of ancient texts was accelerated after the Fall of Constantinople, in 1453, when many Byzantine scholars had to seek refuge in the West, particularly Italy. The invention of printing was to have great effect on European society: the facilitated dissemination of the printed word democratized learning and allowed a faster propagation of new ideas. Sometime around 1450, mathematician Georg Purbach (1423–1461) began a series of lectures on astronomy at the University of Vienna. Regiomontanus (1436–1476), who was then one of his students, collected his notes on the lecture and later published them as *Theoricae novae planetarum* in the 1470s. This "New *Theorica*" replaced the older *theorica* as the textbook of advanced astronomy.

The last major event in Renaissance astronomy is the work of Nicolaus Copernicus (1473–1543). He was among the first generation of Christian astronomers to be trained with the *Theoricae novae* and the *Epitome*. Shortly before 1514, he began to revive Aristarchus's Hellenistic idea that the Earth revolves around the Sun. Copernicus spent the rest of his life attempting a mathematical proof of heliocentrism. When *De revolutionibus orbium coelestium* was finally published in 1543, Copernicus was on his deathbed. A comparison of his work with the *Almagest* shows that Copernicus was in many ways a Renaissance scientist, rather than a revolutionary, because he followed Ptolemy's methods and even his order of presentation. In astronomy, the Renaissance of science can be said to have ended with the works of Johannes Kepler (1571–1630) and Galileo Galilei (1564–1642).

With the Renaissance came an increase in experimental investigation, principally in the field of dissection and body examination, thus advancing our knowledge of human anatomy. The development of modern neurology began in the 16th century with Vesalius, who described the anatomy of the brain and other organs; he had little knowledge of the brain's function, thinking that it resided mainly in the ventricles. William Harvey provided a refined and complete description of the circulatory system. The most useful tomes in medicine, used both by students and expert physicians, were *materiae medicae* and *pharmacopoeiae*. Andreas Vesalius (1514–1564) William Harvey (1578–1657

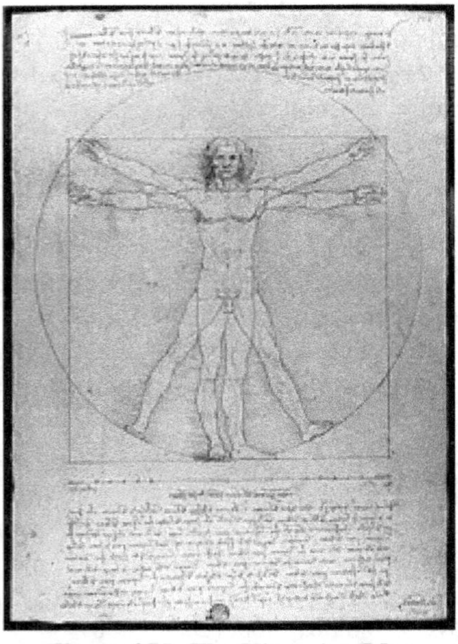

Figure 375: The Vitruvian Man

Leonardo da Vinci

Leonardo da Vinci the <u>*Vitruvian Man*</u>, is a study of the proportions of the human body, linking art and science in a single work that has come to represent Renaissance Humanism. April 15, 1452 Vinci, Italy. He died on May 2, 1519 (aged 67)

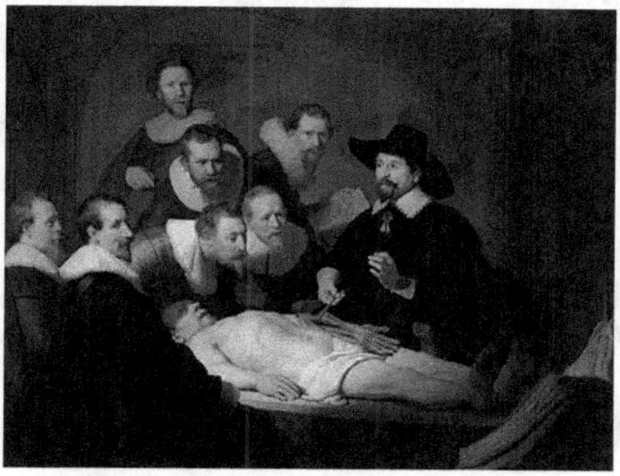

Figure 376: Anatomy Lesson of Dr. Nicolaes Tulp by Rembrandt van Rijn, 1632

Figure 377: Astrolabe of Jean Fusoris, made in Paris, 1400

(Right) Figure 378: The front cover illustration of *De Humani Corporis Fabrica* (On the Fabric of the Human Body), written by Andreas Vesalius, showing a public dissection being carried out by Vesalius himself, c.1550

Figure 379: (Above) Pumhart von Steyr, a 15th-century supergun

Francis Bacon

Figure 380

According to English philosopher Francis Bacon (1561–1626) in *Novum Organum*: "Printing, gunpowder and the compass: These three have changed the whole face and state of things throughout the world; the first in literature, the second in warfare, the third in navigation; whence have followed innumerable changes, in so much that no empire, no sect, no star seems to have exerted greater power and influence in human affairs than these mechanical discoveries."

From the time he had reached adulthood, Bacon was determined to alter the face of natural philosophy. He strove to create a new outline for the sciences, with a focus on empirical scientific methods—methods that depended on tangible proof—while developing the basis of applied science. Unlike the doctrines of Aristotle and Plato, Bacon's approach placed an emphasis on experimentation and interaction, culminating in "the commerce of the mind with things." Bacon's new scientific method involved gathering data, prudently analyzing it and performing experiments to observe nature's truths in an organized way. He believed that when approached this way, science could become a tool for the betterment of humankind

Science

"Science is based on the perception of the external world using the biology-based physical senses." (Lambrechts)

"Science is the systematic way of acquiring knowledge through observation and experimentation." (Tahir)

"Science is the highest personification of the nation because that nation will remain the first which carries the furthest the works of thought and intelligence." (Pasteur)

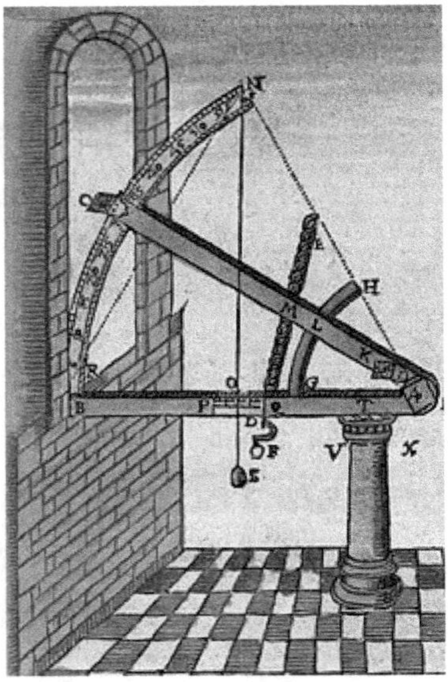

(Right) Figure 381: Astronomy became more accurate after Tycho Brahe (c. 1590) devised his scientific instruments for measuring angles between two celestial bodies, this before the invention of the telescope. Brahe's observations were the basis of Kepler's laws (Johannes Kepler, 1571–1630).

Galileo, 1564–1642

Figure 382

"Though Scripture cannot err, its expounders and interpreters are liable to err in many ways; and one error in particular would be most grave and most frequent, if we always stopped short at the literal signification of the words." **(Galileo 1613)**

"Copernicus never discusses matters of religion or faith, nor does he use arguments that depend in any way upon the authority of sacred writings which he might have interpreted erroneously ... He did not ignore the Bible, but he knew very well that if his doctrine were proved, then it could not contradict the Scriptures when they were rightly understood." **(Galileo 1615)**

"Philosophy is written in this grand book, which stands continually open before our eyes (I say the 'Universe'), but can not be understood without first learning to comprehend the language and know the characters as it is written. It is written in mathematical language, and its characters are triangles, circles and other geometric figures, without which it is impossible to humanly understand a word; without these one is wandering in a dark labyrinth." **(Saggiatore 1623)**

"Of such are the mathematical sciences alone; that is, geometry and arithmetic, in which the Divine intellect indeed knows infinitely more propositions, since it knows all. But with regard to those few which the human intellect does understand, I believe its knowledge equals the Divine in objective certainty, for here it succeeds in understanding necessity, beyond which there can be no greater sureness." [*The Dialogue Concerning the Two Chief World Systems]*

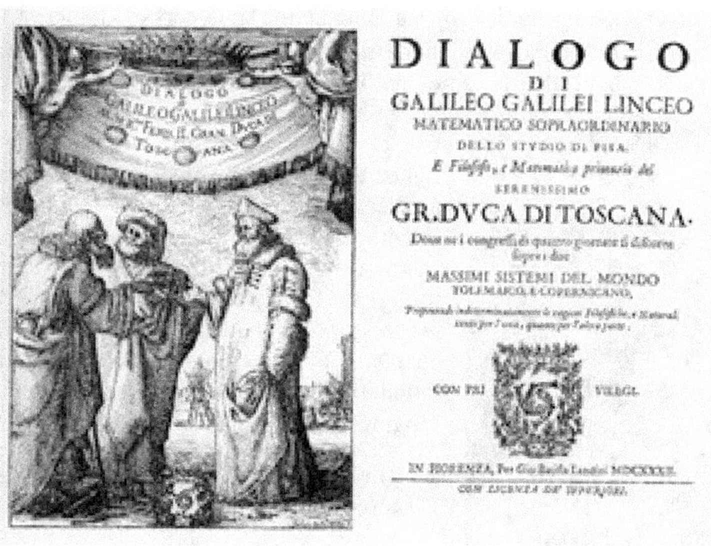

Figure 383: Frontispiece and title page of the *Dialogue* 1632

Isaac Newton, 1642–1727

(Left) Figure 384: *Newton*, by William Blake; here, Newton is depicted critically as a "divine geometer."

"Truth is ever to be found in simplicity, and not in the multiplicity and confusion of things."

"Gravity may put the planets into motion, but without the divine Power, it could never put them into such a circulating motion as they have about the Sun; and therefore, for this as well as other reasons, I am compelled to ascribe the frame of this System to an intelligent Agent."

"The proper method for inquiring after the properties of things is to deduce them from experiments."

"Fidelity and allegiance sworn to the King is only such a fidelity and obedience as is due to him by the law of the land; for were that faith and allegiance more than what the law requires, we would swear ourselves slaves and the King absolute; whereas, by the law, we are free men, notwithstanding those oaths."

"If I have seen further it is by standing on the shoulders of Giants." (1675)

It was Newton's conception of the universe, based upon natural and rationally understandable laws, that became one of the seeds for Enlightenment ideology. Locke and Voltaire applied concepts of natural law to political systems, advocating intrinsic rights; the French physiocrats and Adam Smith applied natural conceptions of psychology and self-interest to economic systems; and sociologists criticized the current social order for trying to fit history into natural models of progress. Monboddo and Samuel Clarke resisted elements of Newton's work, but eventually rationalized it to conform with their strong religious views of nature.

(Left) Figure 385: Isaac Newton in old age in 1712, portrait by Sir James Thornhill

The Enlightenment

(Left) Figure 386: *William and Mary Presenting the Cap of Liberty to Europe*, 1716, Sir James Thornhill. Enthroned in heaven with the Virtues behind them are the royals William III and Mary II who trample on arbitrary power and hands the red cap of liberty to Europe where, unlike Britain, absolute monarchy stayed the normal form of power execution. Below William is the French king Louis XIV.

The Enlightenment was a European intellectual movement of the late 17th and 18th centuries, emphasizing reason and individualism rather than tradition. It was heavily influenced by 17th-century philosophers such as Descartes, Locke, and then Newton, and its prominent exponents include Kant, Goethe, Voltaire, Rousseau, and Adam Smith.

There was no single, unified Enlightenment. Instead, it is possible to speak of the French Enlightenment, the Scottish Enlightenment and the English, German, Swiss or American Enlightenment. Individual Enlightenment thinkers often had very different approaches. Locke differed from Hume, Rousseau from Voltaire, Jefferson from Frederick the Great. Their differences and disagreements, though, emerged out of the common Enlightenment themes of rational questioning and belief in progress through dialogue.

The Enlightenment included a range of ideas centered on reason as the primary source of knowledge, and advanced ideals such as liberty, progress, toleration, fraternity, constitutional government and separation of church and state. The central doctrines of the Enlightenment philosophers were individual liberty and religious tolerance, in opposition to an absolute monarchy and the fixed dogmas of the Roman Catholic Church. The Enlightenment was marked by an emphasis on the scientific method and reductionism, along with increased questioning of religious orthodoxy—an attitude captured by the phrase *Sapere aude* (dare to know).

Rene Descartes (1596–1650)

Descartes, the rationalist philosopher, laid the foundation for enlightenment thinking. His attempt to construct the sciences on a secure metaphysical foundation was not as successful as his method of doubt, applied in philosophic areas, leading to a dualistic doctrine of mind and matter. His skepticism was refined by John Locke's *Essay Concerning Human Understanding* (1690) and David Hume's writings in the 1740s. His dualism was challenged by Spinoza's uncompromising assertion of the unity of matter in his *Tractatus* (1670) and *Ethics* (1677).

These laid down two distinct lines of Enlightenment thought: first, the moderate variety, following Descartes, Locke and Christian Wolff, which sought accommodation between reform and the traditional systems of power and faith; and second, the radical enlightenment, inspired by the philosophy of Spinoza, advocating democracy, individual liberty, freedom of expression and eradication of religious authority. The moderate variety tended to be deistic, whereas the radical tendency separated the basis of morality entirely from theology.

Industrial revolution

The precise start and end of the Industrial Revolution is still debated among historians, as is the pace of economic and social changes. Eric Hobsbawm held that the Industrial Revolution began in Britain in the 1780s and was not fully felt until the 1830s or 1840s, while T.S. Ashton held that it occurred roughly between 1760 and 1830. Rapid industrialization first began in Britain, starting with mechanized spinning in the 1780s, with high rates of growth in steam power and iron production occurring after 1800. Mechanized textile production spread from Great Britain to continental Europe and the United States in the early 19th century, with important centers of textiles, iron and coal emerging in Belgium and the United States, and later, textiles in France.

Population

According to Robert Hughes in *The Fatal Shore* (1987), the population of England and Wales, which had remained steady at six million from 1700 to 1740, rose dramatically after 1740. The population of England then continued to more than double from 8.3 million in 1801, to 16.8 million in 1850 and, by 1901, had nearly doubled again to 30.5 million. Another view sees improved health and economic conditions leading to the population of Britain increasing from c. 10 million to c. 30 million in the 1800s. Europe's population increased from about 100 million in 1700 to 400 million by 1900.

Share of total world manufacturing output (percentage)

	1750	1800	1860	1880	1900
Europe	23.2	28.1	53.2	61.3	62.0
United States	0.1	0.8	7.2	14.7	23.6
Japan	3.8	3.5	2.6	2.4	2.4
The Rest of the World	73.0	67.7	36.6	20.9	11.0

Clearly, the figures for 1900 are most critical for understanding what was happening around the world in terms of modernization, the role of science for war, human demography and health.

China was the world's most technologically advanced country for many centuries; however, China stagnated economically and technologically and was surpassed by Western Europe before the Age of Discovery, by which time China had banned imports and denied entry to foreigners. China was also a totalitarian society. It heavily taxed transported goods, and in general fought political and technological progress.

Modern estimates of per capita income in Western Europe in the late 18th century are of roughly 1,500 dollars in purchasing power parity (and Britain had a per capita income of nearly 2,000 dollars), whereas China, by comparison, had only 450 dollars. India was essentially feudal, politically fragmented and certainly not as economically advanced as Western Europe.

19th Century

The 19th century saw great social change; slavery was abolished, the path of industrialization led to massive urbanization and much higher levels of productivity, profit and prosperity. European imperialism and power brought much of Asia and almost all of Africa under colonial rule.

The first electronics appeared in the 19th century, with the introduction of the electric relay in 1835, the telegraph and its Morse code protocol in 1837, the first telephone call in 1876 and the first functional light bulb in 1878.

The 19th century was an era of rapidly accelerating scientific discovery and invention, with significant developments in the fields of mathematics, physics, chemistry, biology, electricity, and metallurgy that laid the groundwork for the technological advances of the 20th century. The Industrial Revolution that had begun in Great Britain had quickly spread to continental Europe, North America and Japan.

The Victorian era was notorious for the employment of young children in factories and mines, as well as strict social norms regarding modesty and gender roles. Japan embarked on a program of rapid modernization, following the Meiji Restoration (from 1868), before defeating China, under the Qing Dynasty in the First Sino-Japanese War (1895), over control of Korea; China lost.

Advances in medicine and the understanding of human anatomy and disease prevention took place in the 19th century, and were partly responsible for rapidly accelerating population growth in the western world. As noted above, the doubling of Europe's population during the 19th century was echoed around the western world. The introduction of steam boats and railroads provided the first major advancement in sea and land transportation for centuries, changing the way people lived and obtained goods, and fuelled major urbanization movements in countries across the globe.

Numerous cities worldwide surpassed populations of a million or more during this century. London grew to be the world's largest city and became the capital of the British Empire. Its population increased from c. 1 million in 1800 to c. 6.7 million a century later. The last remaining undiscovered landmasses of Earth, including vast expanses of interior Africa and Asia, were explored during this century, and with the exception of the extreme zones of the Arctic and Antarctic, accurate and detailed maps of the globe were available by the 1890s. The press of enlightenment liberalism and socialism became pre-eminent reform movements in Europe.

19th-Century Dynamics of Change

Great 19th-century innovators: Marx, Darwin, Galton, Mendel, Freud, Edison, Beethoven, Wagner, Brahms, Maxwell/Faraday, Pasteur, Keats, Twain, Renoir, Dostoyevsky, Balzac, Daguerre, Cezanne, Debussy, Lincoln, Napoleon, Grant.

Greatest 19th century Innovators?

(Left) Figure 387: Ludwig van Beethoven 1770–1827, portrait by Joseph Karl Stieler, 1820

(Above) Figure 388: Charles Darwin, 1809–1882, c. 1854, when he was preparing Origin of Species for publication

(Left) Figure 389: Karl Marx 1818–1883, photo taken in 1861

(Left) Figure 390: Johann Gregor Mendel 1822–1884

Understanding this process

Look Back: The genocides of the 20th century began at the beginning of WWI, with the Turkish massacre of the Christian Armenian minority. Many believe that the 19th century ended at this time. By the evidence of the optimism of populations in Europe, the inventions, medicine, middle class amenities, the outlook of hope and good expectations, the 19th century was a success. This is especially considering the intellectual, scientific, esthetic creativity it produced in Europe. Yes there were many bloody revolutions, wars.

These were the same people as the Cro-Magnons who invaded Europe in tribal force some 45,000 years earlier. The Cro-Magnons probably came up the Danube River from Anatolia, and spread out throughout Europe very quickly, the climate being a benign interstadial. These people subsequently experienced a demographic explosion—the hunting was very good. That is also why their culture, extending from the Atlantic to the Volga, was so uniform in character. They then and recently shared the same genes.

Several hundred thousand years earlier, their ancestors split off from their blond haired, white skin relatives, probably proto-Neanderthals who stayed north. This presumably tiny group(s) of separatist humans were a different kind of *hominim*. These, few in number, experienced within the proto-Neanderthal or proto-Heidelbergensis types, a process of heterochrony—a different rate of ontogenetic (developmental) maturation.

Their subsequent infantile characteristics carried well into maturity and their entire morphological (structural) and physiological development, and as such, created a very different human being, *Hss*. In all probability, this series of developments did not happen suddenly, but soon enough, it had to have caused a disassociation with the hominin regulars, essentially *Homo erectuse* types. These *Hss* paedomorphs had more energy, aggression, smarts, and even the beginnings of voluntary sound communication (language). At any rate, the secession probably cast small groups adrift to find their place in this universe, and in the process, the wanderers over many millennia shared their genes with other older *hominin* forms that they encountered in their travels around the globe.

The above has to be hypothetical. What is not, is that as a result of these genetic changes and subsequent encounters, we have subsequently identified in history the surviving five racial (sub-species) divisions of humankind, including the Cro-Magnon (European) Caucasoids. All the contemporary racial forms reveal at the least a distinct geographical ancestry.

But since almost none of the other *hominins* that were probably contemporary with the formation of this heterochronic sub-species hundreds of thousands of years ago, then, spoke a declarative "language," the assumption has to be that the proto-Cro-Magnons gave to all of humanity over the millennia, that gift, and genetically.

So, what is this excursion of humanity all about? We have tried to summarize it factually in the foregoing chapters. There is here a fundamental question regarding human nature and our evolutionary destiny. Critically important is the future human status and relationship of highly intelligent, literate human nature, *Hss*. This has to be worldwide! *Nature* has informed us.

The 19th century epitomized this underlying dynamic of the Cro-Magnon creative brain. First, humans want to understand the dynamics of the physical world around them. Second, they want to use this understanding in order to maximize the possibilities for themselves, and their affiliated social/ethnic group to survive, to proliferate with reason. Given the delicacy of human life, having lost the protective behavioral shield of instinct, we need the best of high *Hss* intelligence.

Besides the global desire to understand the ultimate causes of external experience, there are the short run challenges of day to day survival. The millions of years of human voluntary tool making was a basic foundation of this internal desire to work on the external environment, every moment of the day, in order to make the next meal possible.

Transitioning into the workings of larger, in-place urban agricultural communities, there was the development of external symbols of meaning, leading into literate systems of communication. We have at one end of the civilizational curiosity spectrum the gods, making like powerful humans to guide us, celebrate with us, lecture us on our successes and failures, always to memorialize the longevity of the community/civilization.

At the other end of the curiosity spectrum is that day to day exploration of techniques to satisfy the search for the causes of things and the consequent relationship between these causes—clouds and weather, temperature, the angle of the sun, the migration of fish in the rivers, animals semi-annually on their way to search for sustenance—a human system of beliefs, behaviors along with a whole panoply of sensory satisfactions that we desperately need to express: dance, visual arts, love, music, jewelry and human personal beautifications, all extremely powerful energy releases from this new and freakish *hominin* brain.

The modern picture

The Greeks solved part of the general intellectual problem by living in two worlds. Yes, there were the old ethnic gods that guided them out of the northern wilderness and into the solar warmth of the Mediterranean, even along the rocky shores and valleys into which the grains would be planted.

But little by little, their investigations of real, moment to moment life challenges created questions. Nature and other humans created many secular intellectual challenges which the gods and their anthropomorphic supernaturalism could not solve. So, we have here an effluence of brilliant sensory forms, the arts now matched by the discovery of philosophy and science. Then and now, for the Greeks and us, ever integral to this pulsing brain's emotional, intellectual need to know and create the new.

When the entire structure of civilized life seems to crumble in war, corruption and social dissolution, as happened in Imperial Rome, the human mind searched for a more stable and deeper human accounting than what Rome ultimately offered. Here, it was Christianity—indeed, challenging supernaturalism with all the political, external paraphernalia, priests, churches, monasteries, ritual, music, arts, even the intellectual probing of a secular civilization. But above all, there was a moral claim on human behavior.

Consider the migrations, wars, dissolution of the legal structure of the older society a whole new system needed to be developed. This calming down period, the 'dark ages', required half a millennia. Because of the high *Hss* intelligence of the Europeans in this phase of history, intellectual thought was resurrected. The civilizational institutions of the best in the Greco-Roman world came back in different clothing.

Yes, a church, weakened by the usual monarchical surges of the perennial human search for aristocracy: Popes, Kings, Senators, Bureaucrat, Billionaires, Mullahs, Rabbis, Dukes, Counts, Mayors, Presidents, etc.

Underlying this political overcast was the search for a unity of causes in the natural world, the Renaissance, which also unleashed a new and vibrant esthetic element in the *Hss* psyche. Then, arose the Enlightenment, the attempt to rationalize our knowledge of physical nature with the need to order morally and intellectually, the human system of socio-political life and our intrinsic search for freedom of the mind.

19th-century high point

The 19th century unleashed this secular, rational/intellectual thrust of the *Hss* brain, epitomized by the abolition of human slavery and serfdom, the universal assertion of this new civilization in the making of the right of the human mind to study nature and *Homo*, unfettered by religious orthodoxy or ideology. The enormous demographic push upwards seemed to exemplify a new opportunity, a new optimism.

20th century!

It is what came next in the 20th century that suddenly constituted an even newer revolution, perhaps the inexplicable civilizational unraveling of this new brain and this revealed humanity.

Chapter 19

20th-Century Horror

Lies are often more plausible, more appealing to reason, than reality, since the liar has the great advantage of knowing beforehand what the audience wishes or expects to hear.

Hannah Arendt

Achievements

There are two sides to every coin, the saying goes. One hundred years in the history of a species, ours, *Hs*, will soon be seen in context. The tearing passions of loss still in the hearts and minds of so many will fade against the equally passing acknowledgement of both fulfillment as well as unanticipated and unhappy consequence. But the achievements are still there in our encyclopedia of important successes.

Technology—mostly for the better

A Century of Inventions

A change for the ...	Better	Worse	No Diff	Don't Know
	%	%	%	%
Radio	96	1	2	1=100
Automobile	91	6	3	*=100
Computer	87	8	2	3=100
Highway system	84	8	6	2=100
Airline travel	77	8	10	5=100
Television	73	21	4	2=100
Birth control pills	72	10	13	5=100
Space exploration	72	6	17	5=100
Nuclear energy	48	30	14	8=100
Nuclear weapons	19	63	11	7=100

Figure 391: [Pew Research Center]

Civil Rights, the internet, immunization against polio;, penicillin; DNA/genetics, women's rights, working/middle class in developed nations, oligarchic/middle class in indigenous underdeveloped nations; general secularization of nations.

For the Good

Influential people:(Marshall/Nimitz/Spruance), Churchill, J. Dewey, Kafka, Gershwin, A. Fleming, M.L. King. Wright Brothers, N. Bohr, Mandela, Nehru, Einstein, Spearman, Matisse, Balfour, (Zhukov, Rokossovsky, Konev, Vasilievsky) FDR, Truman, Heifetz, Toscanini, Edison..

Perspective: marxism

Karl Marx (1818–1883) planted the seeds of revolution against a new form of inequality—human on human oppression—not by kings and queens, or bishops or cardinals, **but** rather capitalists controlling modern industrial forms of production. The state in his mid-nineteenth century view was an adjunct: a protector of the powerful as well as a receiver of the juices, the surplus value produced but not received by this oppressed proletariat. By the beginning of the 20**th** century, capitalism had both expanded and diversified in its modes of production, along with vast mineral and scientific/technological wealth pouring forth. This late 19**th**-century diversification of economic productivity even surprised the aged Marx. Science and technology had now created great and powerful instrumentalities of power—*military power.*

By some kind of internal human inertial force, instead of class subservience, the masses being born into the 20**th** century world more and more see themselves as nationalists, members of an ethnicity. Consider the welfare, educational benefits larded out in tiny spoonfuls by the state (Prussia, U.SA, Great Britain) in the latter 19**th** century. Consider the languages spoken, the music, literature, poetry, **and** even the national competition for European supremacy **in the** wars first spawned by Napoleon. The early Marxists and revolutionary communists only saw the materialistic side of the working classes. They did not understand, or were **not** willing, to notice the nationalistic/cultural tincture of life that the growing power of the state was emanating.

The revolutionaries of Marx and Engels day saw as their path to equality the capture of the material economic/political state, then to utilize it in order to transform the structure of production, so as to benefit the working class. But in this 20**th**-century transformation, a new generation of revolutionaries began to search out ever new *evil* perpetrators—pseudo members of the capitalist classes, else ethnic or religious minorities that could be accused of being the beneficiaries. They could now endow them through new communication means, as "evil doers." The crime was not only an economic one, **but** rather cultural, religious. **A** new name for the old socialism **is** *fascism.*

In capturing the state and controlling or incorporating into state control the means of production, far from the "dictatorship of the masses," this new revolutionary leadership would not wither away the implicit state. Instead, it would now use its expanded communication/transport/military wherewithal to envelop the masses in unlimited state control, here, a new "revolutionary" elite. Thus came into being a unique horror of the centuries—the state and its bureaucratic structure—a military and secret police with unlimited power, and most importantly, the controlled propaganda means over a vast communication industry to envelop the masses with a new opium.

By 1900

Electricity, dreadnaughts (giant warships), automobiles, telephones, cross continent telegraph, typewriters, oil and natural gas wells, structural steel bridges, light bulbs, phonographs, high speed photography, wireless telegraphy, moving pictures, x-rays, cross Atlantic communication, machine guns, long range artillery, all by the time of Friedrich Engels death in 1895—soon the airplane, 1903.

This vast alteration in the lives of the ever growing masses of Europeans **and** Americans **across** the world had changed the nature of the productive economies of the advanced nations, *Hss*. The dominant modes of industrial production had become blurred as compared to the textile factories and the coal and iron mining, steel production of the steam engine era. Much small-level entrepreneurship, tradesmen, manufacturers, teachers, professors, scientific and technological research, modern farm and mining, a growing governmental bureaucracy and the military were also new tilting factors in this oncoming 20**th** century.

In general, we can argue that technology reshaped the productive profile of the West. Of course, there was a need to create a population which could earn enough to purchase these new technicals and life enhancers, e.g. Henry Ford and the five dollar a day salary of his Michigan workers (1914). The seemingly clear bifurcation in mid-19th century economies between oppressor and oppressed in the sense of supposedly permanent social classes had, by the end of the 19th century, been blurred.

1914: A century of madness begins

Another few years, and this new exemplar "egalitarian" leadership class (new Ottoman Turkey), were exploiting nationalist, Islamic-oriented proletarian sentiment in their own, growing populations. The new "evil" object of their thrust for power was aimed at destroying the Christian Armenian minority, long resident in what was once a tolerant Ottoman Empire. Here, from 1915–1918, they used this modernizing state power to round up and murder well over a million middle-class Armenian innocents, the first of Europe and the world's multifaceted genocides. Here, they raised fictional social class, ethnic, and religious differences as provocations for this modern, governmentally instigated bestiality.

Modernity now produced for the nation state: poison gas, oil and gasoline for transport, attack submarines, big bertha artillery, mines, airplanes, tanks, barbed wire trenches reeking with slime. Estimated human losses in WW1, 1914–1918, are that **there were** thirty eight million casualties, seventeen million deaths—eleven million military and eight million civilian. And yet, the world population did increase. What was needed for this ongoing horror was the need for *explanation.*

With all the new technological advances, **and** industrialized production of the technology, **with** masses of humans coming onto the planet thanks to medical science advances, the *new* was too much for the leadership, from all sides, left and right, to absorb rationally. It was madness on a scale not envisioned by a world that had entered the 20**th** century with an evolving sense of real middle-class comfort and prosperity. Add to this our growing scientific knowledge about the human condition and its history, *Hss* had released potentially vast new socio/political powers for the good of humanity, this elicited from our scientific knowledge of physical reality.

The result of this historical dynamic was the ending of monarchy, the retreat of religion from its determination of human affairs. The Hapsburg, Romanov, Hohenzollern rule amongst many others disappeared, a dangerous vacuum. In 1917, the Russian Revolution broke out. And while the Bolsheviks defeated not only the "white" Russian allies of the monarchy, it also defeated more moderate revolutionary groups attempting to form a more democratic political structure. It also struggled and instigated struggles agains the international powers, but also a great peasant revolt against the Bolshevik *dream.*

The century of horrors had begun: **h**ere is an exemplar. In the beginning, as a committed socialist/communist**,** Benito Mussolini was urging his young hot-blooded radicals to commit to the "necessary bloodbath" the "physical extermination of the bourgeoisie," such that the proletariat could gain "a totality of power." He was against the WW**I**, **and** against arms that would be used against fellow French or German workers. This world war**,** he proclaimed deterministically, "would be a prelude to the revolution**.**"

With the 1920's coming of Adolph Hitler, witness the same antithesis against the established enemy of progress. At first glance, his revolutionary vision turned him toward the radical left.

Adolph Hitler: "There is more that binds us to Bolshevism than separates us from it. There is, above all, revolutionary feeling . . . I have always made allowance for this circumstance, and given orders that former Communists are to be admitted to the Party at once. The petit bourgeois Social Democrat and the trade-union bosses will never be a National Socialist, but the Communist always will." [attributed

20th-century genocide

An overview by R. J. Rummel (1997):

Bosnia-Herzegovina: 1992–1995 — 200,000 Bosnian Deaths.

Rwanda-Burundi: 1994 — 800,000 Tutsi Deaths.

Pol Pot in Cambodia: 1975–1979 — 2,000,000 Deaths **of m**iddle-class eyeglass wearers

German/Nazi/European Holocaust: 1938–1945 — 6,000,000 gas chamber, etc. **Jewish** deaths.

Japanese Rape of Nanking: 1937–1938 — 300,000 Deaths.

Stalin's Forced Famine/Purges: 1932–1935 — 7,000,000 Ukrainian/Russian Deaths.

Saddam Hussein, on the Kurds: c. 1986–89 — c. 100,**000 Kurdish Deaths**

Bangladesh: 1971 — **c.** 1.5–3 million **Deaths**, **via** violence, starvation by Pakistani Army

Mao**:** 1958–62 —38 **M**illion **M**iddle-**C**lass **D**eaths

World War II**:** 60–80 **M**illion **M**ilitary/**C**ivilian **D**eaths.

Rummel—20th-century summary

128,168,000 VICTIMS: THE DEKA-MEGA MURDERERS

- 61,911,000 Murdered: The Soviet Communist Gulag States
- 35,236,000 Murdered: The Communist Chinese Ant Hill
- 20,946,000 Murdered: The Nazi *Holocaust* Genocide State
- **10,214,000 Murdered: The Depraved Chinese Nationalist Regime**

19,178,000 VICTIMS: THE LESSER MEGA-MURDERERS

- 5,964,000 Murdered: Japan's Occupying Savage Military, Asia
- 2,035,000 Murdered: The Khmer Rouge Hell State
- 1,883,000 Murdered: Turkey's Genocidal Purges
- 1,670,000 Murdered: The Vietnamese War State
- 1,585,000 Murdered: Poland's post-World War II, Ethnic Cleansing
- 1,503,000 Murdered: The Pakistani Cut-Throat State
- **1,072,000 Murdered: Tito's Yugoslav Slaughterhouse**

4,145,000 VICTIMS: SUSPECTED MEGAMURDERERS

- 1,663,000 Murdered? Orwellian North Korea
- 1,417,000 Murdered? Barbarous Mexico
- 1,066,000 Murdered? Pre-Revolutionary Feudal Russia

Rummel's analysis

After eight-years and almost daily reading and recording of men, women, and children by the tens of millions being tortured or beaten to death, hung, shot, and buried alive, burned or starved to death, stabbed or chopped into pieces, and murdered in all the other ways creative and imaginative human beings can devise, I have never been so happy to conclude a project. I have not found it easy to read time and time again about the horrors innocent people have been forced to suffer. What has kept me at this was the belief, as preliminary research seemed to suggest, that there was a positive solution to all this killing and a clear course of political action and policy to end it. And the results verify this. The problem is Power. The solution Is Democracy. The course of action is to foster Freedom.

(Rummel, 1997)

20th century: another perspective

The history of the 20th century was basically that of the swath of destruction left across the globe by socialist ideas, from the international socialism of the Bolsheviks and the Soviet Union to the national socialism of Adolf Hitler and his National Socialist German Workers' Party.

"As concerns the 20th Century, the number of people who lived between 1900 and 2002, according to White's calculation under [Worldwide Statistics of Casualties, Massacres, Disasters and Atrocities] was 11,457,490,715. The [Population Reference Bureau's] world population figure for 2002 is 6,215,000,000, so the difference of 5,242,490,715 died between 1900 and 2002—84 deaths for every 100 persons alive in 2002." (Reynolds 2016)

Note: The above wars, genocides, atrocities, etc. were mostly directed by Hss humans against more advanced, fellow Hss sub species colleagues. However, also note that an Hss rationalism in the post war and end of 20th-century political reorganization has turned all of these nations: Germany, Japan, Russia, Austria, Poland, Cambodia, Rwanda, the old Yugolslav confederations into peaceful, at least semi-democratic nations.

World population increase

Many countries in the developing world have experienced extremely rapid population growth since the early 20th century, due to economic development and improvements in public health. China's population rose from approximately 430 million in 1850 to 580 million in 1953, and now stands at over 1.4 billion. The population of the Indian subcontinent, which was about 125 million in 1750, increased to 389 million by 1941.

Today, India, Pakistan and Bangladesh are collectively home to over 2 billion people. Java had about 5 million inhabitants in 1815; its present-day successor, Indonesia, now has a population of over 271 million. In just one hundred years, the population of Brazil decupled (x10), from about 17 million in 1900, or about 1% of the world population in that year, to about 211 million in 2019, or almost 3% of the global population in the very early 21st century. Mexico's population grew from 13.6 million in 1900, to about 128 million in 2019. Between the 1920s and 2019, Kenya's population grew from 2.9 million to 53 million.

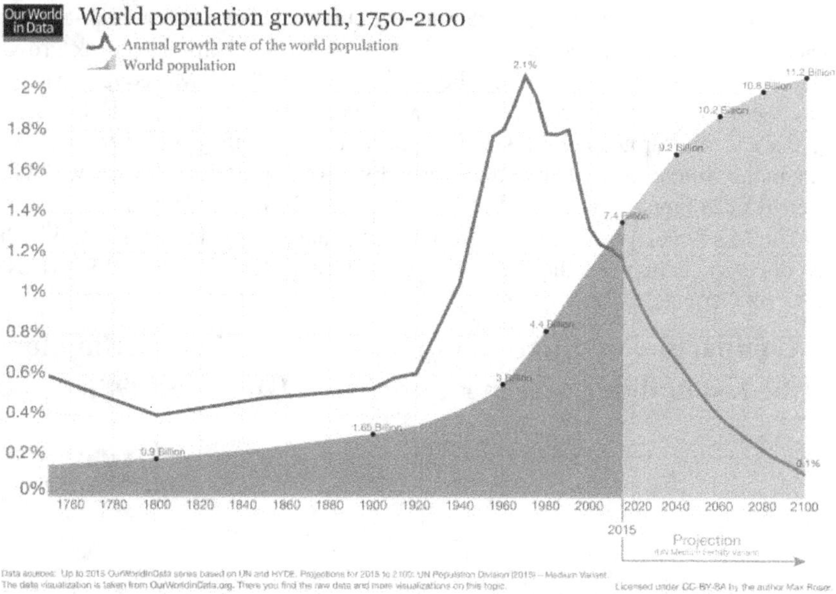

Figure 392: In 2020, the world population was 7.8 billion (Roser and Ortiz-Ospina 2017).

20th-century statistics

(Left) Figure 393: The fastest doubling of the world population happened between 1950 and 1987, where a doubling from 2.5 to 5 billion people occurred in *just 37 years*—the population doubled within a little more than one generation. This period was marked by a peak population growth of 2.1% in 1962. (Roser and Ortiz-Ospina 2017)

While Western Europe's growth rates are currently close to zero, sub-Saharan Africa's rates remain higher than 3%—that is, still higher than the peak growth rates recorded for the world at the beginning of the 1960s. Moreover, in many cases there has been divergence in growth rates. For instance, while India and Nigeria had similar growth rates in 1960 (around 1.7%), they took very different paths in the following years and thus currently have populations that grow at very different rates (about 1.26% for India compared to 2.67% for Nigeria).

The less developed countries' growth rate peaked in the late 1960s and in the least developed countries the population growth rate peaked around 1990. (Roser and Ortiz-Ospina 2017)

Comparison of different estimates of world population for the last million years in Figure below (DeLong 1998).

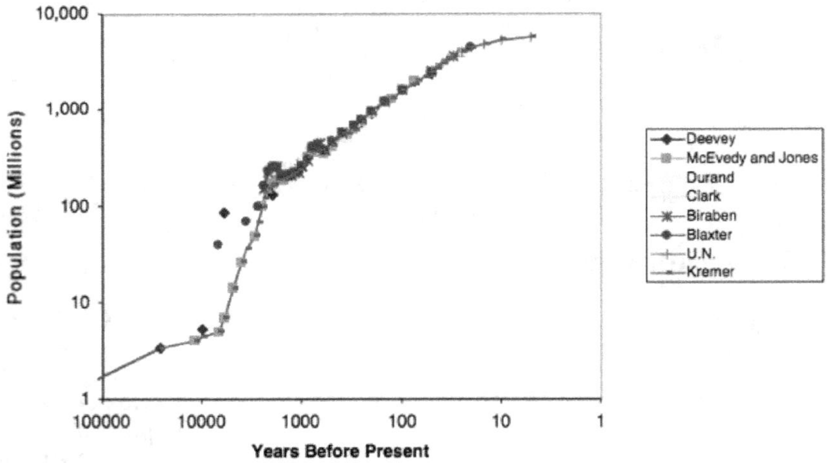

Figure 394

In the year 1900, Belgium and the Philippines had more or less the same population, around 7 million people. By the year 2000, the population of the Western European monarchy had grown to 10 million citizens, while the South East Asian republic at the turn of this century already counted 76 million citizens. The population of Belgium has since then exceeded 11 million citizens, but it is unlikely that this number will rise to 12 million by the year 2050. The population of the Philippines on the other hand, will continue to grow to a staggering 127 million citizens by 2050, according to the demographic projections of the United Nations. (UN 2013)

At the beginning of the 19th century, the world population exceeded 1 billion people for the first time in history. Subsequently, growth accelerated and the number of 2 billion people was already surpassed around 1920. By 1960, another billion had been added, in 40 instead of 120 years' time. It continued to go even faster: 4 billion by 1974, 5 billion by 1987, 6 billion by 1999 and 7 billion in 2011.

In Europe, the modern demographic transition started to take place in the middle of the 18th century. Until then, years of extremely high death rates were quite frequent. Extremely high crisis mortality could be the consequence of epidemic diseases or failed harvests and famine, or a combination of both. As a consequence of better hygiene and a better transportation infrastructure (for one, the canals and roads constructed by Austria in the 18th century), amongst other reasons, crisis mortality became less and less frequent.

Later on in the 19th century, child survival began to improve. Vaccination against smallpox for example, led to an eradication of the disease, with the last European smallpox pandemic in 1871. This way, not only do the years of crisis mortality became less frequent, but the average death rate alos decreased, from an average thirty deaths per 1000 inhabitants in the beginning of the 19th century, to around 15 deaths per 1000 citizens by the beginning of the 20th century. In the meantime, the birth rate, however, stayed at its previous, high level of 30–35 births per 1000 inhabitants.

In the world population debate, the general concerns involve mainly three interconnected consequences of the population explosion:

1) the growing poverty in the world and famine
2) the exhaustion and pollution of natural resources essential to human survival
3) the migration pressure from the poor South to the rich North (Van Bavel, J, 2004), 2013]

Modern medicine and population increase

Vaccines, sugical anesthetic/antisepsis, clean water and sanitation, antibiotics and anti-virals, heart surgery, cardiac care, randomized control trials, radiologic imaging, advancements in childbirth survival, genomic medicine, are all factors contributing to the population increase worldwide.

Population and energy

"The planet could not support the six billion people that exist today without first the commercialisation of coal, then of oil and gas. If these energy sources were necessary for the historically rare and unprecedented population growth that has occurred over the last three hundred years, then this growth might be correlated (and modelled), in some way, after the pattern of consumption of these energy sources." (Zabel 2008)

"In 1750, the world's population was approximately 720 million people. Over the previous 1000 years, this population had been growing very slowly at an average rate of about 0.13%. At this rate population doubles every 500 years and it would have taken over 1500 more years (sometime near the year 3250) to reach our current population of 6 billion people [in 2020, the population is 7.8 billion]. But sometime in the 18th century, circumstances changed and population began growing rapidly." (Zabel 2008)

"The most common explanation for this change in circumstances is that a *mortality revolution* reduced the rate at which people died and that this mortality revolution was brought about by the Industrial Revolution. The Industrial Revolution changed everything. It was an economic revolution, which spawned revolutions in science, technology, transportation, communication and agriculture. As a consequence, humanity began to experience improvements in health, nutrition, food variety, medicine and quality of life. More people survived infancy and childhood and they carried on to live longer lives. Because people were dying less quickly, populations grew more quickly." (Zabel 2008)

Energy, population, health

It has been argued that there is no link between population growth and energy use. Examples of this are in China (in the 1800s) and India and Africa (in the latter half of the 1900s). These regions witnessed explosive population growth without ever having become industrialised. Their use of coal and oil was minimal and was not a significant factor in their population growth.

Karl Marx's critique: Yet, coal powered the trains that criss-cross India. India's extensive transportation network, powered by coal and later oil, became a great distribution network for food. It now enables populations to live in marginal environments (like the deserts of Rahjastan). It probably explains why India has few of the problems Africa has in feeding its population. (Has India experienced many famines?).

In those regions, the modern argument goes, the greatest single factor in population growth was mortality decline. Modern medicine, health, hygiene, education and contraception are all given as reasons for mortality decline. Energy is never mentioned.

However, one position is that every one of the above reasons was a consequence of coal and oil use. The great scientific revolution was driven by men (mostly) who were afforded the leisure to experiment and research because machines and industry released them from the fields and provided the wealth for an educated class.

Vaccinations: In 1855, London sewers were modernised (after a cholera outbreak), and developed. Today, vaccines are manufactured in the first world. They require fuel to be distributed into the third world. Most of sub-Saharan Africa has few oil resources, yet very high population growth rates. How can there be a correlation between energy and population? While these African countries do not have the capital to develop vaccines, they benefit from oil purchased by the first world, that fuels the aeroplanes and trucks that distribute vaccines to them.

The majority of schools, hospitals and health clinics in Africa are built with first world capital, and with first world energy. Money does build the schools and clinics—machines and tools used, designed and built in the developed world, were all produced using fossil fuels.

A tractor, for example, is made of metal, rubber and plastic. Plastic is a direct by-product of oil. Metal is mined and forged using immense quantities of heat and energy. The designer of the tractor uses a computer, office space, lights and heating, drives to and from work in a gasoline-burning car. The tractor is built in highly sophisticated factory that in turn, must be designed and built with metals and plastics, which requires large amounts of energy. The chain of events always has another link, but the ultimate first cause is always inputs of energy.

The introduction of a new energy source affects population growth by:

a) directly lowering mortality rates, through improvements in health and safety, medicine, vaccines, etc.
b) indirectly affecting fertility rates. Fertility rates are directly affected by mortality rates, so a fall in mortality rates as described above will likely by followed by a fall in fertility rates.

A new energy source initially precipitates a dramatic increase in population growth, due to a sharp increase in the population ceiling. Then, after the effects of the new energy source are assimilated into society, growth falls back to some steady state. (Zabel 2008)

Debate

"Energy availability is … of limited usefulness in explaining population growth … Nor is it very helpful to see the rising use of fossil fuels, reflected in better housing, hygiene, and health care, as the key factor driving population growth. Undoubtedly, these changes were of major importance in Europe, but hardly so in contemporary China … Not surprisingly, energy considerations are also of limited help in trying to explain some of the greatest recurrent puzzles of history, the collapse of complex societies." (Smil 1994)

Contra V. Smil: The world's population would not be anything like the 7.8 billion that it is, if not for the discovery, commercialisation and mass use of coal and subsequently oil and gas. Vast inputs of energy into modern society have undergirded and stimulated vast increases in industry, technology, medicine, then the general economy, thus the vast population growth, a powerful example of the unintended consequences of *progress*.

Philanthropy—western *Hss* style

S.S. Hope was a ship converted to sail around the world with a large complement of American medical staff—Project Hope—to help erase disease and famine in undeveloped, mostly "third-world" nations.

"In 2008, [accomplishments completed] after 60 years of bringing HOPE and health to tens of millions of people in more than 100 countries, training over two million medical professionals and providing over $2 billion worth of medicines and supplies. Today, [they] are active in 26 countries, bringing HOPE and health to underserved communities around the world." (Hope Institute)

Here was philanthropic medical aid to many poor people from these nations, along with attempts to teach locals about the need for vaccinations and other death-reducing medical techniques reflective of post WW II scientific achievement in the United States. It did not however, concern itself with either the teaching or the distribution of birth control techniques, and devices to these impoverished nationals. As a consequence, the population increase soared.

While in charitable service from 1958 to 1974, this ship voyaged around the world:

Voyage 1, to Indonesia and South Vietnam, September 1960–September 1961

Voyage 2, to Peru, May 1962–March 1963

Voyage 3, to Ecuador, November 1963–September 1964

Voyage 4, to Guinea, September 1964–September 1965

Voyage 5, to Nicaragua, January 1966–November 1966

Voyage 6, to Colombia, February 1967–December 1967

Voyage 7, to Ceylon (Sri Lanka), February 1968–March 1969

Voyage 8, to Tunisia, August 1969–August 1970

Voyage 9, to the West Indies, January 1971–November 1971

Voyages 10 and 11, to Brazil, February 1972–March 1974

Achievements

One special piece of equipment was a machine called the *Iron Cow*. Using distilled seawater, combined with milk solids and butterfat, it was capable of turning out 2500 gallons of milk daily.

This 15,000-ton ship had a pharmacy, three operating rooms, a radiology department, an isolation ward, and closed-circuit television for viewing operations. The medical crew typically consisted of 150 nurses and 100 doctors, who taught American practices in various medical specialties to colleagues around the world.

In 1950, the population of Nigeria was about 40 million, and the land was almost completely covered with forest. Today, the population of Nigeria is 206 million and almost all the land outside the cities is polluted scrubland with almost no vegetation.

(Left) Figure 395: No animals live here and the topsoil is being washed away. The ability of Nigeria to feed its people is declining fast. (Peterson 2020)

Hss self destruction—ideology

Considering the phase of history the West had fully entered since the 17th century—science, technology, medicine, an enlightened vision of classlessness and democracy—what could have set off this orgy of self-annihilation, basically brothers against brothers? Not merely the butchery, but the poison gas, the gas chambers and crematoria! As luck would have it, America would crown the early stages debacle, the 19th century Civil War slaughterhouse. More recently, the U.S. set off two atomic bombs which, in one stroke, annihilated innocent hundreds of thousands. Yet, it was done in the contemporary context of world-wide horrors. There was also a millennial lesson: *do not start a war unless you are sure that the enemy will not crown your defeat.*

Sadly, the wars continued. If not against fascisms, then we saw capitalist, semi-socialist "democracies" against the 20th century version of communism. Soon to come was Islamic terrorism. One explanation is that it all came to pass too quickly—technology, population surge, a new set of middle class horizons of life, as yet incompletely distributed and acclimated.

There is another dimension to this ongoing 20–21st-century dynamic of social turmoil. The borderless and individualistic laissez faire vision of the 18th-century Enlightenment reflected an economic and demographic openness at the very beginning of the scientific age. While its secularism wounded both church and monarchical state, one could argue that they lived side by side for the relatively short period of intellectual and economic tolerance of the period, 1650–1800. Certainly, the French Revolution gave us insight as to what was to come: terribly bloody times.

The rush of the industrial revolution and the beginning surges of urban populations gave resonance to the idea of Marx and the early socialists. The bloody American Civil War underlined a new concept of human freedom and equality. The Enlightenment had argued for limitless horizons, a future for all under a new secularism. By the mid-19th century, it was clear that economic and social equality was a chimera. The rise of new elites, industrial wealth, now alongside the growing power of the national state and its expanding technological war machine, had substituted the inequalities of agricultural serfdom with a new and mesmerizing economic stratification.

One could say that the grounds for social and political legitimacy once more had the rug pulled out, now from under the urbanizing industrial masses. Karl Marx and his financing colleague, Friedrich Engels, mapped out a theoretical explanation, one consonant with a new conception of humanity in evolutionary nature given by Charles Darwin.

Marx/Engels as we have noted earlier, tried to explain the given industrial system of the mid-19th century. Their complaint was the excessive surplus value profits of the capitalist owners of the system, as well as the governmental alliance which the system fostered. The Marxist/socialist goal was not only to puncture the social evils of the new urban industrial system, but also to plan out a future in which socio-economic equality would reign as the moral foundation of this new world. But also, they identified this ungodly state alliance with capitalist oppression. Marx and Engel's hopeful prediction was that the revolution would lead to *the withering away of the state.*

Come the 20th century, the rhetoric of socio-economic equality resonated across the modernizing world. However, the realities of technological complexity and demographic expansion, pushed the lure of state governance and adjudication to the front of this new economic system. The radicals of the left saw the reins of state control as the key to mastering the competing vectors of power as we transitioned into modernity.

The philosophical underpinnings of 19th-century Marx/Engels were shunted off and degraded by massive propaganda machines emanating from the national ethnic state, a more modern glue to hold things together against the supposed enemies of socio-economic equality. Thus came into being today's ongoing ideology that sets so-called social classes against each other. This ideology has little philosophical/factual legitimacy as it chooses its guilty targets of "oppression;" Jews in Germany and in Europe in general; Ukrainian Kulaks and the Intelligentsia in the Soviet Union; the Landlords of China; Christian Armenians in Muslim Turkey; the Bosnian Muslim middle class; eyeglass-wearing middle class in Cambodia; Igbo-educated Christians in Biafra (Nigeria); Tutsi governing class in Rwanda, etc.

What is important to note, is that the identification and targeting of these groups for genocidal murder because of their supposed leading roles in these nations, should tell us very clearly that this 20th century ideology has left its original "socialist" secular utopian underpinnings for an unconditioned war against high human *Hss* intelligence. The 20–21st-century ideology of the so-called left now demanding socio-economic equality has as its unifying ideological destination, the destruction of the rewards of human creative enterprise, intelligence, political freedom and socio-economic progress. "Promise the masses equality," they propagandize. "Give them their blood rewards for participating in barbaric genocides such as the *Holocaust.*" "Sugarcoat the propaganda" to cover the never ending servitude and enslavement of these self same masses.

French Marxist philosopher Louis Althusser: ideology is "the imagined existence (or idea) of things as it relates to the real conditions of existence."

Where was the *Hss* brain?

We ask, what has happened to this *Hss* brain that created philosophy, science, and the concept, and sometimes reality of *democracy*? What happened to this *Hss* brain that also gave us the beauty and techniques of the Cro-Magnon artists, that of Babylon, Egypt, China, India, even the Medieval; then the Renaissance, Enlightenment and the secular, scientific West? Think of Notre Dame in Paris, Raphael in Rome, Beethoven in Vienna, Shakespeare in London. The art, architecture and music of the western creators was shaped under the influence of the lower brain, the limbic system, often referred to as the "emotional brain," which is found buried within the conceptual cerebrum. Like the cerebellum, evolutionarily, the structure is ancient mammalian. This system contains the thalamus, hypothalamus, amygdala, and hippocampus. Here, below, is a visual representation of this system, from a midsagittal view of the human brain:

(Below) Figure 396: [Serendip Studio]

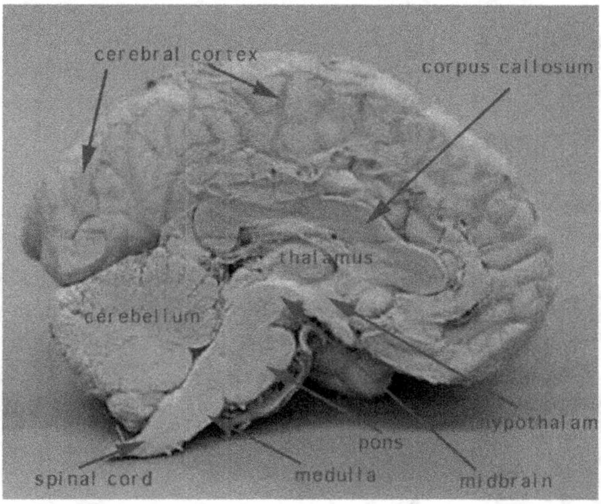

Basically, the limbic system has been passed onto human beings from our emotional mammalian ancestors.

But the key transitional factor was the power stored within the cerebral cortex. The cerebral cortex of our *Hss* human ancestors was subject to selective pressures (Pleistocene ice ages 400–200,000 BP), and survival required predictive thought. The cerebral cortex was called upon to think for life. Positive selective solutions were found in those paedomorphic and ever higher volumes of neuronic grey matter. Darwinian selective expansion brought into play genetic linkages that dragged along a mighty handful of the old mammalian limbic system; think human sexuality, physical aggression and war, love and passion for the person.

In all these aspects of this revolution in brain structure, in order to create conditions of life we call civilizational, the limbic system had to be linked with the *Hss* cerebral cortex to create conditions of mind exemplified in the fine arts, as well as the emotional and cerebral disciplines necessary to create a peaceful community.

The ideological passions thrown up in the horrors of the twentieth century have been precipitated by evolved differences in brain function, which allowed for differences in educational and societal status. In effect, an ancient human morphological/heterochronic revolution created a new human sub-species, *Hss*. But this human sub-species has not as yet been able to create a species- (*Hs*) wide awareness of our evolutionary situation.

Only through politically unencumbered scientific thought will we be able to create a philosophical structure of understanding, such that we humans can fight off the emotional irrationality of demagogic movements. These seek to exploit this as yet unresolved blockade towards the creation of a "panmictic" *Hss* demographic reality.

A panmictic population is one where all individuals are potential partners. This assumes that there are no mating restrictions, neither genetic nor behavioral, upon the population, and that therefore all recombination is possible. This means we will have to create a world of uniformly high intelligence, such that racial and ethnic differences, especially, will disappear, along with social class hierarchies. Then, demagogues will no longer be able release their ideologies upon the unsuspecting. The human race, as panmictic *Hss*, might actually survive in peace. (Solzhenitsyn, 1942)

(Left) Figure 397: An unknown perpetrator (nicknamed the "Death Dealer") at the massacre of local Jews in the Lietūkis garage.

"On the concrete forecourt of the petrol station a blond man of medium height, aged about twenty-five, stood leaning on a wooden club, resting. The club was as thick as his arm and came up to his chest. At his feet lay about fifteen to twenty dead or dying people. Water flowed continuously from a hose washing blood away into the drainage gully. Just a few steps behind this man some twenty men, guarded by armed civilians, stood waiting for their cruel execution in silent submission. In response to a cursory wave the next man stepped forward silently and was beaten to death with the wooden club in the most bestial manner, each blow accompanied by enthusiastic shouts from the audience. Once the mound of the bodies at his feet had reach fifty, the Death Dealer fetched an accordion, climbed to the top of the pile of the corpses, and played the Lithuanian national anthem."

Figure 398: Ottoman Syria, 1915: an Armenian woman kneeling beside a dead child in a field "within sight of help and safety at Aleppo, 1915–1917

Figure 399: France 1942: Jewish deportees in the Drancy transit camp near Paris, France, in 1942, on their last stop before the German concentration camps.

Figure 400: Rwanda 1994: over 5,000 (Tutsi) people seeking refuge in Ntarama church were killed by grenade, machete, rifle, or burnt alive.

Chapter 20

Nature's Revenge: The 21st Century

Seneca's cliff

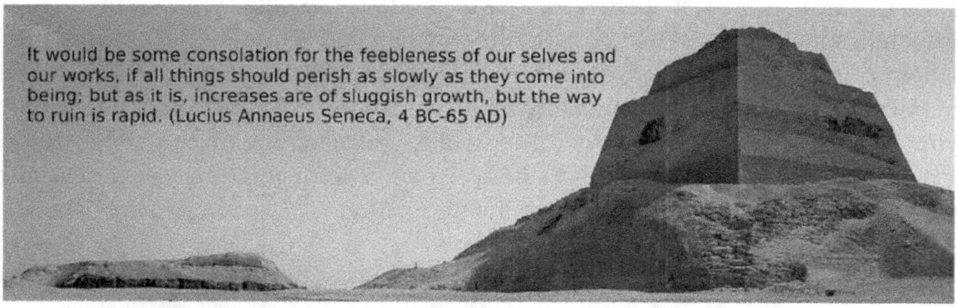

Figure 401: ***Fortune is of sluggish growth, but ruin is rapid***
(Lucius Anneus Seneca, Letters to Lucilius c. 63 CE).

The Seneca effect, or Seneca cliff's, also called Seneca's collapse, is a mathematical model proposed by Ugo Bardi, that addresses a class of problems in nature in which decline is faster than growth. This model is closely related to the work, The Limits to Growth, issued by the Club of Rome in the Seventies. Its main application is to describe various kinds of economics, given the condition of a shortage of fossil fuels, e.g. in relation to the Hubbert curve. However, unlike the Hubbert curve, the Seneca cliff shows asymmetry, which can take into account the delay of effects, such as pollution.

Peak oil. Seneca's Cliff, sudden drying up of oil +condensate+ nat. gas liquids production);

1. Nat Gas price in million of cubic feet, Henry Hub pricing, per million British thermal units

2. Oil= WTI (West Texas Intermediate weight–pricing) or Brent, (heavier than WTI) International pricing

Scientific advances, from the quotidian to the profound, made this human proliferation possible. Before 500 CE, Chinese engineers designed harnesses for plow animals that improved agricultural production, and drilled oil wells that provided fuel for heating and cooking.

The magnetic compass, which came into widespread use after 1300, led to the Age of Discovery and colonial expansion. Flush toilets, invented in 1775, reduced water-borne diseases. Nitrogen-based fertilizer, developed in 1913, helped crops yield more food. Without fertilizer, population analyst Vaclav Smil estimates that two billion fewer people would be alive today. (Yale University)

Population growth: medicine and energy

In 2020, as we write, the human population of the Planet Earth closes in on 8 billion. What does this mean? How did it come about? What of the future?

Typical of many peasant communities in the Third World, the highland Indian population of Nuñoa, Peru, operates close to its capacity for providing members with adequate nutrition. High birth and mortality rates maintain population stability in groups such as this. The introduction of modern medical services could decrease mortality and stimulate population growth, thus upsetting this balance.

The model predicts that decreased mortality would initiate population growth. Some growth would be supported by changes in individual consumption patterns. But, unless decreases in birth rate stabilized the population, it would increase beyond the level sustainable by local resources. Starvation or emigration would cause the population to crash. Thus, introduction of modern medical services could involve "a trade-off between short-term improvements in health and long-term economic hardship for the population." (Blankenship and Thomas 1997, pp. 401–417)

"Medicine has evolved in a very short time span, from the ancient civilization history of healing with the use of natural resources to the conduction of MRI and CTC scans in the modern day world. The evolution of medicine has by and large had a severe impact in the increment of lifespan as well as easing pain and making life a healthier place for most human beings. The increasing level of the average age with a decreasing mortality rates makes me ponder over what the planet would look like in another decade." (Blankenship and Thomas, op. cit.)

"Radical life extensions are under development, and thus growth in population size is inevitable. The planet is said to have already reached its peak of sustainable development, and an increase in the number of souls could be hazardous, and yet we are adding on another 200,000 every single day." (Saraf, 2018)

"For the first time in the history of living things, we are allowing the survival of large numbers of biological misfits, many of whom will become a burden for society … We are allowing the accumulation of defective genes in the human stock by providing a type of medical care that permits those suffering from hereditary disease to live longer and have children … This policy may constitute a step toward racial suicide, however noble it may appear in the light of our religious convictions and present-day ethics." (Dubos 1958)

"The sum-of-energies model assumes that different energy sources dominate during different periods of history. For example, traditional renewables (wood, dung, etc.) were the world's dominant sources of energy until almost 1900. Coal then was the dominant source of energy until the middle of the twentieth century, after which crude oil began to dominate. Oil remains the dominant source of energy to this day, but its share in the energy mix peaked in 1973 and has been declining since. The natural gas share of the energy mix has been steadily increasing and looks set to take over the number one position sometime early this century." (Zabel 2008)

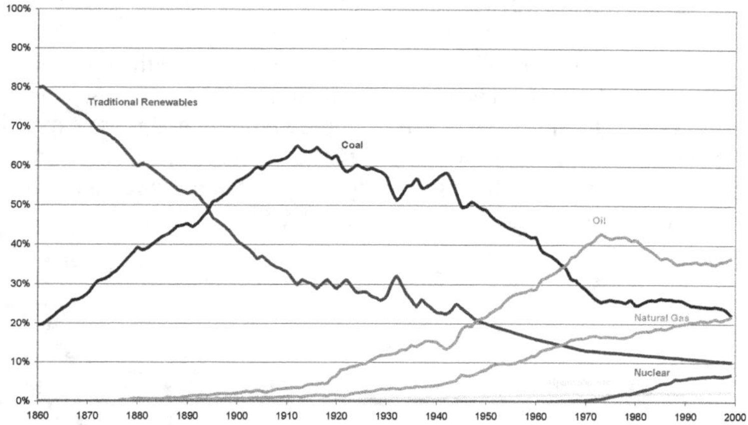

Figure 402: Percentage of total calculated consumption contributed by each energy source [Sources: Jenkins (1989), WEC (1995), BP (2000)]

Determining when the introduction of a new energy source can raise a population's carrying capacity to a new level is not easy and is somewhat arbitrary. One method that seems to fit with empirical observation is that an energy source becomes globally important when it attains a 20% share of the world's energy mix. For coal, this occurred in 1860, oil in 1950, and natural gas attained that level very recently.

	Traditional Renewables	Coal	Oil	Natural Gas
20% of Energy Share	n/a	1860	1948	1990
Energy Share Peaks	pre-1850	1912	1973	n/a
20% off Peak	1860	1940	2000 (?)	n/a

[Energy Share Dynamics]

The newly ascending energy source typically reaches 20% of the energy mix at about the same time as the dominant source has fallen about 20% from its peak. Coal attained 20% of the energy mix in 1860, the same time that traditional renewables had lost 20% of their original 100% share. Oil attained a 20% share in 1948, shortly after 1940, when coal had dropped 20% from its peak of 1912.

This then will be the criterion used to separate the effects of each energy source on the world's population: when a new energy source attains a 20% share of the global energy mix, it has reached a level where it can upwardly shift the population ceiling. (WEC 1995, p. 10)

Peak Oil:

Definition: Permanent highest level of crude oil production on planet Earth.

It seems that the biggest Saudi field is losing its punch.

"Years ago we used to talk a lot about peak oil, the prediction made by M. King Hubbert that the easy oil was going to run out, that it was going to get harder and harder to find the stuff, and it was going to get more and more expensive to get out of the ground. Hubbert wrote in 1948: 'How soon the decline may set in is not possible to say, Nevertheless the higher the peak to which the production curve rises, the sooner and sharper will be the decline.'"

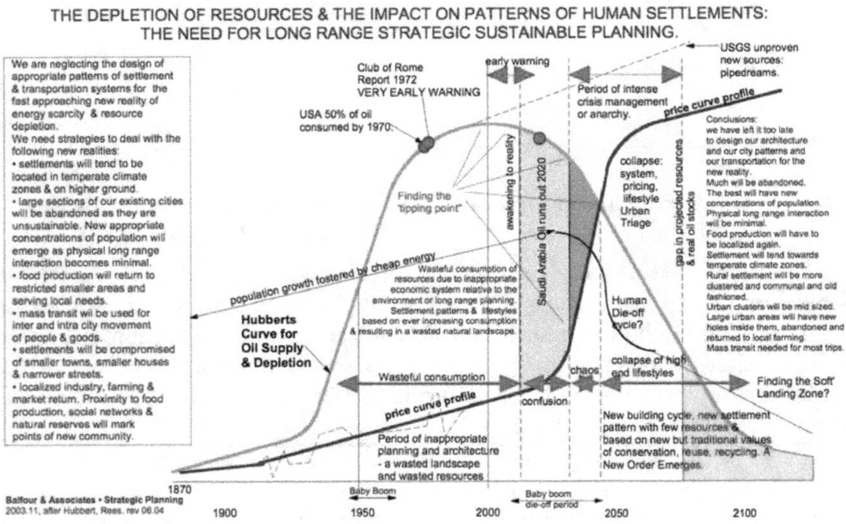

Figure 403: Hubbert peak: Balfour and Associates/ get ready for the great boomer die-off

According to the predictions made back in 2005, right about now, the Saudis are running out and we are smack in the middle of confusion, heading for chaos. Of course, we are not—we are flooded with fossil fuels, thanks to the fracking boom.

But according to Eric Reguly, writing in the Globe and Mail, there is trouble ahead, because that prediction about Saudi oil may not be that far off. He writes that the giant Ghawar field used to produce ten percent of the world's oil, five million barrels a day.

In fact, Ghawar is not as resilient as we were led to believe. We just found out that its output has fallen substantially since Aramco previously came clean on its reserves and production. If Ghawar is losing momentum fast, peak oil—remember that theory?—might be closer than we had thought. And Ghawar is just one of dozens of enormous conventional-oil reservoirs scattered around the planet that are in various stages of decline.

Those include the North Sea, Alaska's Prudhoe Bay, and Reguly reminds us that Mexico's Cantarell reservoir used to supply 2.1 million barrels a day and is now down to 135,000.

"The US Permian shale basin now supplies 4.1 million barrels a day, but fracked wells run out pretty quickly, and the fracking companies are all losing money. Better sell that pickup truck; it may well cost a lot more to fill it. As Reguly concludes, the Ghawar field is indeed in trouble, … and if it does collapse, peak oil will come a bit sooner." (Alter 2019)

Peak demand as issue

Another perspective focuses on the coming of "peak demand," a time when renewable or other sources of energy precipitate a peaking and then a permanent reduction in the demand for both oil and natural gas. In effect, don't worry about "Seneca's Cliff" when it comes to the question of availability and price of oil, gasoline, natural gas—so sayeths BP Corp's specialists on the future.

BP: "The point at which oil demand will peak has long been a focus of debate. BP chief economist Spencer Dale and Bassam Fattouh, director of The Oxford Institute for Energy Studies, argue that this focus seems misplaced."

Much of the popular debate is centred around when oil demand is likely to peak. A cottage industry of oil executives and industry experts has developed trading guesses of when oil demand will peak: 2025, 2035, 2040? This focus on dating the peak in oil demand seems misguided for at least two reasons.

First, no one knows; the range of uncertainty is huge. Small changes in assumptions about the myriad factors determining oil demand, such as GDP growth or the rate of improvement in vehicle efficiency, can generate very different paths.

Second, and more importantly, this focus on the expected timing of the peak attaches significance to this point as if once oil stops growing it is likely to trigger a sharp discontinuity in behaviour: oil consumption will start declining dramatically or investment in new oil production will come to an abrupt halt. But this seems very unlikely. Even after oil demand has peaked, the world is likely to consume substantial quantities of oil for many years to come.

The comparative advantages of oil as an energy source, particularly its energy density when used in the transport system, means it is unlikely to be materially displaced for many decades. And the natural decline in existing oil production means that significant amounts of investment in new oil production is likely to be required for the foreseeable future.

The date at which oil demand is likely to peak is highly uncertain and not particularly interesting. Rather, the importance of 'peak oil demand' is that it signals a break from the paradigm that has dominated oil markets over the past few decades.

In the past, any mention of peak oil would have been interpreted as a reference to peak oil 'supply': the belief that there was a limited supply of oil and that as oil became increasing scarce, its price would tend to rise. This basic belief has had an important influence on oil markets since the 1970s and before. Oil producing countries rationed their oil supplies safe in the belief that if they didn't produce a barrel of oil today they could produce it tomorrow, potentially at a higher price. Oil companies spent huge sums of money exploring and securing oil resources that were expected to become increasingly harder and more expensive to find.

Peak oil demand signals a break from a past dominated by concerns about adequacy of supply. A shift in paradigm: from an age of scarcity (or, more accurately, 'perceived' scarcity) to an age of abundance, with potentially profound implications for global oil markets as they become increasingly competitive, and for major oil producing countries as they reform and adjust their economies for an age in which they can no longer rely on oil revenues for the indefinite future.

One key implication of this paradigm shift is its impact on long-run price trends. The move to oil abundance is indeed likely to herald a more competitive market environment. But the assumption that oil prices will be determined simply by the cost of extracting the marginal barrel of oil risks ignoring an important aspect of global oil production.

Many of the world's major oil producing economies, with some of the largest proven reserves, rely very heavily on oil revenues to finance other aspects of their economies. The current structure of these economies would be unsustainable if oil prices were set close to the cost of extraction. Many oil producers would be forced to run large and persistent fiscal deficits or to cut back sharply on social provisions, which, in turn, would likely have knock on implications for global oil production and prices.

The argument is not that large oil producers cannot change the structure of their economies: the age of abundance means that structural reform to reduce oil dependency is more important than ever. But history has shown that economic reform and diversification can be a long and challenging process. As such, the pace and extent of that reform process is likely to have an important bearing on oil prices over the next 20 or 30 years. It is not enough simply to consider the marginal cost of extraction, developments in these 'social costs' of production are also likely to have an important bearing on oil prices over the foreseeable future.

Summary: The prospect that global oil demand will gradually slow and eventually peak has created a cottage industry of executives and commentators trying to predict the point at which demand will peak. This focus seems misplaced. The date at which oil demand will stop growing is highly uncertain and small changes in assumptions can lead to vastly different estimates. More importantly, there is little reason to believe that once it does peak, that oil demand will fall sharply. The world is likely to demand large quantities of oil for many decades to come.

Rather, the significance of peak oil is that it signals a shift in paradigm – from an age of (perceived) scarcity to an age of abundance – and with it is likely to herald a shift to a more competitive market environment. This change in paradigm is also likely to pose material challenges for oil producing economies as they try both to ensure that their oil is produced and consumed, and at the same time diversify their economies fit for a world in which they can no longer rely on oil revenues to provide their main source of revenue for the indefinite future.

"The extent and pace of this diversification is likely to have an important bearing on oil prices over the next 20 or 30 years. It seems likely that many low-cost producers will delay the pace at which they adopt a more competitive 'higher volume, lower price' strategy until they have made material progress in reforming their economies. More generally, it seems unlikely that oil prices will stabilize around a level in which many of the world's major oil producing economies are running large and persistent fiscal deficits. As such, the average level of oil prices over the next few decades is likely to depend more on developments in the social cost of production across the major oil producing economies than on the physical cost of extraction." (Dale 2019)

Peak demand?

Pandemics, who would have thought. See chapter 24, this study.

Peak oil production

R. Heinberg: "Falling production of conventional oil is pushing prices higher, and high prices are driving demand down. 'Peak demand' is peak oil by another name—de-fanged and de-clawed. Euphemisms don't change reality, except by first changing people's perceptions of reality, and hence their actions."

Here's reality, stripped of niceties. Oil is the lifeblood of industrial society, since nearly all transport fuel is oil-based and transport is essential to trade. Industrial nations have no ready and adequate substitute for petroleum. Therefore, unless preparations are made, the inevitable decline in global conventional petroleum production will gradually strangle the world economy, starting with the most oil-dependent nations.

Production rates of conventional oil will decline more sharply as time goes on, making it harder with each passing year to keep total liquid fuel production steady, harder still to grow it.

"As long as prices remain high, increasingly desperate efforts to slake society's thirst for liquid fuels will lead to the exploitation of more costly substitutes for conventional crude, such as tar sands—but these sources also happen to carry more environmental costs and risks. Problem not solved." (Heinberg)

Fossil energy and the agricultural crisis

J. E. Woods, et al.:

> Modern agriculture is heavily dependent on fossil resources. Both direct energy use for crop management and indirect energy use for fertilizers, pesticides and machinery production have contributed to the major

increases in food production seen since the 1960s. However, the relationship between energy inputs and yields is not linear. Low-energy inputs can lead to lower yields and perversely to higher energy demands per tonne of harvested product. At the other extreme, increasing energy inputs can lead to ever-smaller yield gains. Although fossil fuels remain the dominant source of energy for agriculture, the mix of fuels used differs owing to the different fertilization and cultivation requirements of individual crops. Nitrogen fertilizer production uses large amounts of natural gas and some coal, and can account for more than 50 per cent of total energy use in commercial agriculture. Oil accounts for between 30 and 75 per cent of energy inputs of UK agriculture, depending on the cropping system. While agriculture remains dependent on fossil sources of energy, food prices will couple to fossil energy prices and food production will remain a significant contributor to anthropogenic greenhouse gas emissions. (2010)

Fossil fuels and agriculture

Until recently, it seemed that cheap oil could override constraints imposed on agriculture by lack of land, soil and water. What is the oil connection? Where is oil (or more generally, fossil fuel) used in agriculture?

We'll use the example of corn production in the U.S., focusing only on farm production. That is, we'll exclude energy inputs associated with processing the corn into food products. What are the major energy inputs into corn production?

In decreasing order (by kcal/ha; where kcal = kilocalorie [kilogram calories] or quantity of heat equal to 1000 gram calories), the 'big 8' in US corn production are:

- nitrogen fertilizers
- irrigation
- gas + diesel fuels
- machinery (including energy costs of manufacture)
- drying of harvested corn
- seeds (includes all inputs required to produce the seeds)
- phosphorus fertilizers
- Herbicides

[People,.oregon state.edu/~muirp/fossfuel.htm]

M. Nichols: "While tractors may be the first piece of equipment that comes to mind when farm equipment is mentioned, other energy-consuming machines don't get as much attention. For example, we don't usually associate slaughtering animals with water consumption, but we should. Slaughtering just one animal can use up to 132 gallons of water."

That matters because the water doesn't just appear—it has to be pumped in and out of the slaughterhouse. Those pumps run on electricity, which runs on fossil fuels. When the fact that roughly 55 billion animals are slaughtered each year is considered, this is truly astounding.

Although the amount of fuel it takes to transport food is enormous, even if we only ate local produce and dairy we would still have an area of waste: packaging.

"Just like it takes a large amount of fossil fuels to operate the machinery that pumps water into slaughterhouses for sanitation, it takes a lot of fossil fuel to run the machines that make packaging required to transport food, whether it's across the country or just down the street." (Nichols)

Another agricultural perspective

"The systems that produce the world's food supply are heavily dependent on fossil fuels. Vast amounts of oil and gas are used as raw materials and energy in the manufacture of fertilizers and pesticides, and as cheap and readily available energy at all stages of food production: from planting, irrigation, feeding and harvesting, through to processing, distribution and packaging. In addition, fossil fuels are essential in the construction and the repair of equipment and infrastructure needed to facilitate this industry, including farm machinery, processing facilities, storage, ships, trucks and roads. The industrial food supply system is one of the biggest consumers of fossil fuels and one of the greatest producers of greenhouse gases." (Church 2005)

Population explosion = consequences+psychosocial/political issues

The most common reaction to crowding is stress, particularly over time and in an important space like one's home. For example, when it is crowded, people typically have negative feelings such as anxiety and frustration about restricted behavioral options. Our choices of what, where, and when we do things are constrained. If these restrictions are experienced repeatedly, crowding can also lead to feelings of helplessness wherein we start to question our own ability to effectively manage the environment.

When people experience crowding, their social interactions change. Two results are common: they withdraw from others, creating more psychological space when physical space is limited, and they become more irritable and potentially aggressive. The natural tendency to cope with crowding by social withdrawal may become a characteristic way of interacting with others.

One of the ways researchers mark whether a situation is stressful or not is to use physiological measures like blood pressure or stress hormones (e.g. cortisol, epinephrine). If crowding is a stressor, then it should affect these physiological measures. Both laboratory research, usually with college students, and community studies provide evidence that crowding can cause physiological stress. If you carefully observe yourself or others who are in a crowded situation, you can also see non-verbal indicators of stress. For example, when it's crowded, people will fidget; adjust their clothes, hair, jewelry, and so on, often avoiding eye contact. Next time you are in a very crowded setting (e.g. elevator, train), see if you notice a link between how crowded the setting is and how much these behaviors occur.

[2020 Pandemic Lockdown, see Chapter 24]

"As our city population densities continue to increase, more and more of us will become exposed to some of the psychological impacts of overcrowding:

1) Lack of privacy. Most of us are used to having some degree of alone-time and personal space and privacy. Removing such privileges often results in depression and the exacerbation of other psychological symptoms and illness.
2) Strained relationships. Overcrowding has been found to strain both social and familial relationships. This usually affects caretakers and their dependents most significantly.
3) Increased irritability and aggression. Overcrowding and noise are well known to promote restlessness, irritability and heightened aggression, something that is already having detrimental effects in many of our large cities.
4) Subculture of complaints and discontent. An increase in dissatisfactions and growing social discontent, especially among young adults, often leads to outbreaks of civil unrest." (Winch)

Seven ways seven Billion people affect the planet

G.D. Dobelko: Here are seven ways seven billion people affect the planet, according to recent research:

1. Security: Nearly 90 percent of countries with very young and youthful populations had undemocratic governments at the end of the 20th century. Eighty percent of all new civil conflicts between 1970 and 2007 occurred in countries where at least 60 percent of the population is under age thirty, says demographer Elizabeth Leahy Madsen. According to research by demographer Richard Cincotta, these countries may achieve democracy, but are less likely to sustain it. (Cincotta and Madsen)

2. Climate: The impact of demographic trends on climate change is complex. Aging in industrialized nations could reduce carbon emissions in the long term, while urbanization in developing countries could increase emissions, according to research led by Brian O'Neill of the *National Center for Atmospheric Research*. Overall, slowing population growth by 2050 could meet 16–29 percent of the reductions in carbon emissions necessary to avoid climate change. (O'Neill)

3. Water: By 2025, 1.8 billion people will be living in countries with water scarcity, and fully two-thirds will be living in conditions of water stress. People are using groundwater faster than it can be naturally replenished, putting us in danger of "peak water," says MacArthur "Genius" Fellow Peter Gleick. "We cannot talk about water without also understanding the enormously important role of population dynamics and population growth." (Gleick)

4. Food: Population growth has forced more than twenty water and/or cropland-scarce countries to import grain, making them vulnerable to food-price volatility in the international marketplace. To meet the demands of the future, we must double world food supplies by 2050, if not sooner, all while reducing our impact on the environment, according to Jon Foley of the University of Minnesota.

5. Forests: The growing demand for energy has helped devastate tropical forests, as more than two billion people depend on wood for cooking and heating, particularly in developing countries. Projects in Indonesia, Nepal, and Uganda are fighting deforestation by providing alternative energy and incomes along with health and family planning services.

6. Biodiversity: Population density is connected to the loss of biodiversity in many regions. Data from the Apache Highlands along the U.S.-Mexico border indicate that biodiversity tends to drop off at population densities of more than 10 people per square kilometer, according to research by Richard Gorenflo at Penn State University.

7a. Future Growth: By 2050, the UN says global population could range anywhere from 8 billion to 11 billion—and where it ends up depends in large part on the status of women in developing countries. "Even if fertility rates remain constant at current levels (which is unlikely), developing regions would grow from 5.7 billion in 2010 to 9.7 billion in 2050, but the total population of developed countries would remain essentially unchanged," writes Elizabeth Madsen.

7b. Population growth is not just a planet's problem. Women in developing countries with high fertility rates are more likely to suffer from poor health and low literacy. It is a vicious cycle: More children, inadequate healthcare, and less education make it harder for women to help their families adapt to scarce supplies of food, water, and energy. (Dabelko 2011)

More peak oil/demography consequences

Some of the most obvious consequences of hitting peak oil are directly related to the economy. A drop in oil supplies will lead to a sharp spike in prices. And because many industries rely on crude oil and related products, other facets of the economy will see drastic changes. Major sectors like agriculture—which are heavily dependent on the oil industry for pesticides, fertilizers and fuel—could see a steep decline. But the ripple effect could continue to transportation and even the food industry, which could see increases in prices. In a worst-case scenario, large areas of the world could experience famine because of higher food prices.

People rely very heavily on crude oil and its many byproducts. That means that any decrease in oil production may result in a change in our culture and technology. Because of the reliance on fuel for transportation, a drop in oil supplies may make it unsustainable for people to live in metropolitan areas unless they increase the use of

> alternative means of transport. The majority of the impact of peak oil would likely be felt in the lower to middle income families.
>
> (Kenton 2018)

Peak oil is a reality with serious consequences

Bear in mind the huge scale of the industry and the production [and refining, transport] infrastructure required. The vast bulk of production is coming from conventional oilfields, the majority of which are past peak and whose production is in decline. A consideration of the discoveries waiting to be developed and the timescale to put them into production reveals a significant gap, apparent even on close consideration of the work of the International Energy Agency (IEA), which masks this gap as production that will come from as yet unidentified, undiscovered fields. It is totally unrealistic to anticipate future discoveries on the scale required to fill this gap, given the historical record (especially this century) and the fact that most promising oil provinces have already been well explored and developed.

The potential of a future boost in production from Iraq and Iran can be assessed, as can that from unconventional sources including Canadian Oil Sands and US tight oil (shale), but all of this combined is insufficient to fill a large demand-supply gap when considering the horizon 2020–30. What is actually important is not the date of peak global oil production, but the decades of decline that follow. They imply a continuous global economic contraction until the underlying basis of the global economy—liquid hydrocarbons for everything that moves—is substantially changed.

"Furthermore, rapid growth in the domestic consumption of the key oil-exporting countries means less will be available for export—oil exports from key exporting countries will decline 1much faster than their production declines (the so-called "export land effect"). Peak oil and its consequences are as serious for the global economy as climate change." (Tobin 2015)

Contra peak oil

In 2009, Dr. Christoph Rühl, chief economist of BP, argued against the peak oil hypothesis:

> Physical peak oil, which I have no reason to accept as a valid statement either on theoretical, scientific or ideological grounds, would be insensitive to prices ... In fact the whole hypothesis of peak oil—which is that there is a certain amount of oil in the ground, consumed at a certain rate, and then it's finished—does not react to anything ... Therefore there will never be a moment when the world runs out of oil because there will always be a price at which the last drop of oil can clear the market. And you can turn anything into oil if you are willing to pay the financial and environmental price ... (Global Warming) is likely to be more of a natural

limit than all these peak oil theories combined ... Peak oil has been predicted for 150 years. It has never happened, and it will stay this way.

Summary warning—21st-century Seneca's Cliff

Vast overpopulation, ruled by techno/political/economic elites dominate the declining levels of communication, educational intellectual abilities—Hollywood, mass media and cultural barbarism, all act as reform blockade.

Consequential events, *tipping points* (Seneca Cliff)—Peak Oil, Peak Natural Gas—cannot be avoided. There will still be coal, and some emergency nuclear minerals for energy production. When will this happen, 2025–2030? What will be the consequences?

Scenarios: When the price of crude oil (at 2020 dollar level) reaches a range of $150-200 per barrel, WTI (West Texas Intermediate) and dry Natural Gas reaches a price of $10-15 per million cubic feet, Henry Hub—BTU (British Thermal Unit}s;--spiraling inflation begins; bond price interest rates rise, mortgages, cost of money much higher. GDP falls, wages/profits decline, prices rise, Weimar Republic like, high inflation and economic decline.

What happens, for example, if India cannot raise the funds to supply their agriculture with necessary fertilizers/pesticides for their human billions? Expensive commodities, metals, rare minerals = expensive food. The cost of internally produced fossil fuels can be regulated by government price controls. But fossil fuels are an international market: the U.S, and most other countries, except Gulf States, Saudis, Russia, and perhaps a few others, have to import fossil fuels to balance their own productivity. They can import only if they have their own resources and have the cash or exportable products to balance payments.

Thus, there are problems necessitating price controls. Government taxation hit by private sector deflation, hit by higher energy prices even with general price controls, fewer tax receipts. , Here we confront serious inflation: higher interest rates as governments start to print money.==great unemployment in the millions in private sector, inevitable also in governmental sector, except for police and armed services.

(Note: U.S. printing of money in 2020, virus pandemic. Will this lead to general governmental control of private sector and result in fascism?)

What about the third or developing world, those nations with vast, impoverished populations? China and India, with their great populations and meager internal fossil energy resources, will be in desperate straits. Russia and their former nationals in mid Asia, like Kazakhstan, have these resources. What about Japan and the Koreas, they without?

What alliances or wars will develop? What about the billions incapable of fighting to get needy resources? Take the wealthy Saudis, Gulf states, etc, are they ripe pickings for powerful northerners in Europe or Asia who need Middle Eastern fossil energy?

Triage: to preserve international stability.

Could the international community rise up to try to create order in this millennial civilizational collapse, by deciding which populations could not be saved, and which might survive with minimal international resource allocations? We are a world in disintegration, like a species undergoing negative biological selection. Only here, we have the cortical decision-making capacity possibly by *Hss* elites with military power. They have the unlimited capacity to decide the future of humankind. Here, expect the salvation of some people, nations, ethnicities, races, and the turning away from those which are destined to disappear.

Two Alternatives to the Crisis

1. A long slog of penury in those nations and people, overshooting the natural resources necessary for modern prosperity because of demographic bloat—bare maintenance in a time of desperation. Testing of human socio-psychological tolerance for mass urban life, no matter the accompanying barbarism.

2. Else, Seneca's Cliff, a sudden collapse of the system within a generation, 20–25 years, unleashing a whirlwind of uncontrollable, socio-economic forces attempting rapid self-correction. Vast and sudden changes with the aim of stabilizing western civilization.

Make a wager. Will precious batteries, even with their vanishing lithium, cobalt, nickel commodities, reappear to save us?

Pessimism

The expected evolutionary victory of *Hss* is not necessarily the outcome determined by the sociobiology of the human species. Remember the *Holocaust*, the French Revolution. There is also the reality of the limbic system and ideology. The human brain, at least in relatively short cut-outs of time (a century or two), can negate the long-standing dynamics of primate evolution—advancing cortical brain power.

Remember also, the Zanj Rebellion, a major and terribly bloody uprising against the high cultured Abbasid Caliphate, which took place from 869 until 883 CE. Beginning near the city of Basra in present-day southern Iraq and led by one Ali ibn Muhammad, the insurrection is believed to have involved enslaved Bantu-speaking people (Zanj), who had originally been captured from the coast of East Africa, then enslaved and transported to the Middle East.

Ideology and the lower brain

The affluent "in the warm uplift of the all-encompassing swamp," the supposed *Hss* elites, can always be overthrown in leadership by their own "quislings," manipulating emotional, if real underclass grievances and their consequent mass movements. The result for humanity is a long escalator slog downward. Here, in the early 21st century, we already see the portents of the decline of western civilization since its hopeful apogee before 1914.

Further, we should always recall the origin of Soviet genocides. Lenin overthrew the Czarist autocracy by means of a supposed alliance with the masses, ostensibly to create a socio-political utopia! He died from a liberal's gunshot and his successor, Stalin, created a brutal genocidal dictatorship for 65 years, 1925–1990. Fortunately for humankind, it finally disintegrated at the behest of these self-same masses. Dissolved, a modest neo-fascist autocracy came into leadership, fortunately, without genocidal programs.

Will the 21st century bequeath humanity a peaceful and cognitive civilization?

Chapter 21

Significance of *Homo sapiens sapiens*

Prelude

The ancestors of modern bacteria were unicellular microorganisms that were amongst the first forms of life to appear on Earth, about 4 billion years ago. For close to 3 billion years, most organisms were microscopic, and viruses, bacteria and archaea were the dominant forms of life. In July 2018, scientists reported that the earliest life on land may have been bacteria living on land, 3.22 billion years ago.

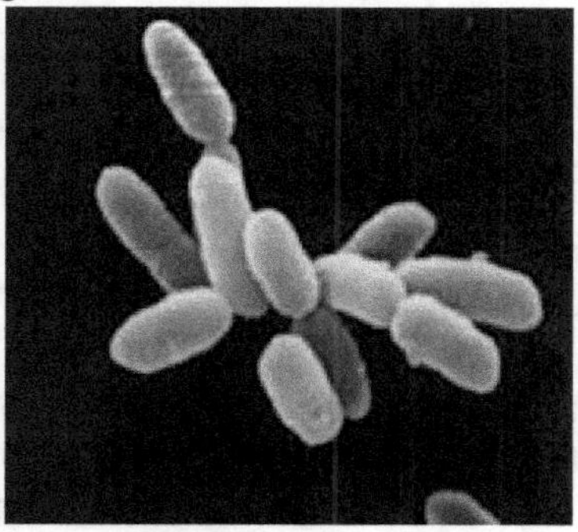

Figure 404: Archaea prokaryote microbes

Archaea, prokaroytic microbes, were first found in extreme environments, such as hydrothermal vents. Bacteria were also involved in the second great evolutionary divergence, that of the archaea and eukaryotes. Here, eukaryotes resulted from the entering of ancient bacteria into endosymbiotic associations with the ancestors of eukaryotic cells, which were themselves possibly related to the archaea.

Eukaryotic cells—cells with internal "organs" (known as organelles)—came into being, c. 2 bya. One key organelle is the nucleus: the control center of the cell, in which the genes are stored in the form of DNA. Eukaryotic cells evolved when one simple cell engulfed another, and the two lived together, more or less amicably—an example of "endosymbiosis." The engulfed bacteria eventually became mitochondria, which provide eukaryotic cells with energy. The last common ancestor of all eukaryotic cells had mitochondria—and had also developed sexual reproduction (protists).

The eukaryotes divided into three multi-cellular groups, c. 1.5 bya. The ancestors of modern plants, fungi and animals split into separate lineages, and evolved separately. We do not know in what order the three groups broke with each other. At that time, they were probably all still single-celled organisms.

A eukaryote cell had the advantage over bacteria and archaea in that they could maintain their physical integrity in a greater breadth of environments, in contrast to the two wildly mutating uni-cellular biological domains. In their new sexually reproducing amplitude, they were a great step forward in going against the second law of thermodynamics, and not melting into the physical environment. The evolution into multicellular forms between 1.5 and 1 bya added on to this resistance of amplitude and ordered mutational variation, again exemplified by the three great forms of life, animal, plant and fungi, formed from one-celled eukaryotic protists (amoeba-like).

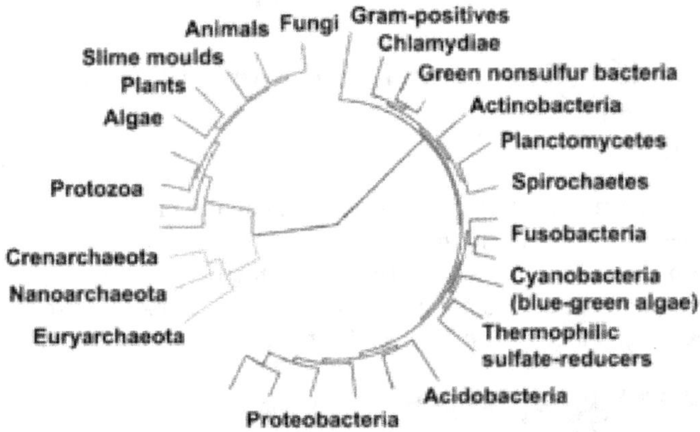

Figure 405: Evolutionary tree showing the divergence of modern species from their common ancestor in the center. The three domains are colored, with bacteria blue, archaea green, and eukaryotes red.

Dynamic ancestor

The separation of the archaea and the eukaryote ancestors of animals, and thus we humans, is such that we are in constant protective opposition from the bacteria and viruses in terms of disease. Their wildly mutating character makes it difficult for modern medicine to keep up with an as yet unknown bacterial/virus outlander who could be immune to medicines' own deadly efforts.

It is puzzling that despite being one of the most numerous and ubiquitous life forms on earth, no member of the domain ***Archaea*** has been described as a human pathogen.

What the eukaryote forms of bio-chemical life can teach humans today is the adaptive nature of living forms, especially with animal movement to attempt to avoid energic dissolution by altering behavior within the life term of the individual biological unit, here a living human being.

Some animal forms secured their destiny by being fixed, sessile filter-feeding absorbers of food as they moved with the currents of the waters (e.g. oysters). Others were destined to adapt with a more aggressive motile and predatory profile. The vertebrates thus created a wholly new and progressive option, first in the seas and then on land, 600–400 mya.

It is important to note in terms of evolutionary options, that those morphological and behavioral adaptations that seem to herald the coming of the primates and then humans were long existing adaptive innovations from the earliest stages of both vertebrate life in the waters and on land—warm bloodedness, placental live births, large brains, long distant migrations, highly predatory relations with life neighbors. Here fish, skates, snakes and especially the sharks and sting rays stand out: Hammerhead viviparity, c. 60 mya, is especially impressive, considering their early soft boned cartilaginous structure formed well before the amphibians and reptiles (dinosaurs) appeared on the planet. We have yet to find such a fully modern morphology in a fossil mammal before c. 90 mya.

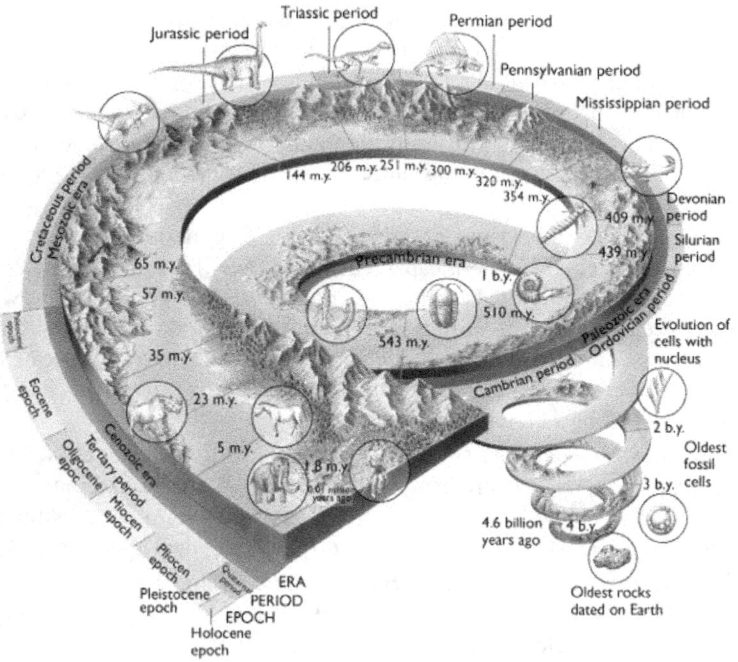

Figure 406: The Swirl of life

Achieving sapiency

In the explosion and diversification of the mammals after 90 mya, especially after the demise of the dinosaurs, c. 65 mya, we witness the seemingly sudden eruption of a wide variety of apparently recessive genes, the accumulation of a billion years of evolution, all in the adaptive niche that our planet had prepared for life. From then, we see the gradual dominance of mammalian, mainly placental reproduction.

Our guys and gals arose from a somewhat conservative line of creatures, at first emphasizing defensive adaptations, moving in light and dark, mostly frugivorous, with an omnivorous backup. The important genetic adaptations consisted of staying out of the way of rodents and predatory cats, wolves, and panther types, those on the make, but also mostly living alongside a variety of specialized vegetarians, they still providing animal protein milk for their young.

The primates successfully brought their young to reproductive maturity and gradually spread into a variety of adaptive niches—monkeys, lemurs to apes and finally *hominins*. This took place over many tens of millions of years. Today, it is recognized that the precursors of the hominins were not proto-chimps living 6 mya. As we have described in earlier chapters, we now have a wide variety of c. 10 mya fossils that point to a separation of an ape/pre-hominid primate of at least twenty million years past. This latter claim is mine, written decades ago. (Itzkoff 2000)

The genes for success needed time to express themselves selectively, pruned by environmental happenstance, and an increasingly important symphony orchestra of brain/behavior adaptations. And they came. Smartness, the ability to predict danger and opportunity worked in terms of reproductive success. The brain, some three to two million years ago, began to throw light on the opportunities for hunting/gathering success in the fabrication of tools for a variety of "industrial" uses.

There is much enthusiasm among the authorities in defining these tool-using *hominins* of fairly enlarged skull capacities as *sapiens*. True, there were enlarged skulls, belonging to individuals still using primitive technologies, such as with the Caucasoid Neanderthals in Europe and their contemporary *hominin* forms in Africa and Asia. None of these earlier *hominins* give evidence of the use of spoken language as we now recognize it. Thus, the appellation *sapiens* for these fossil anticipators might best be withheld.

As we have noted in an earlier chapter, the controversy over Neanderthal language capacities is still presently in recess. We have no external evidence. Their Mousterian culture, somewhat more in advance than other fossil brethren, shows no sign of significant abilities towards objective, declarative expression.

The current hypothesis is that *Hss* originated some four to two hundred thousand years ago, probably in Eur-Asia, the Caucasus? *Hss* was a product of radical rate gene developmental changes, which allowed for the growth of that vast overblown excrescence, the human brain, and that delicate bone plate that covered it. Some researchers see *H. Heidelbergensis* in Eur/Asia as the common ancestor of both the Neanderthals and *Hss*, some 600,000 BP.

At the c. 90,000 BP range, we have found tools and harpoons in south central African sites, home of the Bantu, and they are still using these artifacts. The Cro-Magnon Magdalenian culture in Europe reveals similar fabrication of such fishing harpoons, at the c. 18,000 BP level. The assumption is that these Cro-Magnons wanderers at 90,000 BP, had much of the technology in hand, as when they arrived in Europe, c. 45,000 BP.

Recent discoveries on Mt Carmel in Israel (teeth and jaws) argue for a *Hss* presence in that area from about 180,000 BP. The Jebel Irhoud fossils from the Atlantic coast of Morocco, c. 300,000 BP argues for an earlier modernity seemingly closer to *Hss* than the Neanderthals. (Nature 2018; 2017)

We note the beginning of the use of color, in Africa, in the wall paintings of humans now seen as the ancestors of the Bushman—Capoids, Boskopoids—also at c. 90,000 BP. The spread of their very early and diversified click languages, their preoccupation with art, large endocranial capacities and dominant spread of this race over Africa, north and south, as well as into southern Spain, argues for the genes of *Hss* to have begun its domino impact and the extensive spread of *Hs* humans over the planet, starting at the latest, c. 100,000 BP.

Even in far off Tasmania, Australia, New Guinea, these genes of the Cro-Magnons must have had their long-distance domino impact in moving the various forms of *Homo* over the sapiens line; giving the acquisition of language in all humans, the real taxonomic standard of sapiency.

Perspective

The great morphological and thus behavioral revolution within the *hominin* clade began between 400–200,000 years ago. Given its historical perspective, it was a northerly fragment that broke off from the well-established hyper-erectine pathway that the other main line Caucasoid group, the specialized Neanderthals, were evolving from.

Our claim is that these new, radical, baby-faced humans created their nascent culture and adaptive thrust into modernity, certainly by 100,000 BP. The hybrid Mousterian culture of Skhul and Qafzeh, found in today's Israel, date to that period. The morphology of these large, non-Neanderthal skulls hint at this hybridization of the two groups. Sixty thousand years later, in what is now Moravia, the Czech Republic, Mladec show similar levels of morphological hybridization, yet now surrounded by Cro-Magnon cultural artifacts.

The remnant racial differences that we see around the world in recent historical perspective should tell us of the ongoing hybridization of the *hominins*. By now this an old tradition, at the least from 100,000 BP. In the last several hundred years, it has speeded up exponentially, most recently under new communicative and transport unification (immigration) dynamics. This social development has been precipitated by the spread of *Hss*, the Cro-Magnon genetics of intellectual behavior.

The selective evolutionary dynamic of ever-larger brains goes back some 500 million years to the dawn of vertebrate evolution. At times, in the seas and upon land, selective conditions have speeded up the growth of the brain and the adaptive potency of the vertebrates, here even harking back to one of the most ancient vertebrates, the sharks and rays.

As long as the external environments within which the vertebrate line has been forced to readapt continue their rhythmic pulsations, this heritage of changing external dynamics—geological, climatic, biological—will continue. The press for the natural selection of higher intelligence in the vertebrates has been fulfilled in the explosive emergence of *Hss*.

Human population explosion

1800 = 1 billion; 1900 = 1,65 billion; 2000 = 6.2 billion; 2020 = 7.8 billion

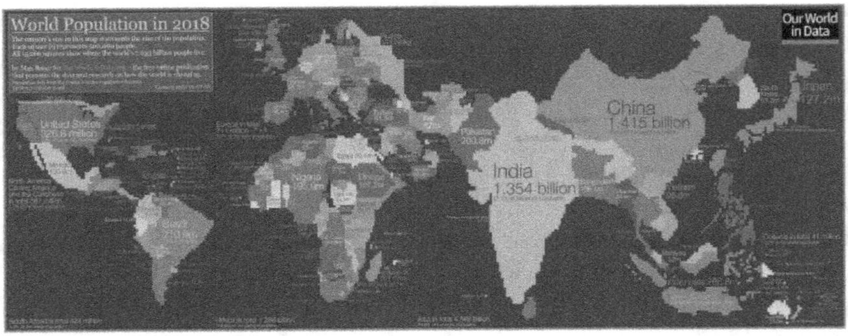

Figure 407: Population densities, 2018

That century, the 20th, as we have noted in chapter 19, was the most horrific experience for humanity in historical times. It was accompanied by a great change in the structure of social life, this precipitated by modern scientific thinking, technology, medicine, fossil energy. Natural death rates diminished, and new scientific advances in mobility and communication brought the world of humanity together. This was all led by the ongoing dominance of the Europeans, descendants of the Cro-Magnon *Hss*.

Accompanying these advances in middle class life, undergirded, of course, by our mining of nature's great fossil fuel heritage, the ensuing wars made ever more deadly by modern science, the genocides unleashed by new human perceptions of their social selves added up to this vast slaughter even amidst growing wealth, and an explosion of humans, both *Hss* and *Hs*, onto this planet.

In terms of the significance and impact of this horde of humans, in 2020, now approaching 8 billion, an almost incomprehensible multiplication from the population of humans on this Earth in 1900: 1.65 billion. Can the resources of the planet provide for this mass? Can the socio-psychic balance of humans, packed into greatly expanding urban congestion be managed by their leadership elites? This is the question that should be front and center of those who wish to fulfill the civilizational traditions of *Hss* as we look to the future when planning our daily disciplines.

The above question brings forth the need to understand what has happened to the seemingly placid civilizational transitions of the past. Consider some of the glories of the human *Hss* intellect that undergirded this *hominin* leap into overwhelming power. Then, there has occurred the even more important tragedy for so many accompanying life forms, which we have destroyed. Here, we do not underestimate the accompanying and ongoing human tragedy of recent centuries.

Lost instinct

The theory of signs is a good open door to the attempt to understand our contemporary human dilemma. Signs are communicative inputs external to a living creature, acting as a tripping point for behavioral action. We speak about two sub-classes of signs, signals and symbols. A signal can be interpreted as a fixed informational element, such as a red or green light. Throughout the world, its connotation is fixed. Yes, of course in the human domain, we can by choice change the behavioral meaning of a signal.

In the animal world, these signals, a shadow that a mouse perceives while crunching a piece of corn, alerts the flee and hide behavioral nexus built into an instinct. The higher you go up the phylogenetic scale in terms of brain size and complexity, animals have more room to "decide" the behavioral meaning of signs/signals in terms of "what to do." Nature has perfected these instinctual reactions to long-term selective specializations in behavior. Usually, animals have little choice. Nature has told their neurological system how to "understand" these signals and translate them in to instinctual responses.

The function of an ever larger brain, selectively, is to give more choice within the life term of an animal to make alternative decisions. This is where the defensive primate brain held its own. It was not as rigidly programmed as that of a wolf or a reindeer. And as we go up the time line of primate progress towards humanity, we see this advantage taking place in the enlarging hominid brain with ever fewer instinctual restraints on their sign/signal behavior.

With the coming of that revolutionary hetero-chronic transformation of the *hominin* line, the instinctual hold over animal behavior has seemingly disappeared. A vast protuberant cortex has now disciplined three billion years of life into a thinking machine of behavioral alternatives, here and now. We humans, *Hs* and *Hss*, now live in the symbolic world of sign choices and thought projections of the future.

"Man has, as it were, discovered a new method of adapting himself to his environment. Between the receptor system and the effector system, which are to be found in all animal species, we find in man a third link which we may describe as the '*symbolic system.*' This new acquisition transforms the whole of human life. As compared with the other animals, man lives not merely in a broader reality, he lives, so to speak, in a new *dimension* of reality. There is an unmistakable difference between organic reactions and human responses. In the first case a direct and immediate answer is give to an outward stimulus, in the second case the answer is delayed. It is interrupted and retarded by a slow and complicated process of thought. At first sight, such a delay may appear to be a very questionable gain. Many philosophers have warned man against this pretended progress. 'L'home qui medite,' says Rousseau, 'est un animal deprave': it is not an improvement but a deterioration of human nature to exceed the boundaries of organic life." (Cassirer 1962, p. 24)

But of course, the advantage that lay in this new morphology was this brain's capacity to organize experience into predictive systems of social action. Building on the gift of spoken language, this new *Hss* creature soon advanced his group into a civilization of abstract meanings, eventually by the Sumerians, written language then throughout the migrating, hybridizing Cro-Magnon world, mathematics and the physical accoutrements that fascinate us about the ancient *Hss* civilizations, these we have surveyed above.

Abstract world of *Hss*

Abstraction: "The ability to think about objects, principles, and ideas that are not physically present. It is related to symbolic *thinking*, which uses the substitution of a symbol for a concrete object or idea."

"Abstract thinkers think about how everything relates to the bigger picture. They are always looking for the deeper meaning or the underlying patterns in things. Abstract thinkers want to understand how everything relates to everything else … ***They enjoy subjects that use a high degree of abstract thought, which includes subjects as varied as astrophysics and poetry*.**" (Learning Mind)

"Someone who thinks more abstractly can do better on intelligence tests. Abstract thinkers are better with language, can see what's not there, and understand concepts that are important to humanity's development. Most of our modern inventions would not exist if there were no abstract thinking. Meanwhile, someone who is not an abstract thinker may do poorly on an intelligence test." (Thomas 2018)

With the advances in science and technology towards the end of the 19th century, it became clear in the European nations descendant from the Cro-Magnons, that the need for universal literacy and mathematical competence required educational testing to assist the young in gaining a position in the modern national societies of the future. Individuals such Francis Galton in England and researchers Alfred Binet and Theodore Simon in France, attempted to devise testing instruments which could predict future intellectual competencies through the developing universal educational systems of all the nations of the world.

The expansion of universal education and the awareness of the correlation of academic abilities and subsequent achievements in the higher intellectual area of a nation, and thus its national power, military and economic amongst its peers, set off a great scholarly/psychological search for the meaning of academic and intellectual abilities, early on, especially amongst the European and North American nations.

The existence of the *g* factor was originally proposed by the English psychologist Charles Spearman in the early years of the 20th century, starting in 1904. He observed that children's performance ratings across seemingly unrelated school subjects, were positively correlated, and reasoned that these correlations reflected the influence of an underlying general mental ability that entered into performance on all kinds of mental tests.

Spearman suggested that all mental performance could be conceptualized in terms of a single general ability factor, which he labeled *g*, plus a large number of narrow task-specific ability elements After studying general intelligence, Spearman concluded that intelligence is a cognitive ability that can be measured and numerically expressed. Thus came the phenomenal growth of IQ and other scholastic and ability/talent tests.

The "g" factor in the analysis of human intellectual performance whether academic or in specialized areas, law, medicine, aviation, etc. has engendered over a century of controversy and research. A different world demography has entered our space in the century plus. We are no longer researching the Caucasoid descendants of Cro-Magnon.

Probably the most influential of modern researchers and exponents of the concept of "g" as a definitive measure of human intelligence are Hans Eysenck and Arthur Jensen.

Gradually, the study of genetics has moved from the early plant researches of the mid-19th century priest Gregor Mendel to an international array of researchers. Their work and countless others have made the study of human genetics the most powerful tool of both psychological and medical research. Thomas Hunt Morgan, Ronald Fisher, Francis Crick, James Watson, G.H. Hardy, Robert Plomin, Wilhelm Weinberg, Walter Fiers, Francis Collins, are a few of the many who have broken open this field to almost unlimited horizons. It is fast approaching integration with psychological research on human intelligence variation and behavior, ultimately the issues of *Hs* and *Hss* civilizational capacities. The research of Robert Plomin and the impact of DNA research on understanding the power of genetics in explaining human intellectual behavioral differences exemplifies this newest scientific horizon. (Plomin 2018)

Genetic key to social behavior

Volkmar Weiss was born in 1944, in the eastern part of Germany and there to be educated. This young student showed talents in biological science and thence in genetics. Surprisingly, the communist East Germans were interested in genetics and did not send their scientists in this field off to imprisonment or death as did their tutelary Soviets to the east. The Soviet Union hated Darwinism and the science of genetics, favoring "socio-political creationism" that sent so many of their geneticist off to the gulag. Sadly, this interest of the East Germans and their secret police, the STASI, was an interesting reaction ro the Nazi madness in pursuit of the master race.

Hitler, insane, but in some ways practical, sent his officers back home on short term leave to create future heroes and heroines of the "Aryan Race." The East Germans inherited this theme. They wanted to arise from the ideological dysgenic ashes of one political system by stimulating the births of the intellectual classes that yet remained within their domain. They wanted at the least to propel births of future Olympic greats to legitimize this terror state.

In 1969, Hans Grimm head of the Humboldt University (Berlin-East) Physical Anthropology Institute suggested to Weiss, then a research student in genetics, to initiate studies of family backgrounds of talented mathematicians. This was to be based on the results of 2,800,000 East German students on international mathematics competitions.

The East German politicos were wise enough to allow Weiss in his studies to have full access to the ongoing research in the West, as to genetics and associated fields, and of course to utilize his growing competency in the languages of the nations of the West. This he did and succeeded with especial success, even to the point of becoming acquainted with some of the West's most famous anti-genetics collaborators with the STASI, during this Cold War period. Weiss published in western journals, achieving surprising status with his agile feet in both worlds. Clearly, he was quietly ever more addicted to the freedom of expression of colleagues across the Channel and Atlantic.

In a paper he published in an American journal after the walls came down (walls—1989, paper—1992), he summarized his long developing research and conclusions concerning the genetic element in human intelligence variation. Indeed, these were German and Austrian subjects. But there was also a larger significance for the international world of the variation in capacities revealed in the intellectual labor of these experimental populations. His paper was titled "Major Genes of General Intelligence" in *Personality and Individual Differences*. In this paper, he reveals his full acquaintance with world wide psychological/genetic research, both historical and contemporary.

By this time, he was consumed in an attempt to find the major gene or genes responsible for high "g" as well as lower forms of "general" or abstract intelligence. His research pursuit was to find the underlying biochemical code, the enzymes in brain metabolism which would highlight these model genes. As with most earlier and more contemporary researchers, this "major" gene complex has yet to be isolated.

Using mathematical achievement levels correlated with I.Q. testing, thence correlating these results with family social status profiles, Weiss believed he had achieved a functional view of the as yet undiscovered genes for intelligence, high, mid-level and lower, here using East German subjects. We must note that East Germany along with West Germany had in the previous generation lost a vast mine of possible subjects to their sociological profile in the war, especially of the Jews. What he was testing were the children of the survivors, including Panzer officer corp., Luftwaffe pilots, and Submarine crews.

Even if Weiss' sociological research was relevant, it was a different genetic profile of a population then had existed in Germany at the beginning of the 20th century. One can picture the result, the genetic information in a parallel 19th-century controversy. This was the observed orbital trembling of the planet Uranus as it made it way around the sun. The wonder that these calculations presented led to many hypotheses. Finally, the prediction of an extra planetary object causing gravitational perturbations in Uranus led to the observation of Neptune and its gravitational impact on Uranus.

But out of Weiss' contemporary researches as extrapolated back into a broader sociological panorama, came a hypothesis of the existence in modern society of three distinctive genetic structures. These genetic differences led determinatively, according to Weiss, to three educational achievement levels and thence to a wider separation in vocation and thence social class.

One might say the hypothetical realities of these mathematical genetic distinctions had been arrived at; M1M1 = 130 IQ; M1M2 = 112 1.Q and M2M2 = proportionally, below the higher means. So too the sociology of these genetic profiles showed up in the kinds of life achievement of the families of these children—physicians, engineers, hair dressers, truck drivers and down the line of abstract and practical social skills, to the generational unskilled.

It should be said that the predictive achievement of the children, given their mathematical achievement, the wider I.Q., were supported by the subsequent professional achievements of families. Weiss's work has been long validated by the American research of L.M Terman and M.H. Oden (1948, "Genetic Studies of Genius," Vol.4.) a twenty-five-year study of gifted children growing up.

Note that the traditional mean I.Q. of Caucasoids, Europeans and North Americas is c. 100 I.Q. Weiss did believe that Northeast Asian Mongoloids had similar I.Q.s as the European and North American Caucasoids. However, Weiss made no recognition that the Northeast Asian Mongoloids were tri-racial hybrids, much Caucasoid and with some Australid genetic elements.

While Weiss saw these assumptions to be characteristic of stable national environments, perhaps East Germany, 1945–1989, he said that the mixture of genotypes in marriage M1M2 might, by skipping a generation, consonant with Gregor Mendel's 19th century research, produce a proportion of M1M1 progeny. One would, by example note the Darwin family in England, late 18th to 19th century and the Mendelssohn family in Germany mid 18th to early 19th century show interesting skip a generation production of great talents (Erasmus Darwin's grandsons, Charles Darwin and Francis Galton; one of Moses Mendelssohn's grandchildren was Felix Mendelssohn

Weiss agreed with the results of a number of researchers that showed that gifted mathematicians of I.Q.s 170–196 showed no greater professional, economic or sociological achievement profiles than those with an I.Q of 135. Weiss, "… this should follow from major genes, where beyond I.Q. of 130 all individuals should be M1M1." (Weiss 1992, p.18)

Weiss's presumption of a stable political/sociological environment and the endogamaus interbreeding of intellectual/personality types, led him to hypothesize an eventual "balanced polymorphism for human intelligence and its concomitant biogenetic social structure."

The 1992, publication of his seminal paper indicates skepticism of the universal socialist goals for the eventual achievement of a "classless society." This was the political dream within which he had grown up. However, his conclusion postulates that differences in the vocational needs for various levels of abstract intelligence will create social stability, but also a permanent biogenetic hierarchical society, archetype of the future.

The origin of this permanent human condition is given in his exemplification of the Upper Paleolithic Cro-Magnon hunter of mammoths, 20,000 BP. He envisions a band of hunters: twelve men, six, mm2/mm2 with I.Q.s of 94; five, mm1/mm2; hunters .of IQ 112; they, all led by one alpha male mm1/mm1, of 130 IQ. He states "Any hierarchical society can tolerate only a very small percentage of high I.Q. individuals in relatively humble social positions, because such individuals represent always, in view of the limited number of leading positions, a potential threat to the ruling elite." (Weiss op cit, p.22)

Note: They are still at it.

"Genome-wide association meta-analysis of 78,308 individuals identifies new loci and genes influencing human intelligence" 52 **candidate genes** linked to intelligence

"The researchers discovered that the genes that were the strongest linked to intelligence are ones involved in pathways that play a part in the regulation of the nervous system's development and apoptosis (a normal form of cell death that is needed in development). The most significant SNP (Single nucleotide polymorphisms, frequently called SNPs [pronounced "snips"], are the most common type of genetic variation among people. Each SNP represents a difference in a single DNA building block, called a nucleotide) was found within FOXO3, a gene involved in insulin signaling that might trigger apoptosis (**a form of programmed cell death that occurs in multicellular organisms).** The strongest associated gene was *CSE1L*, a gene involved in apoptosis and cell proliferation." (*Nature Genetics* 2017, p. 1107–1112)

Hss: the puzzle

It is almost a generation since Volkmar Weiss wrote his important paper. His view and hope for a stable social situation has by now been radically outdistanced by advances in technology, medicine and genetics. Hidden behind recent revelations of massive research into the genomic character of populations, especially in Russia and China, has frightened many in the West, especially in the United States.

While the report below focuses only on the limited military impact by genetic knowledge applied to making a "better soldier," we can be sure that hidden beyond the façade of political correctness is an active research and policy endeavor to face the issue of intellectual variance in the military as well as in the general national population profile. There is here a sharply focused awareness that high intelligence in all phases of national life and work may well be the key to national dominance, even survival.

A new front if opening in the intensifying economic and diplomatic competition between the United States and China: human DNA.

"I think we're just seeing the leading edge of it," retired Lt. Gen. Thomas Spoehr, director of the Heritage Foundation's Center for National Defense, told the Washington Examiner. "Everybody knows that we ought not to be sharing technical data on hypersonics, directed energy, things like that. This is one of those things that people haven't thought of."

"They're thinking about it now. FBI officials are wary of letting scientists with ties to the Chinese government study the DNA data that American researchers use to develop cutting-edge medicines. Their skepticism has prompted a pair of top Republican senators to probe whether the Department of Health and Human Services is giving an unintentional boost to China's genetic research industry at the expense of U.S. economic and national security. 'China can already compel the genetic information of over a billion of its own people,' Sen. Marco Rubio told the Washington Examiner. 'And if it can acquire—through companies it controls or through cyber theft—genetic information on millions more, [the risk] can range from unfair advantage in the development of life-altering cures to … the creation of biological agents that only harm people of a certain genetic makeup.'"

"The idea of biological weapons uniquely suited to kill Americans 'sounds like something out of science fiction' the Florida Republican acknowledged. But he's not the only one thinking along those lines. 'I'm not an expert on genetic medicine, but I can tell you that people that are agents are stealing information about medicines that only harm people of a certain genetic makeup," he said.

The Russian government wants to study genetic information to develop 'the so-called genetic passport of a soldier,' as one of Moscow's top scientists put it recently. Military officials [Chinese] hope to tighten surveillance on Uighur Muslims, a repressed ethnic and religious minority. And last year, a Chinese doctor aided the birth of 'the world's first known gene-edited babies,' an operation that drew international condemnation because he had modified the DNA in their embryos 'to try making them resistant to their father's HIV infection,' as the Wall Street Journal explained.

"It's theoretically possible to combine those concepts 'to make people super-soldiers,' according to Spoehr, a former commandant at the Army's Chemical, Biological, Radiological and Nuclear School. If researchers can identify parts of DNA that make some people better athletes than others, for example, and it's possible—though widely condemned as unethical—to edit DNA to encourage or suppress certain qualities, governments around the world might consider engineering biological weapons and soldiers to resist their effects."

"Genomic data is also the next phase of breakthrough medicines via gene therapy/editing, a Senate Republican aide told the Washington Examiner. If China has cures to cancers or Alzheimer's, it could give them leverage in larger issues [emphasis added]. We've seen them use this strategy in the ongoing tariff debates, since we're dependent on China for many traditional drugs." (Gehrke 2019)

We are still policy-wise, loathe to consider that the issue of intellectual variance in our demographic imprint is critical to our future, the USA, as well as that of our world at large. We must put our minds to the need to increase this international intellectual profile and not be satisfied with a stable and hierarchic political/economic structure, as Weiss suggested in his paper, and that of many others who disdain even thinking about the subject.

Scientist Weiss believes in the inevitability of a hierarchical society. This, he concludes from the diverse genetic levels of intelligence in European society, a major puzzle as to the nature of intellectual variability even within *Hss*.

His exemplars are the Cro-Magnons of Eur-Asia on their anti-extinction of mammoth hunts for their own survival. For this, he writes, they need a planning leader of IQ of at least 130 (MM1). Then they back the leader up with five men of IQ 112 (M1M2), perhaps to hone in for the kill. Finally, to cut up the slain mammoth and to drag it back to camp, Weiss postulates the need of six more men of IQ 94.

Even as Germany was united in a democratic structure of political life, 1990, Weiss is skeptical, since nature argues against egalitarian democracy. Too many people of IQ 130 and above equals constant violent struggles for power to gain the fruits of leadership. Stated in more colloquial terms, "If we have too many smart people, who is going to work at McDonalds or clean the rooms of the Waldorf Astoria?"

But is this the human experience of nations or ethnic groups having a lack of entrenched social classes, and presumably enjoying equable high intelligence. Certainly, the Jews long before the *Holocaust*, lived peacefully in their segregated communities, even as there were differentiated and hierarchical skills needed to be fulfilled within their communities. Psychologist Hans Eysenck estimated Jewish Ashkenazi (European) intelligence mean at about I.Q. 118, compared to the European mean of I.Q. 100.

The Scandinavian nations, before welcoming migrants from other continents to their shores, created highly developed civilizational norms of social life. In modern times, they lived peacefully amongst themselves. The truth is that high human intelligence expresses itself in a number of ways, unrelated necessarily to the thrust for power and wealth. There have been many men and women of religion who live and function on meager financial tailings. We have exemplars such as Leonardo, Shakespeare, Beethoven who envisioned a life different from Napoleon, MacArthur, Buffet, Gates.

In fact, what creates a civilization is this diversity which yet emanates out of the minds of those with I.Q. higher than 130. Perhaps the reason why those with I.Q.s higher than 150 do not receive higher financial or professional emoluments as compared with those who hover around 130, is that their pursuits (intellectual, creative, communal) satisfy their mental and physiological needs of life.

Another way of looking at the almost 50% of Weiss's hunters at 94 I.Q. can precipitate some good arguments. How would a nation with 50% of its citizens being M2/M2at I.Q. c. 94l 25% M1/M2 at 112 IQ; 25% at IQ c.130 compete in war and economics with a nation where 50% of its citizens are M1/M1 at 130 IQ, 25% at M1/M2= IQ112 and 25 % at M2.M2, IQ averaging=c.94. Which of these nations would be destined to become a subservient colony of the other?

The *Hss* brain and behavior

Our contemporary view of the structure of intelligence and its variation across the planet tells us a somewhat different story than the picture of a pure "g" factored intelligence, and thus a determined I.Q. destiny. Here, of course, mathematical ability might give us a nicely simple predictive picture of an individual's destiny. Early I.Q. researchers had to add a verbal "sister" to IQ and other exams of intelligence and education.

Why verbal tests? One can argue that the acquisition of spoken language and finally amongst the *Hss*, written language, cleared the way for mathematical thinking, there being an underlying relationship even if they show independent strengths in testing and in life. One of the explanatory hints for these relatively independent skills of mathematical and verbal ability lies in a brain which has several or more independent profile variables in strength and weakness.

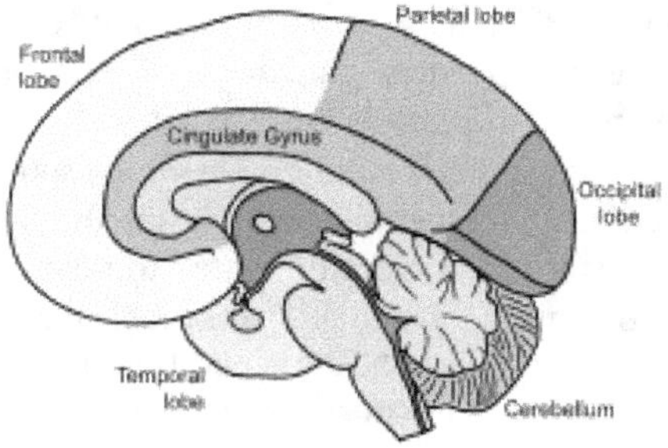

Figure 408: Lobes of the brain

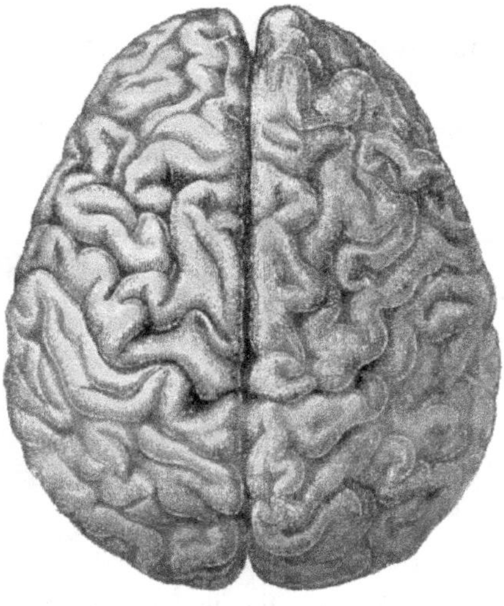

Figure 409: Frontal lobes=pink; parietal lobe=green; occipital lobe= blue

The cerebrum, consisting of the cerebral hemispheres, forms the largest part of the brain and overlies the other brain structures ... Each hemisphere is divided into four main lobes—the frontal lobe, parietal lobe, temporal lobe, and occipital lobe

For example, the silent purrings of the grey mantle of our brain reveals the parietal and occipital lobes as mainly specialized for abstract thought. The frontal lobes of the brain integrated into thought but specialized in ordering the executive/ambitious drives of humans, the so-called "p" or postponement/ planning factor that shapes the uses of the temporal and parietal's fluid reasoning elements.

Of course, there are the regulation of motor controls, sensation, homeostasis (internal), language, the amygdale; this latter transforms the mammalian emotional structural heritage in powerfully thrusting up and integrating with the other brain functions, importantly the cerebral cognitive abilities of humans.

So we speak about humans and their diverse abilities, emotions, personalities, as deriving from this great integrative assortment of structures and functions of a brain that creates individuality as well as intellectuality in an individual. One mysterious hint at the independent functions of our brain that create the institutions of culture and civilization is the "savant."

The so-called "idiot savant" is a condition in which someone with significant mental disabilities demonstrates certain abilities far in excess of average. The skills at which savants excel are generally related to memory. This may include rapid calculation, artistic ability, map making, or musical ability. Usually just one special skill is present. Those with the condition generally have a neuro-developmental disorder such as autism spectrum disorder or have a brain injury.

Savant syndrome

Other names	Autistic savant, idiot savant

Figure 410

Kim Peek, the savant who was the inspiration for the main character in the movie *Rain Man*

Specialty	Neurology, psychiatry
Symptoms	General mental disability with certain abilities far in excess of average
Types	Congenital, acquired
Causes	Neurodevelopmental disorder such as autism spectrum disorder, brain injury
Frequency	c. 1 in a million people

The hint here is that these preternatural abilities are separable from the cognitive, “g” factor. It tells us that the integrative qualities of the brain can, in humans, be separated out or joined with “g” in human functioning, giving us a mathematical genius such as Richard Feynman or an autistic savant of extremely low “g” such as Kim Peek.

Figure 411: 1918-1988

Functional Areas of the Human Brain

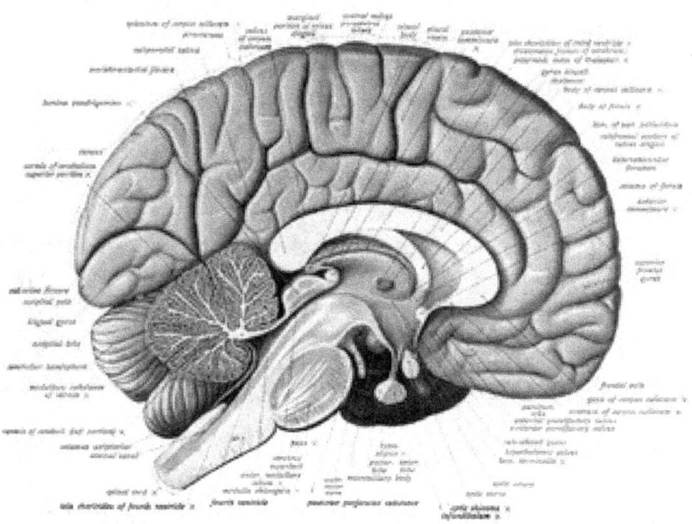

Figure 412: Human brain bisected in the sagittal plane, showing the white matter of the corpus callosum

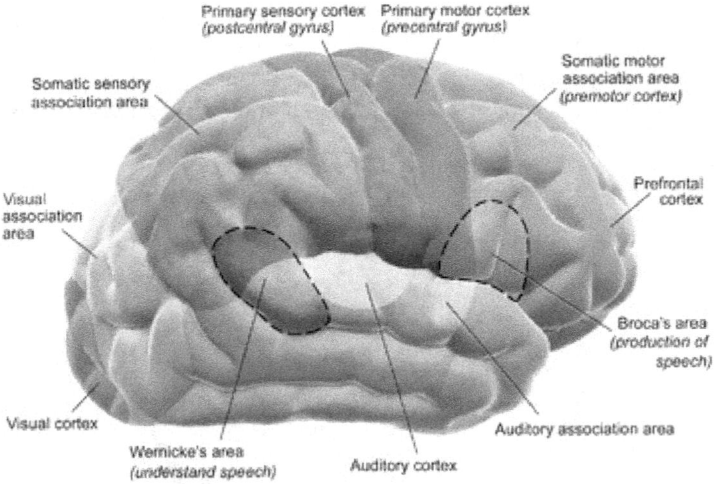

Figure 413: Human Brain

American education ethnic profiles

Throughout the advanced scientific societies of our planet there is a recognition that the talent for mathematics and physical sciences achievements can be early identified in young children, with special schools being established to nurture this talent element of "g."

In NYC, this tradition was established alongside schools for the arts where early talents also need to be recognized and educationally nourished. In the area of mathematics and science, a number of high schools for 14–18 year olds was established, Stuyvesant High School being the most selective.

2010 U.S Census

Self-identified race	Percent of population
White	72.4%
Hispanic and Latino Americans (of any race)	16.3%
Black or African American	12.6%
Asian	4.8%
Native Americans and Alaska Natives	0.9%
Native Hawaiians and Other Pacific Islanders	0.2%
Two or more races	2.9%
Other	6.2%

The first United States Census in 1790 classed residents as "free white" people (divided by age and sex), "all other free persons" (reported by sex and color), and "slaves." The 2000 Census officially recognized six racial categories.

Preliminary 2019 NYC demographic/racial profile

White: 42.78%
Black or African American: 24.32%
Other race: 15.12%
Asian: 14.00%
Two or more races: 3.33%
Native American: 0.40%
Native Hawaiian or Pacific Islander: 0.05%
(World Population Review)

Figure 414: Stuyvesant H.S. students, NYC (*The Atlantic* 2019)

A push to diversify New York City's most elite public high schools continues to face backlash from groups that argue the plan would discriminate against Asian American youth.

Of the 895 current eighth graders who secured a spot in next year's 2019 Stuyvesant freshman class, just seven identify as African American. (These seven could be of mixed race such as is Barack Obama, Kamala Harris [Senator from California], Cory Booker [Senator from NJ].)

Roughly three in four students currently enrolled at Stuyvesant High School identify as Asian American, for example, dramatically eclipsing all other racial groups; their African American and Latino peers account for an almost non-existent slice of the school's population. (Lennihan)

In 2017, 22 percent of Harvard's admitted class identified as Asian American, compared with 5 percent of the country at large. In New York City, residents who are identified with this racial group make up a larger but still modest 15 percent of residents, see above statistics

The population referred to sweepingly as "Asian American" comprises more than 30 nationalities and ethnic groups: from Chinese and Cambodian to Burmese and Bangladeshi. Close to three-fourths of Indian Americans aged 25 or older had a bachelor's degree in 2015, according to Pew data. An analysis of 2010 census data by an NYU-based urban-policy think tank found that the city's majority-Asian neighborhoods are more economically depressed than those of most other racial distributions: their average household income of less than $52,000, for example, was less than that of majority-black neighborhoods.

According to Statistical Atlas, a website that analyzes data from the U.S. Census Bureau, California's ethnic population break down as of July 2017 was 39.7 percent white, 37.4 percent Hispanic, 13.3 percent Asian, 6 percent African American, 2.6 percent mixed and 1.0 percent other.

Cal. Polytech University, San Luis Obispo, a Public University of modest academic/science selectivity. Ethnicity of students:

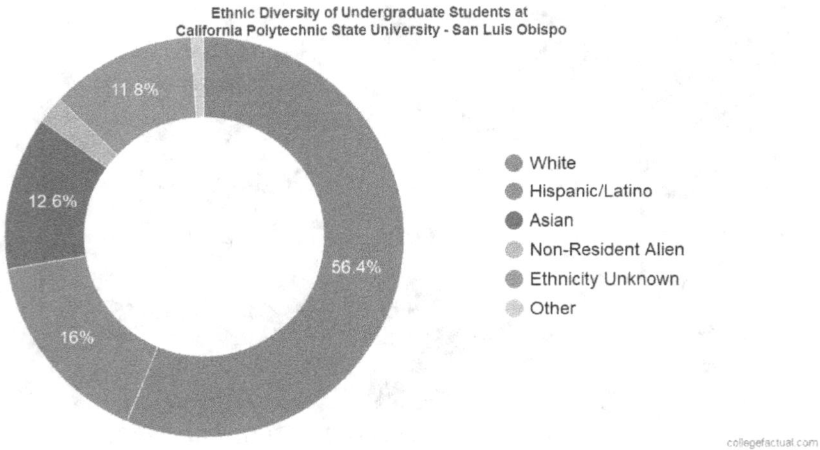

Figure 415

Looking at the California State University system as a whole, Mexican American students make up the largest ethnic group at 31.8 percent, followed by white students (23.5 percent) and Asian students (11.6 percent). Cal Poly's proportion of white students is more than double the CSU average.

California Institute of Technology
This institution is a small, highly selective non-state school specializing in mathematics and the hard sciences
Demographics of Caltech student body (2018–19)

	Undergraduate	Graduate
White American	27%	34%
Asian American	40%	10%
Underrepresented minority	18%	5%
International	9%	47%
Two or more races / Unknown	5%	4%

On the University of California side (UCLA, Berkeley, etc.), Asian students comprise the largest ethnic group at 33.7 percent, followed by Latino students (24.2 percent) and white students (22.5). African Americans make up 0.7 percent.

None of the 10 University of California schools has a white population larger than 34 percent, including universities comparable to Cal Poly such as UC Davis (24.7) and UC Santa Barbara (31.8).

The racial demographics of the school, one of the most intellectually selective in the United States, substantially differ from those of the nation as a whole.

Today, black students make up only 0.9 percent of the student body at CalTech. In 1995, there were no black students in the freshman class. For the current freshman class, there were 30 black applicants to Cal-Tech, 9 of whom were accepted. There are two black faculty staff out of a faculty of 285. Cal Tech has a private endowment with a total value of $3.2 billion.

Ethnicity and the verbal dimension

Figure 416: Spelling bee winners

This year, 2019, 562 spellers competed in the Scripps National Spelling Bee, and after a grueling 20 rounds, the eight left at the end were named co-champions.

Just a cursory scan across their faces is an insight into a steely culture of excellence that's been dominating American spelling contests.

"Seven of the eight winners this year are of Indian descent. And no, they're not born great—they're made great through a culture that instills a drive for achievement and mastery of the English language." (CNN)

And of course, these dark skinned achievers do carry within their genoplasm the memory of Cro-Magnon, *Hss* (Dravidian, c. 5000 BCE; Indo-Aryan, c. 2000 BCE). These once light-skinned invaders from the north and west genetically intermixed with the existing indigenous Paleolithic Australid dark-skinned peoples to create a modern hybridized ethnicity, both Hindu and Islamic in religious orientation.

Conclusion

The lesson learned from these vignettes is simple and clear. The phenotype of any particular ethnicity/race on this planet, c. 2020, is no sure guide to knowing their quantity and depth of the Cro-Magnon, high "g" intelligence potential. We are in a new phase of human evolution wherein the importance, if not the mystery of the nature of *Hss* geno/phenotype demographic penetration and power on our planet, is yet to be learned.

Chapter 22

Does *Hss* have a Future?

Bad News

Genesis 41: "Then Joseph said to Pharaoh: 'The dreams of Pharaoh are one and the same. God has revealed to Pharaoh what he is about to do. [26] The seven good cows are seven years, and the seven good heads of grain are seven years; it is one and the same dream. [27] The seven lean, ugly cows that came up afterward are seven years, and so are the seven worthless heads of grain scorched by the east wind: They are seven years of famine.[28] It is just as I said to Pharaoh: God has shown Pharaoh what he is about to do. [29] Seven years of great abundance are coming throughout the land of Egypt, [30] but seven years of famine will follow them. Then all the abundance in Egypt will be forgotten, and the famine will ravage the land. [31] The abundance in the land will not be remembered, because the famine that follows it will be so severe. [32] The reason the dream was given to Pharaoh in two forms is that the matter has been firmly decided by God, and God will do it soon.'"

Wisdom: nothing human lasts forever

"Our reality is a moving stream."

Heraclitus

The wider human species has enjoyed over 200 years of economic, scientific, technological growth, much new wisdom, but alas, as in the 20th-century, unbelievable human carnage.

So how does this biblical novella speak to us? For the cows and grains that Pharaoh dreams about, we dream of endless cheap fossil energy and scientific medicine. In reality, these will cure the ills of today and tomorrow. But wait, there is the other dream which ends in famine. We humans of the total *Homo sapiens* clade, today should think peak oil and plague. Many of us hold a trembling looking glass that sees a less than halcyonic future for all of humankind. The overstuffed elites turn their blind eye. After all, which social class shapes our sense of reality?

Here are some of our pet words: "economic growth, low interest rates, under populated nations, higher minimum wages, middle class communities." Dream a bit more: bankruptcies, unemployment, depression, homelessness, shrinking GDP, addiction, etc. Why? An unannounced environmental cataclysm has begun to cloud over our optimism. It is not global warming or climate change, and not even terrorism, even though we may yet enjoy a plate full of this scourge.

First the *Pandemic*, then *Peak Oil* when it arrives in our consciousness, sometime in the next half dozen years, 2020–2026, will add to the shaking of the current *weltanschauung* gifted to humankind by the power people. Then, as reality begins to establish itself in the minds of those masses affected by this new dream world, there will be the inevitable international attempt to come to terms with the consequences of *pandemic fear* and the consequences of fossil fuel demand that soars beyond fossil fuel production (especially that of crude oil). As we have delineated in an earlier chapter, the consequences for the international economy, both first world and third world, will penetrate deeply into every institutional dimension of our lives, from grocery shopping to schooling.

At the moment, the world population is closing in on eight billion. Have no fear, it will begin to shrink. Climate change rhetoric will be barely a whisper in the mass media. The price of practically everything we purchase will begin to hit the roof. Most fearfully to this writer, is that the human detritus packed into urban sink holes throughout the world will be terribly vulnerable. The scientific structure devoted to the discovery of new antibiotic reactions against the ever mutating viral/bacterial enemies of humanity will be stretched thin to deal with this blackest of clouds over our future.

Draught, floods, and earthquakes—minor physical disruptions of our lives in comparison with the sudden spread of one or more new infectious and contagious diseases? All this, a product of an animal species which has enjoyed unprecedented evolutionary success in producing a cancerous demographic bulge, the ecological destruction of animal and plant life. Our gift to the evolution of life on earth, the rampant desiccation of nature's natural resources, all within this 200-year-old banquet. The understandably irresponsible elites and the rest of us sapiens, in our ignorance, have been pushed to the edge of the cliff.

Just think about the socio-psychic impact on those populations, once happily engorged in the prosperous *Hss* nations of our world. Is it too late? Is there any hope? Don't think that our season of riots, vandalism, and hatred-spewing is over, or that fascism will not raise its ugly head.

International struggle for stasis

Within several years of awareness of this long-term challenge, the big guys of the international community, plus the smaller nations still with significant production and reserves, will get together to try to oversee the production and distribution of oil and gas and their derivative liquids, such as gasoline, propane, etc. Take Venezuela as an example. This nation has enormous amounts of thick, muddy oil reserves which can be liquefied with the heavy investment of capital. Other currently unstable, but potentially large producers like Libya and Iraq will also have to go under the international thumb. There will be an international fund for this purpose, and also, most importantly, some kind of "divvy up our survival" fund, with the idea of avoiding massive military expeditions and war to grab what one or two nations are able to absorb, considering their economic and military profiles. One suspects that serious consideration will be given to the question of "triage."

The natural mineral and agricultural resources of the world, given the increasing scarcity of these resources—front and center fossil fuels—cannot reasonably support a modern lifestyle with eight plus billion humans on this planet. Certainly, uranium for nuclear energy will balance out with natural gas and coal for electricity. But it will be expensive and temporary. The energy potential of solar and wind are so weak for the needs of the vast urban centers around the world, and their energy cost is also soaring; they will fade into the ecology history books, along with batteries.

China and the U.S are very powerful. North America, especially Canada, has large oil reserves, c. 2025. They do not have the reserves to support the population of the U.S., Mexico and beyond. Canada will have to give. How far can this go, to also include Europe? The United States, China, the EU and the Americas will all want to share Venezuela's resources. South America will get some of this, here excluding Colombia, Brazil, Ecuador, who have some oil, but hardly enough to maintain in power the current European-origin oligarchic ruling class.

China, with 1.4 billion humans and importing ever more oil, will be a big problem. So too India, with its own 1.4 billion. One guess is that China, the Koreas, Japan, and probably Vietnam will sign up with Russia and its developable resources to share technology and military acumen. The scientifically undeveloped oil nations will have to transform themselves into modestly have not wealth nations. Why? The Russians still have economic and military power, reasonable demographics, and are centrally located to also exert this power south into the Middle East as well as into all of Asia. Expect this alliance to include the former Asian Soviet "republics," with their as yet modest demographics and relatively heavy natural resource potential.

The Chinese and Japanese have always looked at the vastness and potential of Siberia, still so far from Moscow. It is empty of humans and there is so much of need and value under its frozen soil. Note the only recent historical land invasion of Russia from the East, is by Japan in 1939. A Soviet army under a younger General Zhukov almost annihilated a large Japanese army from Manchuria, invading the then Soviet Mongolian territories. Better for Russia now to join the East than have to contest the West, including the U.S.A. Alright, things are different today, and the East has also learned many new things.

This brings up the question of South Asia. The bet here is that India, Pakistan and Bangladesh will forget their religious antagonisms when it comes to a basic standard of living survival, at the least for some percentage of their populations. India in 2020 is moving rapidly—technologically and militarily. Its agriculture depends on cheap oil. All three nations, heavily overpopulated, will in hunger probably look west toward Iran and possibly Iraq and the Gulf States. At least that will be their negotiating card in the international attempt to avoid catastrophic wars, and to prevent a cascade of national starvations and disease.

Saudi Arabia and the Emirates would remain a luscious gravitational attraction. By 2025, the Saudis will not even have the memory of that sea of oil that was Ghawar. But oil states will still have "more than they need" even for a Western standard of living's basic needs, given that they can turn their multitude of royalty back to work, and return the emirate Disney Lands back to functional survivability.

Russia is closely north and could only land this southern prize at the price of war. The Europeans, now the beneficiaries of the vast Russian natural gas reservoir pouring west, might wake up to see it now pouring east. Even a scaled-down European military would be in the Middle East in an eye blink with the assistance of American long range missiles, probably with the addition of the Hungarian, Polish, Ukrainian, Rumanian military to get this Middle Eastern "black gold." In the end, it will probably be better to negotiate than risk the fatality of a war between great military powers.

There was in early 2020, a world wide production of close to 100 million barrels of crude oil a day. Demand stood at c. 105 million consumed every day. Prosperity demands fossil fuels, so production and demand with hope, will return in a few years.

Remember, every year after peak oil, 2020–2025, we will see less productivity of this vital resource. Guess how much higher the price of gasoline at the pump would be beyond the predicted $10 per gallon, sometime between 2022–2025, both in Europe and the USA?

Pathways to a new culture of modernity

Note the ineffectuality of both the League of Nations and the still in-place United Nations, the latter inept in stopping the continuing terror, the ideological barriers between nations and sadly the genocides of the 21st century. Now, the *Hss* Christians are in the hands of the *medieval* barbarism of recidivistic Muslim gangs (Apologies here to the great Islamic Middle Ages).

The agreements to share natural resources by predicted international alliances of convenience will be exactly that—temporary convenience by the existing elites and ruling bureaucracies. In terms of longevity, it will probably not be in place as long as the Congress of Vienna, which did calm things down for a good piece of the 19th century.

Great demographic changes have since occurred, and also science, medicine, as well as a dream of universal classless prosperity—but also the memory of the horrors of the 20th century and the increasing terrorist chaos of the 21st. There is plenty of contemporary evidence today for both international and internal national hints of social dissolution—both demographic and class-oriented. This is the result of the combination of great wealth for the few and the experience of the masses of economic stagnancy. How long will the media propaganda for both the pseudo symbols of socialist and free market models of human progress succeed?

The pandemic and the oncoming fossil fuel crisis will clearly support the aphorism, "*when the tide recedes we will quickly know who was bathing in the nude.*" The clearest comparison and deviation from the sociological condition of world society 1914 as compared to c. 2025 lies in the nature of the earlier disintegration, the calamity of world wars, genocides, and the *Holocaust.* History will likely define the 20th century in terms of these horrific events more so than the real advances in technology and medicine. How will we define the 21st century from the perspective of 2025–30?

It is important to point out that this twentieth century horror was a war of *Hss* v *Hss*, guided by ideological frameworks, the end of the monarchical establishment and church power. Also, the tremendous instability propelled by advancing science and technology, the power to kill, and the growing mass, percentage-wise of heretofore scattered *Hs* populations in Asia/Africa/ South and Central America. The growth and instigation of communist and fascist ideologies promoted the philosophical rationale for a resurgent nationalism and helped to promote the wars and the genocidal enthusiasms.

The struggle amongst the *Hss* populations of Europe and Asia was characterized by the vast wealth and power unleashed by the scientific and medical discoveries of the later 19th and 20th centuries. It is completely understandable, as the human world transformed itself into an industrialized urbanized international culture, experiencing a flood of technological discoveries, all of which literally transformed the lifestyle of the expanding billions on the face of the planet.

As Marx predicted, the change from rural agrarianism (modes of production) and village cultural endogamy to this new reality of a connected world with highly altered behavioral norms, could do nothing but upset this civilizational apple cart. The new modes of economic production disrupted all previous truces, the agreements for peace in place.

Indications are, in late 2020, that a new set of international behavioral norms will recive its sudden dynamic from the proverbial shift in our "modes of production."

The question of race

Species: Technically, a species is a population or groups of populations that can potentially interbreed freely within and among themselves. This is a naturally-defined concept, something which exists by itself. Subspecies, on the other hand, are subgroups within a species that have different traits and are defined by scientists. (Andrei 2019)

Race is a grouping of life forms, here humans, based on shared physical or social qualities into categories generally viewed as distinct by society. The term was first used to refer to speakers of a common language and then to denote national affiliations.

The placing of the different historical races of humans as *subspecies* is controversial.

Figure 417: Lookalikes! A new study has shown that giraffes might actually be four different subspecies. Photo by Luca Galuzzi

The distinction here proposed between Hs humans and Hss humans is a psycho-social one, given that there are very few distinctive racial groups in the 21st century, the cause of which is long and ongoing hybridizations of Homo. The definition of all existing human groups, pseudo races and ethnicities, as members of the species Homo sapiens is secure. However, the ongoing socio-cultural separation of humans by their intelligence, vocational abilities, independence of economic competence, will define humanity over the next several hundred years. Thus the proposed subspecies appellation of *Hs* and *Hss* humans.

Conflict

The only recent *Hss* v *Hs* war/genocide was between the Tutsi rulers of Rwanda and Burundi and the indigenous Hutu. The Tutsi were descended from northern, cattle-raising hybrid (Negroid/Caucasoid ethnicities) Watusi, who had migrated south in earlier generations. This Hutu majority ethnicity of this area staged a revolt against the Belgian endowed leadership (Tutsi), 1994. Over the generations there had been much Interbreeding between the two groups. But the Tutsi's still held on to their cultural traditions and probably practiced selective eugenics. A great genocide of the Tutsi civilian population ensued. The Tutsi military, part of which escaped to Uganda, reconstituted itself after the genocide and returned to re-conquer both nations.

Beyond this terrible African event, *Hs* and *Hss* populations either lived apart geographically, else were absorbed into the general demography of the larger nations, 20-21st centuries.

Identification

Our genus, taxonomically *Homo*, recognizable several million years back, example *Homo habilis*, Africa, c. 2 mya; also *Homo neanderthalis* (interbred with Cro-Magnon, but now not thought to be an *Hs*), Eur-Asia, c. 40,000 BP. Today *Homo sapiens sapiens*, originated about c. 400–200,000 BP, genus *Homo* with spoken language and a power mentality. At the same period, from c. 400,000 BP, the genetics for language was gradually spread over the continents amongst the various sub-species (racial) groups of *Homo*. This must have marked, along with the origination of language probably in Eur-Asia, the genetic and then cultural transformation of these many peripheral lines of *Homo*. We are thus on solid ground in enabling the taxonomic label of *Homo sapiens* for all historical humans. Still, there was the taxonomic case for *Homo sapiens sapiens*, the revolutionary originating *heterochronic/paedomorphic* human form, from, c. 400,000 BP.

Small groups wandered widely, probably engaging within itself of a natural selective/eugenic shaping of its cognitive character: symbolic skills of prediction, communication, art, written signs/symbols (such as with the Cro-Magnons). Our modern world today consists of one generic form, *Homo sapiens*, which has most probably gained its speech specialization from the genetics of *Hss* (this the freakish product of heterochrony/neoteny), *Hss*, has continued to pass on its highly selective genetics for several hundred thousands of years to a variety of *Homo* types in Africa, Asia, Europe, the Americas.

The numerically larger ethnicities/sub-species of *Hs*, which grace our planet today surely have samples of *Hss* genetics in their corporate genoplasm, even in the New Guinea highlands, jungles of Philippines, India, islands throughout the world. In the long run of evolutionary time, these genetic trickles of *Hss* will assert themselves gradually, and build up in all ethnicities. This is because the selective trend in the *hominins*, then *Homo*, has been to empower cognitive thought with the ability to destroy the lesser ranks of brain/cerebral cortex size and complexity, *hominins*.

The scientific term subspecies/*race* is applicable in taxonomic analysis throughout all forms of life. As applied to *Homo sapiens*, the term and concept *race* should be considered history. At one time in the history of our species it was correctly and historically applied to humans (Carleton Coon). Thus, at one time in the evolutionary history of humans there was enough identifiable similarities in phenotypic structure within geographical populations such that we could recognize this sub-species' taxonomic appellation: the five identified geographical races of *Homo sapiens*: Negroids, Capoids (ancient Bushman), Australids, Mongoloids, Caucasoids.

An *Hss* world to come?

Recent Hybridizations: The beginning of the 20th century mass killings of *Hss* by their peers began in WWI, 1914. Simultaneously, a revolution was taking place in the old Ottoman Empire. A revolutionary group had taken over in 1908.

As WWI began, so too, in 1915, did the genocide of their Christian Armenian ethnic populations begin. These Armenians constituted an advanced socio-economic group of tradesmen, civil servants, and well-off farmers, but also they inhabited large enclaves of poor-but-able peasant communities. The minority Kurdish Muslim community of Turkey, now in internal opposition to the Turkish establishment, has in retrospect admitted complicity in the total killing of as many as 1.5 million Christian Armenians.

Today, the Kurds of Iraq, they, Sunni Islam descendants of the Indo-European Medes of the ancient world, have established a liberal secular enclave in Iraq. They protect the various remnants of the ancient Christian communities in Iraq, even while the surviving Armenians have their own secular Christian national identity. This is not to say that the Turkish people will always be ruled by a pseudo democracy of extreme nationalist policies. Without Turkish militarism and with recent Iraqi Kurdish plenitude, the political conditions into the future are delicate.

As with the Kurds and Turks of the Asiatic West, and extending all over the world, issues of race have less and less scientific solidity, whereas ethnicity looms large. As the levels of fossil fuel production decrease, the reality of the possible energy substitution of renewals will be seen as a chimerical dream. The universal press will be concentrated on the economics of survival. To maintain our scientific, medical, and technological civilization will become the underlying subject matter of the contemporary *Hss* mind. Question: who are these *Hss* in this large and inchoate sub-species? In any case, we can no longer use race.

In the U.S. for example, while the majority of citizens who identify themselves as Afro-Americans are poor, many are excluded because of poor education or discrimination from the job and enterprise markets. But there are many millions of completely competitive Afro-Americans who are independent and successful sociologically and economically when compared to their "White" or "Asiatic" peers.

This situation of hybridization and cultural amalgamation is world-wide. However, we should be aware that already in 2020, the gap in educational and intellectual capacities amongst the indigenous national populations across the globe is growing. Science and technology have now created jobs for computers and robots. They may make many human hands irrelevant.

Intellectual test

Many of those who today are losing out to automation may never find work, especially as we enter this new age of fossil fuel shortages, pandemic threats, great inflation and the consequent international depression which will afflict all of the near eight billion humans of this world. Many of these billion have barely had the generational time to acclimate to this new scientific culture. Humans do not easily and voluntarily make themselves obsolete. But, in the transition from plenitude to rationing, they may, by their desperation precipitate a not so quiet internal national and international war for economic survival, no less middle class status.

If we are not to again descend into international wars for economic and social survival, else within the nations precipitate anarchy to decide who gets what, then we will have to establish rational rules for the transition. Here, from a "racially" and ethnically divided world competing for power and survival the slow/fast path into a panmictic species of similar bio-genetic intellectual capacities becomes paramount. The evolutionary goal will be the coming into being of a rational, secular philosophical plan for *Hss* humanity.

What we should plan for is a condition of humanity in unity. Heck, without pressure for a *tabula rasa* of appearance homogeneity, a new *laissez faire* of reason would allow for pre-existing phenotypic diversity of appearance, our "racial" residue will be of the first order. We don't have to look alike to be socio-politically equal.

We are also going to be forced to move our species, hopefully *Hss* in intelligence and capability, into a balance with nature's supply of resources. This means gradually deflating this massive population bulge. It has brought the world to the abyss.

Unifying the species: criteria

There is an intellectual and practical criterion for membership in the greater sub-species, *Hss*. This is that all individuals have intellectual and personal qualities that allows them to earn his/her own living consonant with being able to independently, as a family, raise and educate their children for the middle-class *Hss* culture in the making. Note that we are here supporting the existence of a variety of ethnic interpretations of this higher intellectual culture. We need to give communities of thinking and living alike families the opportunity to stay together to gradually form a variety of ethnic sub-divisions of our international civilization. See Alabama, Vermont, Texas, California, etc., with interesting places and people. The concept is "*cultural pluralism*" amidst universal laws for humanity.

How do we evaluate and grade these qualifications for the stimulation and perseverance of such a civilization? One might use I.Q. tests for job interviews and evaluations. In theory, the mathematical brilliance of the Stuyvesant High students early on in their math lives reveals this variability of ethnicities/races of high *Hss* intelligence. We do know that high math achievement is a cue for future high socio-economic positions in society. Researchers added the verbal element to I.Q. tests because verbal excellence touches on a great variety of talents and abilities.

Without the discovery and development of written language by the Mesopotamian Sumerians some five thousand years ago, there probably would not be a mathematical vocabulary that theoretically transcends and opens up a new world of intellectual/scientific reality. So we need both skills in varying combinations to do the work and creative breakthroughs of our evolving civilizations. *E Pluribus Unum.*

Scientist Volkmar Weiss, as we discussed his views in the preceding chapter, used an I.Q. score of 130 to denote special leadership talents at the base of the hierarchy of mental wizardry. Run of the mill MD's average out at about 115 I.Q. The great breakthrough creators of medicine certainly were and are persons of I.Q. ranges of 130–160. Beyond that line of demarcation, we find only a few extraordinary I.Q.s. The evidence seems to be that these phenomenal humans tail off in their proportional civilizational achievements. And we have to recognize that those "geniuses" often come with little "p" (postponement/drive) factors.

Are they *Hss* too? One guess is that if they knew they would lose the privileges of working alongside other *Hss* types, they would get back on their heels and at least land jobs sorting cars and scooters in a parking lot. The evidence is that higher intelligence humans are more efficient and reliable in a low-skilled job than individuals of lower intelligence. But the smart ones have to want to do the job.

There will be plenty of testing, besides I.Q. The hope, if we are not to engage in physical conflict, is that the adjudication process will become an easily ascertainable one as more work in the depressed economic state of the world becomes automated. The result of any rational population policy must be the uplifting of the intellectual and vocational level of populations from the *Hs* state of unskilled dependency, below Weiss's 94 I.Q. level. What any utopian vision requires is that all national populations can live within the law, and also work to be economically independent. This is the democratic vision of "the diminished state," social life ever within a forward-looking dynamic world.

This dream of a competitive playing field that is flat must include those of the higher levels of education, knowledge and power who will lobby hard for their children. This, of course is classical nepotism. The favored few may yet have to acquiesce to seeing their children descend to a lower socio-economic level. Unfortunately for those high on the economic totem pole, "Mother Nature" does tell us that we cannot avoid that universal law of *regression* towards the mean. Few parents of 130–140 I.Q. have children whose I.Q.s are equal to their parents. With luck it will only skip a *mendelian* generation, as the grandchildren might shine.

Fear of Ideology

"Ideology is a collection of normative beliefs and values that an individual or group holds. In other words, these rely on basic assumptions about reality that may or may not have any factual basis. The term is especially used to describe systems of ideas and ideals which form the basis of economic or political theories and resultant policies. In these there are tenuous causal links between policies and outcomes owing to the large numbers of variables available, so that many key assumptions have to be made."

Ideology is what destroyed the 20th century. These beliefs about the nature of modern society, and its sharp separation of social classes, was due to a misapprehension of the nature of the evolving industrial/technological breakthrough of *Hss* Europeans and North Americans. Add the complexities and the enormous value-added wealth that industrialism/technology produced for the leading creators and exploiters of this new knowledge and the natural resource utilizations that undergirded this accumulation of wealth.

So we had ferocious nationalisms, Muslims destroying Christian Armenians (Turkey 1915); communism/socialism, Soviet Union, c. 1919, spreading death by action and inspiration throughout Europe and Asia; Fascism, as expressed in Germany and most of Europe, 1930–45. The theme was the same, the divide between wealth and labor. Advertised by the controlled technological media to the masses, were the many examples of terribly unjust and unnatural socio-political exploitations. Supposedly, they could only be cured by the extermination of the identified exploiters. So here they were, Jews and other minorities of *Hss* education and intellectual skills, as well as the upper/middle educated classes of a myriad of societies, China and Cambodia being several 20th century exemplifications. Give us power the ideologues enunciated and we will right these "wrongs," by any means.

In the case of Japan, we see a long arc of 20th century nationalistic invasions against not only racial/ethnic nationalistic rivals such as the Russians, 1905, and British and Americans, 1940–45. Certainly, fellow racial/ethnic neighbors, the Chinese and Koreans, from 1910, were added to their aggressive mix. The Japanese, while in hatred against their "putrid odor" Caucasoid rivals as well as their fellow Asiatics, reflected an ancient militaristic expansionism. Thus, in the 20th century, they made war, armed by their great industrial/technological leap forward and egoistic pride in their ethnic power to conquer and expand. This ideological, militaristic optimism should have early on been shadowed in that decisive military defeat by the Soviets in 1939. Finally, there was peace after the ultimate nationalistic humiliation, by the USA in 1945.

The terrible *Hss* "civil war" of 1914–18 was similar to the Japanese nationalistic aggressions. However, in contrast, it concluded by pushing religio/monarchist ideological domination into history. Both democratic/socialist and ideological fascist/communist politico-military entities quickly filled the political blanks. The horrors of WW II and beyond should have taught historians that smart ideological *Hss* types can always destroy their own kind. This lesson could have been learned from the Americans, the Civil War 1861–65, else the Greeks, the Peloponnesian War 431–404 BCE.

The great ideological misconception was caused by the nature of a changing social structure. This change, undergirded by vast industrial-technological advances in killing, vast and cheap fossil fuel energy, scientific disease control medicine, had changed the nature of leadership trends in political power accretion. This historical dynamic made it possible for unscrupulous *Hss* ideologues to create, through use of the mass media, a mythos of vulnerable pseudo targets, *glorious revenge* by the supposed exploited masses.

Our ideological problem: today and tomorrow

Consider the short term challenges for our world 2020–2025: clean water; clean air; food production (fertilizer, pesticides, insecticides), cheap fossil energy, cheap electricity, slowing the demographic glut, medicinal resources world-wide; rational political leadership; informative media; informed citizens. Decisions will have to be made.

The closest threat to the *Hss* elite's present self-satisfied power contentment is fossil fuel energy, severe shortages, possible exception, coal, and for a short while uranium. Think $150–200 a barrel crude oil; think $10–15 mbtu, natural gas. Consequences: Inflation of all resource elements and food; bankruptcies in tourist and entertainment industries and many others; unemployment on a vast scale; governmental shrinkage of services, reduced tax revenues, unemployment for bloated national, state, local bureaucrats, private debts (rising interest rates. In effect a great international emergency. Hopefully, desperate dialogues amongst the nations of the world to share in that fossil energy remnant which at one time undergirded the great civilizational expansion, *plus* the wars and genocides.

What happens on the human scale of social class divisions? People, leaders start rationalizing and attempting to explain the great shrinkage of wealth and the destitution of vast populations. One asks how the taxonomic distinction between *Hs* and *Hss* computes, considering as we have noted above, that for any job in a free economy, from cleaning hotel rooms, to directing parking at a football stadium, the more intelligent always win out, *if they are willing and have the law.*

Libertarians intellectually push for the diminished state, small government, to enhance the free exercise of life needs by the citizenry. Even under the benign leadership of an FDR during the depression of the thirties, "big government" came to the rescue with a series of federal institutional financial interventions which over time have grown enormous budgets and are forever to be on the books, *Merci Beaucoup.*

The evolutionary question: When undergoing a great economic/social crisis, the screams of "I will save thee" will have a powerful magnetic and seductive allure. Will the populist powers that are sure to gain political control be able to resist the dysfunctional call for *equity*, not from "each according to his ability to each according to his need," but to each, equally. Throw down the billionaires, millionaires, rich MDs, research scientists, etc. They can still function with less. When people are starving or in great need for medicine, clothing, housing, will the populists be able to reward high achievement? Who will be able to raise and educate children independently, as governments hand out pittances, equally for all children.

Economic survival by government

Below are a few statistics both of the U.S and Europe where the possibilities for survival, economically, socially, scientifically are most probable in such an emergency, but at the same time already under stress.

What this all means is easy to understand by just looking around the western world, including the United States.

United States

Figure 418: [Getty Images]

The federal government estimated that in 2019, it would receive $3.645 trillion in revenue. Most of the taxes are paid by legal residents, either through income or payroll taxes. The government expected to spend $4.746 trillion. Almost 60% of the revenue pays for mandated benefits such as Social Security, Medicare, and Medicaid.

Interest on the U.S. debt will be $479 billion. The U.S. Treasury must pay it to avoid a U.S. debt default. Interest on the approximately early 2019 $23 trillion existing debt is already the fastest-growing federal expense. The remaining $1.4 trillion pays for everything else. It's called discretionary spending. The U.S. Congress changes this amount each year. It uses the President's budget as a starting point.

The discretionary budget will be $1.426 trillion. More than half goes toward military spending, including the Department of Veterans Affairs and other defense-related departments. The rest must pay for ***all*** other domestic programs. The largest are Health and Human Services, Education, and Housing and Urban Development. They can be considered transfer payments for the needy (*Hs*).

There is an emergency fund of $200.1 billion. Most of it goes to Overseas Contingency Operations to pay for wars. A growing portion is set aside for disaster relief for hurricanes, floods, and wildfires, etc. The 2019 budget deficit is estimated at $1.101 trillion. That's the difference between $3.645 trillion in revenue and $4.746 trillion in spending.

Federal Debt = c. $22 trillion. As of December 31, 2018, debt held by the public was $16.1 trillion and intra-governmental holdings were $5.87 trillion, for a total of $21.97 trillion. Debt held by the public was approximately 77% of GDP in 2017, and the U.S. debt ranked 43rd highest out of 207 countries.

State and local government debt: Q1 2019: 3,052.579 (+ more) 3 trillion, 52 billion, 579 million

Updated: Jun 6, 2019

Apr 5, 2018: in Federal and State Welfare spending of the more than $1.1 trillion spent in the 2016 financial year, federal expenditures accounted for $829 billion (74 percent), and state expenditures accounted for $297 billion (26 percent). Most state spending ($213 billion) occurs in a single program: Medicaid. (Rector)

"We are spending more than enough money to fight poverty but not spending it in *ways* that actually reduce poverty." (Tanner 2012)

A policy analysis released by the Cato Institute in April 2012, shows that despite nearly $15 trillion in total welfare spending since President Lyndon Johnson began the "war on poverty" in 1964, the poverty rate in the United States has remained relatively constant. The cumulative cost of Great Society programs, $22 trillion by 2014 is now above our current federal debt. (Heritage Foundation)

In 1964, the national poverty rate was around 19 percent and falling, while the poverty rate in 2012 was around 15.1 percent and climbing. It is now at the lowest level in nearly a decade. 2019 = 10.5%, child poverty = c. 20%. (Census. Gov)

Note: The above is for a *normal* year. 2020 is another story, *predictive* in this model of the future

Europe

Figure 419: Public spending / GDP in Europe. Legend: maroon > 55%, red 50–55%, orange 45–50%, yellow 40–45%, green 35–40%, blue 30–35%

Approximate populations of non-European origin in Europe (about 20–30 millions, or 3–4% depending on the definition of non-European origin), out of a total population of approx. 831 million.

"Western Germany: 42 percent of children under six come from migration backgrounds" [//www.breitbart.com/europe/2018]]

"The number of migrant-background residents in Germany has reached a new all-time high at 25.5 percent of the population. The new statistics come from a report released by the German Federal Statistical Office this week which states there are now 20.8 million German residents who come from migrant backgrounds, defined by having at least one parent born overseas." (***Kronen Zeitung*** 2019)

"When I was in high school, there were around 50,000 people with a non-Western background in Denmark. Today, there are almost half a million. In one generation, our country has changed." (Rasmussen 2019)

Just how great are the problems was revealed in a recent survey, *Integrations Barometer 2018*, published by the municipality of Copenhagen. The Integration Barometer—which measures the degree of integration in the municipality among young people with a non-Western background—showed that almost one third of 18–29 year-olds (31%) believe that "religious and cultural laws must be adhered to, even though they may be contrary to [Danish] law." The issue, then, is whether these young people believe that Islamic sharia law should take precedence over Danish law. The statistic represents an increase from 2016, when a similar survey showed that 24% wanted sharia law. Additionally, the number of youths who view democracy in a positive light has fallen from 86% in 2016 to 79% in 2018. (Bergman 2019)

Welfare states differ quite dramatically in the size of the budgets devoted to social protection and redistribution, with net social spending (2011, after taxes, tax breaks and social benefits are taken into account) ranging from a low 14.2% of Gross Domestic Product (GDP) in Estonia to a high 31.3% of GDP in France (OECD 2013). Further comparisons of social/welfare spending percentage of GDP: Japan = 21.9; United Kingdom = 20.6; U.S. = 18.7; France = 31.2; Germany = 25.1. (Van Kersbergen)

Ethnicity: a deeper look

Humans think, communicate and philosophize, using symbols, and meanings. The brain feeds on these made up meanings. Meanings elicit understanding, then action to fulfill the predictions that the brain/symbols create about the dynamics of things. Symbolic thought craves dynamic stimulation. Meanings can involve religious interpretations of the universe, a dramatic caste of gods in action. Our symbolic needs can also focus on the various secular activities in the material universe around us. But always, there is language to interpret the gods or the cost of lumber. Here, we start to recognize our ethnic nature. Our language, and even our gods can be different than neighbors down river, all while the price of a bushel of wheat is the same.

Of course it is the cultural and thus the ethnic differences between social groups, peoples, that turn blazingly hot. One can negotiate the price of oil but not the value of one's language or religious heritage. But not always, and that is the historical puzzle. We live in a world of self-created cultural symbols. Symbols can fade away in meaning as our religious sensibilities change, and our language of course is in a perpetual dynamic. Thus, humans can give up their gothic languages to become a Roman, and speak pigeon French, or Spanish or whatever. You can fight for X, and then shove X away for a more interesting Y.

Historian Toynbee saw the rise and fall of civilizations as having a religious dynamic of outlier innovation. This is true to the extent that the specific religion mobilized the dynamic cultural efforts and elements of the particular ethnicity and strove to enlarge its meanings and physical boundaries of belief. The key which he may have overlooked lay in the embedded symbolic powers of the civilization, its deepest bondings. Did the symbols of ethnic identity energize, have continued creative momentum, let the smartest re-creators have their way in shifting meaning and thus changing the symbolic hold of the old religion/culture?

The great civilizations had this quality of bringing diverse people into its symbolic world. It could be the Roman military or/and its law. It could be a deeper super material view of one's identity and future which Christianity then presented to a declining and deadening symbolic universe that once was western Rome. The secular West, eventually created out of the concomitant Christian contradictions between the secular and the theological, experienced a powerful, ethnically pluralistic surge into modern science and enterprise. This new civilization exploded in its dynamic realization of the many symbolic cultural powers of the human mind. From the 15th to the 19th century, a great revelation of the potentialities of human creativity took place.

Today, we look forward in apprehension at the fate of the products of this surge. Our encompassing materialistic demographic expansion is only one of the signs that are blatantly telling us that the forward thrust is slowing. The meanings of this civilization as it was shaped from the later medieval to the beginning of the 20th century are yet not clearly appreciated in retrospect. Can this now, 2020, universal civilization survive the crash of poverty?

The world wars, the genocides, and the *Holocaust*, instigated by a powerful, modern and largely *Hss* nation, Germany, and its many European followers surely has to tell us that even modern, sophisticated, seemingly secular elements of our civilization can contain self-annihilating elements. The dynamic unifications that this great new civilization's ideas and practices introduced philosophically dissolved the old monarchical and theistic glue. Can it now self-dissolve in chaos on its own merits?

Note how the ideological unity of Soviet-style communism evaporated in 1989–90, luckily without too many bullets flying. The press of armament rivalries with the still-functioning West impoverished this declining, universal, tyrannical, symbolic vision. In its stead arose a plurality of ethnic states. Likewise, the vision of a Europe united, EU, in its economic and social vision, now without war or genocide, is itself beginning to fall apart before our own eyes. The supposed international economic dynamics have begun to fade. Ethnic symbols seem to be giving more life and meaning to these respective populations. And of course, as with the League of Nations, the United Nations is still hostage to the diverse ethnic and ideological nationalisms which bind individual nations internally as potentially stable cultural/political units. And, to hell with the outside world.

Hss survival?

This is the short version of a topic which should become an important academic enterprise for economic and political thinkers.

Some nations such as with the former Soviet Union, once a conglomerate of ethnicities, will not survive the ideological siren calls of the "identity," ethnic demagogues demanding their front and center place in the symbolic food lines of a number of current political entities. Those nations will surely continue to breakup into their respective ever creative ethnicities. There will be many foci for division, political, economic, ethnic/religious. In a time of socio-symbolic contraction national ethnic political unity becomes a virtue. The lure of the universal symbol no longer excites meaning.

Russia, now largely Slavic and with modest demographic pressure given its geography and great natural resource potentiality, will stay whole. It needs to build up its intellectual capital, which was horribly destroyed during too many 20th century wars.

China will need to modify or dissolve its ideological tyranny. Then, having an unusual unity in diversity culture, several hundred spoken languages in addition to official spoken and written Mandarin, the latter intelligible to all educated Chinese, it will still break up. It is hard to believe that a China in economic and military retrenchment, in the 21st century, can stay politically unified.

India, with a large Muslim minority, could still function. This Islamic minority is comprised of largely capable, already secularly functioning *Hss*. There is linguistic diversity here and thus a political/ethnic separation, north/south might occur, given the already existing poverty nationwide. However, India as a whole, along with Bengali-speaking, secular Muslim Bangladesh, might survive. It is serious about its transformation into a secular, scientific democracy.

Europe, let us hope, has completed its millennial preoccupation of killing its high *Hss* cousins. Its well intended experiment with the EU will probably educate them in the importance for each nation of their ethnic symbolic identities. And they each will have to suffer an independent national fate, with Norway and possibly Russia fluttering to the top of the economic totem pole.

For the United States, the future is questionable. The massive invitation to the world for legal or illegal immigration to further inflate the welfare state, following on the yet insoluble consequences of the enslavement of Africans, is reflective of the ideological breakdown of its democratic governance. The USA was once a great and pristine ecological/geographical, political experiment. But as equality and economic growth falter, the variance in intelligence, education, the *Hs-Hss* divide, will surely as it does now, announce the entrance of a renewed army of ethnic demagogues.

The key question: can each political entity reduce its material needs equitably within its borders, and at the same time allow for the Marxian admonition: “from each according to his abilities to each according to his needs,” to take hold?

For *Hss* humans, survival requires that their nation must make room for the most valuable members of society to function and raise intelligent educable children. Even under very debilitating social and economic conditions, communities of *Hss* education and mentalities could come together to preserve scientific research, technological advance, medical protection and the necessary ethnic and cortical emoluments of the practical and creative arts. Together, with an association of philosophically like-minded *Hss* nationalities, a new civilizational dynamic could be born. Down does not necessarily mean “out.”

Chapter 23

Summation: Human Nature/Evolution's Message

Human Nature

Issues:

1. Expansionist limits of a genus/species line; limits are both ecological, geographical. geological, and biological—food sources, reproductive competition, over adaptive specialization. Considering the hundreds of millions of years of multi-cellular evolution, hundreds of millions of species have undergone expansion in numbers and in geographical dominance, then and inevitably, disappearance/extinction. There were causes: over specialization, then changes in the external environment for which they are not adaptively prepared; else competition from other biological forms which are more efficient in gaining resources and reproductive dominance.
2. **A revolution amongst the primates/anthropoids, hominids/ hominins, c. 400–200,000; a new reconstruction of the hominins, focused on the millions of years of adaptive and selective potency of the mammalian/primate/hominid brain. This brain potency characterized by greater ability to organize sensory experience for long term adaptive, reproductive, selective behavior; these skills based on anticipating events, the continuous reception and adaptive organization of sensory information.**

Figure 420: Hominin Dinner Party

Hss and *Hs* Hominins today—visually indistinguishable

Recall that the taxonomic status of *Homo sapiens* requires the human ability to speak with objectivity: "Look, clouds are coming, it will rain." While all historic humans have this ability, it is doubtful that the Caucasoid and large, primitive-brained Neanderthals were able to do so. Thus the taxonomic label *Homo neanderthalis* seems appropriate. It is probable that the absorption on of Cro-Magnon genetics (origins some 400–200,000 BP) in the vicinity of 100,000 BP, the ancestors of the Bushman, the large skulled Boskopoids and their very ancient click languages may have crossed the *Hss* taxonomic line. The closure evidence, their vivid art expression, argues that they were already, at the least *Hs*.

The test of *Hss* today must remain a purely practical and social element. Can an individual and his/her ethnic community perform, survive socio-economically, as independent actors at any subsequent stage of cultural advancement? The modern *Hs*/*Hss* brain is characterized by morphological changes described as neoteny, paedomorphosis. Neoteny, also called juvenilization, is revealed in the delaying or slowing of the physiological development of an organism, typically an animal. Neoteny is found in modern humans. Neoteny and paedomorphism, are similar terms, specifying heterochrony, developmental changes in the rhythms of ontogenetic (individual) development.

This evolutionary phenomenon has had wide spread recurrence. It seems to occur when a line of creatures is under great selective pressure, thus allowing bio-adaptive (heterochronic) outliers to become positively selected and thus thrive on the basis of those often unrevealed sets of mutations

As part of this heterochronic reconstruction, maintaining infantile characteristics late into sexual maturity, the lower, emotional brain, and the limbic system was pulled along in genetic linkage with the growth of this increasingly adaptive cortex. The primary structures within the limbic system include the amygdala, hippocampus, thalamus, hypothalamus, basal ganglia, and cingulate gyrus. The amygdala is the emotion center of the brain, while the hippocampus plays an essential role in the formation of new memories about past experiences.

It is the limbic system which has shaped the enormous sexual drives of humans, and forms a partial explanation of the aggressive and war-like proclivities of cultures, but always under possible neo-cortex inhibition, celibacy, and voluntary poverty (monks-philosophical pacifism).

The two major functions of the new human brain can be summarized as the (a) cortical—thinking in universal causal terms; (b) the emotional plural dimensions of cultural thought, the arts, language, ethnicity. While all cultures have a dimension of these two elements in human behavior, the difference between primitive, stagnant cultural traditions and the dynamic complex abstract dimensions of the great civilizations, as noted in earlier chapters, reveals the embellishment of the emotional valences, the arts, language, religion, ethnicity with dynamic elements contributed by cortical thought. The result is what we call high intelligence, the core of the taxonomic appellation *Hss*, and its response and adaptation to change.

Here, we produce as consequence the two critical cultural elements, the *universal* and the *plural*. Humans and their cultures have a variable capacity to cortically balance the interactions between these two trends in human thought. Any endogamous social group, even the size of a nation of three hundred million can root its arts expression in the amplifications of a rock band, a string quartet by a Beethoven, or an opera by Verdi. So too, it can reveal its technological "know how" in the creation of a wooden plow, else the atomic bomb.

The universal mode of thinking is sourced in humanity's search for the causes of things so as to predict "what's next." So we create universal principles linguistically (philosophy), else mathematically Newton/Einstein, that will mentally predict the principles of "everywhere" and "always." The effective source of these universal principles can be laid at the powers of "gods," else Aristotelian or Galilean principles.

The universal motives in human culture and cognitive civilizations are supposed to give us a predictive guide to the future. Most often in human history, these principles become laden with emotional resonances from our limbic system and turned into ideologies. Sadly, these ideologies utilize grist from socio-economic and military differences between individuals, ethnicities, religions and national power houses, which too often can lead to serious physical conflicts.

Our evolutionary path: two new sub species

The evolution of life could have halted with the Virus and bacteria, they, still around, after c. 4 billion years. But there was innovation in life's challenge to Sadi Carnot's "second law." Living forms fought against energy degradation, pulling in more "food" for energy production than spewing out waste into the ether. As a result, our planet has been able to build multi-cellular forms, here, creatures with backbones, and recently, animals with large brains.

A not rare but still freakish event in an evolutionary line of large brained *hominins,* as we have preached, occurred some 400–200,000 BP. This baby-faced, thin boned creature surfaced with an enormous skull protuberance in its process of maturation. Within this thin bone covering, a new adaptively dynamic mentation was born.

In terms of evolutionary time scales, we humans who write and think about these matters, have only been at this puzzlement for the above short evolutionary moment. Evolutionary history tells us that a recognizable "species" hangs around for about five million years on average. We human products are part of a new and extended species now consisting of two sub-species, *Hs* and *Hss*. The criteria of identification of these two forms of humans do not lie in their external phenotypic differences (old races and ethnicities). These existing differences are a heritage from an ever-distancing past.

The real reshaping of the species is taking place interior to that abundant skull structure. It is the capacity to deal intellectually with an external world, both socially created and physically challenged by "Mother Nature." The criteria by which we can judge these differences lies in the individual's capacity to deal intellectually through education and experience with this ever reshaping of human cultural existence by fellow high "g"- (general intelligence) factored humans, now taxonomically identified as *Hss*.

Can a person be educated to function independently in their respective societies whenever confronting more abstract demands on his/her educated intelligence? Considering the different levels of socio-economic conditions lived in by different ethnic groups and nations throughout the human world, it is also clear that the historical heritage of ever-more abstract cultural traditions, and the new scientific dynamics will make their demands clear to all human groups in the oncoming decades.

Nature informs us

Given the current view that the heterochronic revolution that produced *Hss* took place some 400–200,000 BP, it is understandable that the adaptive and selective forces that made this group of *hominins* ever more powerful, was their ability to predict events critical to their survival. Here was a natural selective dynamic better than other associated *hominins* traversing the northern Eurasian ice age terrains. But reproductive and thus demographic success was only one part of this revelation to themselves, and the external world at large, about their unique human nature.

At the time level of 100–90,000 BP in South Africa, the precursor Capoid racial/ethnic clade was already experimenting in the use of colored mineral and vegetative materials to paint and illustrate the world around them. Eventually, this artistic need found its way throughout Africa and into earliest historic times. There are tools from this time frame of history both in Africa and the Far East which resemble the later Cro-Magnon artistic tool kit (this discussed in earlier chapters). The genes were traveling.

About c. 45,000 BP, we find the Cro-Magnons entering Europe with a fully evolved cultural profile, the *Aurignacian*. In the thirty or so thousand-year occupation of Europe and West Asia, the world also witnessed the extinction of the Neanderthals. Here is exemplified the psycho-social power of this new competitive presence, and hints at the adaptive/selective power of high intelligence. And yet, the evidence of the socio-cultural lives of the Cro-Magnons places their art on a passion level higher than the practical hunting gathering economic preoccupations of life itself.

The above considerations bring us back to the question of the large endo-cranial capacity of the South African Capoid Boskapoids. Their body type may reflect an early hybridization with the wandering "baby faced" paedomorphs of the north, else as Carleton Coon hypothesized an early relationship with the wandering proto-Mongoloids. Indeed, the inherited delicate morphology of the now disappearing Bushman ethnicity reflects this ancient infusion. The few in numbers were the northern wanderers of the earlier stages of the *Hss* revolution,

400–200,000 BP. At this later stage, 90,000 BP they apparently were adeptly spreading their genes around. It is this genetic transformation of *Homo* which we have historically discovered throughout the planet. These humans at the least, can speak discursively. And it argues for the genetic transmission by these ancestors of Cro-Magnon to a world wide taxonomic uplift into *Homo sapiens.*

But what about the meaning of their transformative preoccupation with the arts? The production of food resources is certainly central to the positive selection of any human or animal group. Obviously, the Cro-Magnons were very successful hunters. But they also carried sculpted hand sized artistic/sexual evocations of their "girls back at the farm," clearly to keep their minds on the objects of their economic efforts. Throughout the Eur-Asian lands, these Cro-Magnon bands were puzzlingly uniform in their artistic passions. Their descendants in Europe, Africa and Asia were likewise involved in selective/survival efforts but also able to indulge their vital interest in the arts and other seemingly non-practical limbic/intellectual interests.

The abstract mind

Homo Ludens, originally published in 1938 by Dutch historian and cultural theorist, Johan Huizinga, discussed the importance of the play element in the history of culture. As noted in our earlier discussion, Huizinga suggested that play is central to the generation of culture. Play, however should be thought of as an element in the far more general nature of human abstract thought.

The coming together of the intellectual creation of symbols to guide our thought and behavior, combined with the joining of the emotional power of the limbic system to infuse juice into our thinking, creates the possibility for all the game playing that humans engage in. Universally, all *Hs* humans aspire to distance themselves from the dross of matter—the drudgery, one might say, of making a living through physical and mental labor.

We play at every opportunity. Play gives us meaning, even in the win or loss to a competitor. There is the energic juice of mentation in a Beethoven string quartet, a football game, a sonnet by Keats, and a stock market speculation. All such activities are laden with new symbolic possibilities, an abstract evocation of the ever chancy in our life experience. Our five senses receive the info, then the juices of our mammalian limbic system resonate to the meanings that our abstract system of cortical thought recognizes. Human life is permeated by this symbolic game of mining new emotional and cognitive evocations of our intellectual need to know.

We find this dynamic in all the societies and tribes of *Hs*. But it is the *Hss* mentality that takes this symbolic world deeply into the cortex and creates abstract systems—mathematical, physical, chemical, musical, artistic—that coordinate the sensory. This occurs only in civilizations that rise in symbolic meaning far beyond the thump of the beat as expressed not only in the jungles but in the cacophony of a Broadway. Talk to a mathematician, and you will hear the same language of form and emotion as with J. S. Bach.

The Cro-Magnons, for over thirty thousands of years, were ecologically bound to a mostly unchanging economy of hunting/gathering, certainly with some variations—the herding of reindeer, taming horses and wolves (dogs). The bloodiness and dependency of this hunting economy of need certainly had to impact their psychies. Why else were these animals, especially the horse, ritualized esthetically in the paintings within the caves. In addition to a steady diet of steak, the impacts on their minds however, were and are to stimulate them to objectify esthetically and abstractly the latent beauty of those creatures. These inner mental dynamics and their ongoing technological "utensils" represent in nutrio what all the other great historical civilizations quested for, the abstract vision.

Evolutionary message

There is a message here for those of us living at the end of the 20th and beginning of the 21st centuries. There is soon and inevitably to come an era of ecological restriction, as we have over-indulged in our natural resources, that have made possible the wars, genocides, the consumer culture of Walmart, and the world wide communication advances that gives us a Hollywood and the mass media.

No matter how advanced technologically, we are still prone to a pandemic accident. In one moment, we are sophisticated, wealthy. In the next, we are soon fated to undergo that Biblical discipline of scarcity.

Even those in power, indulging in their billions, their momentary unchallenged material fruits, will not command the future.

Eventually, a wise generation will have to come forth to decide in which direction this human planet will evolve. The lessons of the evolutionary past, the Irish Elk, Mr. Brontosaurus and all the other heralds of the end of the line, giganticism of size or numbers should help to clue us in about end of civilizations.

The Cro-Magnons, even the sadly disappearing Bushman descendants of the artsy Boskopoids, tell us that as a species we can survive in small numbers. As we recognize that our natural resources will no longer undergird this particular human setting of either eight or nine billion mixed sub-species humans, there still will be policy choices that can lead to civilizational wonderment.

Ecologies of the creative mind

Small Is Beautiful: A Study of Economics as if People Mattered, 1973, is a collection of essays by German born British economist E. F. Schumacher. The phrase "small is beautiful" came from a phrase by his teacher Leopold Kohr. It is often used to champion small, appropriate technologies that are believed to empower people to live better, in contrast with phrases such as "more and bigger is better."

Schumacher argued that the modern economy is unsustainable. Natural resources (like fossil fuels), are treated as expendable income, when in fact they should be treated as capital, since they are not renewable, and thus subject to eventual depletion. He further argues that nature's resistance to pollution is limited, and can and will break down. He concludes that government effort must be concentrated on sustainable development, because relatively minor improvements, for example, technology transfer to Third World countries, will not solve the underlying problem of an unsustainable economy.

Schumacher's philosophy is one of "enoughness," appreciating both human needs and limitations, and appropriate use of technology. It grew out of his study of village-based economics, which he later termed Buddhist economics, and is the subject of the book's fourth chapter.

Clearly, Schumacher was onto something. His concern with the anomie of our evolving urban dystopia has been realized as the dynamics of modern industrial/technological expansion proceeds. It has been facilitated by a huge dollop of "consumers" ensured by modern medicines' miracles, here subsidized by the West. The seemingly endless glut of cheap natural resources fuelled this "bigness," clearly in a world unconcerned with his Buddhist religio/philosophical inspiration. In 1973, when his book was published, the world population was 3.9 billion. Today it is about 7.9 billion.

Perhaps the greatest concern of this clearly indigestible demographic surge has been the cultural landscape of this new planet-wide culture. As we have learned from the time of the Boskopoids and the Cro-Magnons, a highly abstract culture uniting both limbic/cortical dimensions of our intellectual/symbolic needs has characterized what we still recognize as the great creative civilizations of history.

Classical demographies

Thoughtful writers have long discussed the social contexts within which the great abstract cultural and national traditions have developed, to leave at the least an impression of human intellectual creativity. The demographic issue is important to note. If we claim that eight billion humans of a wide variety of *Hs* and *Hss* intellectual levels, considering the natural resource issues alone that cannot long be maintained, we have to ask what is possible so as to live under decent physical conditions as well as enjoy an interesting society.

The pandemic alone warns us that even very advanced nations, can catapult the world into trauma, death, and impoverization. Just a few dozen air flights out of the country into the international arena, and humanity's "civilization" is shut down behind closed doors. For how long?

In ancient Mesopotamia, in Sumer, the first true city was Uruk. At its peak, around 2800 BCE, the city had a population between 40,000 and 80,000 people living between its six miles of defensive walls, making it a contender then for being the largest city in the world.

Aristotle insisted on the existence of a minimum population (from *Politics, VII*), as well as a maximum size, in both cases without specific numbers. In treating size, Aristotle gives emphasis on the public function of cities: "It is vital that the citizens know one another." He was also worried about the problems of security when cities become too large: "Foreigners and half-breeds usurp without difficulty the rights of citizens because it is easy for them to escape notice owing to the size of the population."

In *Laws, V. 74*, Plato states that the ideal republic would have 5,040 citizens, i.e., heads of households. This figure implies an optimum size population of about 20,000 people. He linked his optimum size of city to the need for communications among citizens, stressing that "the city must remain sufficiently small to permit the holding of public meetings with all of the citizens present."

Greek city-states of the ancient world did in fact remain limited in size. Athens (Attiki) was the largest Greek city-state, approaching a population of approximately 100,000 by 500–450 BCE.

"The other Greek city-states rarely had populations as many as 40,000 people. As a general rule, as soon as a city approached 20,000, its people emigrated to found new cities." (Leonidas C. Polopolus)

The following statistics mark out the West (not to negate the creative civilizations and contexts of Southern Asia, or Northeast Asia), societies, cities which in the time frame suggested here, were pregnant with or had given birth to important advances: Florence 1425, population 60,000. (Annenberg Learner); Paris at the end of the 15th century, with 150,000–180,000 inhabitants; Venice was the second largest city in Europe after Paris and probably the richest in the world; 1500, Venice city, c. 150,000-200,000; London, 1500= 50,000-100,000; 1600=c. 200,000. Alexandria, Egypt, Greek founded, c. 331 BCE, population c. 200 BCE=c. 200,000.

Consider for example, the conditions of life of the inhabitants of Egypt, c. 100 CE, with a population, c. 6 million and the condition of Egypt's people today with a population of c. 100 million. Then consider the leadership of the United States c. 1800, population c. 5,309,000 (894,000 slaves) and in 2020, c. 340,000,000 people.

Completing our biological revolution

Time's Passage: The dynamic of ever more intelligent hominins from at least 10 mya is current scholarly orthodoxy. The biological revolution which probably occurred somewhere in western Asia, some 400–200,000 BP stimulated the human brain to create esthetic technologies and pictorial art. These first-to-be-noticed gene transfers showed up in Western Asia and South Africa some 90,000 BP. Then, in an eventful migration and demographic explosion into Europe and east into Western Asia, a new civilization (Cro-Magnon) was created between the climatic interstadials, from c. 45,000 BP.

In small doses, these genes then and earlier kept migrating to the four corners of the Earth. In the process, all the peripherals *hominins* became *Homo sapiens*. Yet, the genetic/cultural achievements of the Boskopoid Capoids throughout Africa, and the Caucasoid Cro-Magnons throughout Eur-Asia and then throughout the world, saw the triumph of sapiency and radically reshaped the face of millions-of-years-old *hominin* profiles and culture. The so-called five races which historians and anthropologists have discovered, have continued this process of hybridization such that today, *Homo sapiens sapiens* genes are deposited throughout the human world and are doing their slow work of modernization. A new transition of the *hominins* is in the cards. Sadly, through the crush of crisis, these to now be revealed, ongoing for how many decades, centuries?

The crisis is manifold. The current pandemic lockdown of the world will soon be followed by an oncoming shortage of fossil energy which has over the past 150 years undergirded the demographic explosion. Scientific medicine's ability to counter the grip of disease over our sprawling populations may be limited. Certainly climate change will as usual create challenges to life, the greatest danger being a parallel crisis in water availability for the eight billion or so humans of the mid 21st century. This cliff which we are on the cusp of experiencing will create economic/social/political challenges never before experienced.

Will rationality, intelligent prescriptions of policy, save the human species from horrors of bestiality such as those that characterized the 20th century?

Transitions: The reshaping of our human world from our historical racial divisions into populations of *Hs* and *Hss* intellectual capabilities is already well under way. The scientific and technological prowess of the *Hss* mind is already pushing the *Hs* component into an evolutionary corner. Automation, robotization, AI (artificial intelligence) will be exponentially exported throughout the world as we go from the current plentiful availability of fossil energy, water, and natural resources, into a world of physical and cultural miniaturization/scarcity, where the wherewithal for financing human endeavor is shifted to the machine, the computer, the silent worlds of electronic information and communication. But do not forget the criminal lessons of political tyranny today exhibited by the followers of General Secretary Xi.

If our proposed taxonomic and socio/political perspective of an oncoming division between *Hs* and *Hss* is a real description of our present and future evolutionary revolution, we must reiterate its concrete bite. The jobs will disappear. There will be pressure in some nations, US included, to enforce paid employment for all hands regardless of merit appropriateness. Other nations of more homogeneous *Hss* profiles will have to squeeze the most "money" from each employable, thus stringent competency testing. For the ideological nations, U.S. included, the international competition will deliver serious challenges. Hopefully, covert rationality will surface and quietly enable a nation's economic solubility. How many billionaires will the masses tolerate world-wide?

Along with all attempts at making work lead to social equality, the question is bound to surface: "what about the reduction in population?" How to persuade the clearly *Hs* unemployables to lead the way in those nations which are quite variable in intellectual and education/vocational skills? Here again will surface the memory of the *Hss* versus *Hss*, wars, genocides and the *Holocaust* of the 20th century.

Today, entrance into the taxonomy of *Hss* stands at a verifiable high school diploma, U.S.A. the I.Q. educational level, averaging 90–100. Beyond the free market of employment, even the governmental bureaucracies will have to meet functionality by laying off millions. Where will the taxes receipts come from now, pandemic threats and costs, crude oil $150–200 a barrel, natural gas and even coal equally spiraling up in price? One can conceive that the I.Q. and equivalent educational/vocational skill levels will quickly demand workers in the 115–130 I.Q. range, medical doctors today averaging, c. 115 I.Q. [see discussion by Volkmar Weiss in Ch. 22]

Considering that in this oncoming worldwide and probably centuries-long crisis, vast populations will be supported on some kind of version of the dole. In the case of national chaos and possible violent anarchy, portions of the globe could again experience genocide, or triage, possibly requiring international peace-keeping interventions. Consider the consequences for the planet's humanity if plague again hits the world community. With economies in systemic default, there will be no short-term solutions. Our species may have reached an evolutionary tipping point, auguring an opportunity for plans looking to long-term sustainability, else a deeply unfocused set of ideologies threatening the extinction of the genus/species itself.

A message from the Emergency Senate: "Dear friend, we have welfare subsistence aid for you and a person of concern. The ongoing receipt of this aid depends on your not having any children. We will provide what is necessary to avoid pregnancy for yourself or a beloved one. We do not suggest celibacy."

Note that the Chinese "one child" policy of Chairman Mao was a national response to an economically overpopulated nation. Now, with high flying exports and industrialization, there are jobs for all, even with two or three children per Chinese family. Could Mao's policy be re-introduced under new medical threats, energy and other natural resource depletion conditions? *You bet.*

The word amongst some historians is that the Athenian Assembly, consisting of 500 male citizens, in the late 400's BCE, and during the costly Peloponnesian war with Sparta and its allies, voted to require all virile men to acquire a second wife to have children, considering the terrible manpower losses Athens suffered in the war. Socrates, a retired sixty-year-old hoplite, took on a young and second wife. She presented him with two sons. This family witnessed his execution in 399 BCE, when Socrates was aged seventy years old, his guilt voted by the *hoi polloi* majority.

Prospect

The dominance of the genes for ever higher *hominin* intelligence led to the great revolution that occurred some 400–200,000 BP. In terms of the creation and longevity of this new sub-species by "mother nature," we are yet babes in this evolutionary wood. The question now relates to our ability to surmount this important blip in our inevitable progression towards a "pan mictic" world species of ever more intelligent *Hss*. First, we need to leave *Hs* types behind, in the gentle dust of evolutionary history. This we have done repeatedly to lesser intelligent hominins over the past 10 or so million years. The fossils are clear about this.

There must be groups of wise humans quietly discussing and planning in theory how to redirect this technological civilization. Presumably, they also understand the nature of the oncoming darkness of natural resource scarcity, and our constipated world civilization's terrible vulnerability to the seductions of the ideologists.

Stalin had his five year plans, and industrialization—perhaps the only wise look into the future that he had. It allowed his countrymen to defeat the Germans and their European and Japanese allies in World War II. Chairman Xi, today, has his twenty-five-year plan to make China the world leader "in everything," especially autocracy. Will it have worked?

One day, not too long into the future, we all will wake up and discover a new, ongoing set of facts that will falsify this make-believe, ideological civilization that has "educated" us. With some wise planning, we could ameliorate the inevitable pain and traverse a downward path—"small is beautiful"—with the least of potential conflicts, human despair, and barbarism. It would inevitably be a several-century exercise in disciplined rationality, leading forward to a time when the species could again experience its inner civilizational dynamics—remember the unruly but creative 19th century.

If evolution has taught us anything, it is that the pathway for *Homo* is ever toward the realization of the promise that *Hss* represents to the world of life on planet Earth.

Some of these are creative, disciplined, symbolic advances in all physical/technological areas of life; an ethnic division of nationalities to explore modest and possible cultural divergencies and interactions. Think 19th-century classical music and literature. The utopian vision of economic and social equality of national populations, where humans can interact rationally to confront the inevitable challenges and questions of the future, is at the top of our list. Then, a must is a world government that shares a unified, scientific, philosophical perspective and supports democratic political forms of adjudication.

This is probably not going to happen peacefully, given the enormous philosophical, ideological and economic differences we have today between *Hss* and *Hs* populations. Even within the ranks of majority *Hss* educated citizenry, there is little agreement about the pathway into the future. The only thing we can be modestly assured of, is the ultimate resiliency of the *Hss* mentality and the energetic thrust of its sensory, limbic and cortical capabilities to look ahead predictively.

Beyond these expectations for the long term future of the species, we of the living generation can only wish the young of the next, *good luck*!

Chapter 24

Pandemic: Perspective on Our 21st-Century Apocalypse

In a dark time, the eye begins to see.
American poet Theodore Roethke (1908–1963 CE)

Nature's autonomy

The global infection of the Corona Virus gives sharp factuality to the warning described in Chapter 20 of this writing; a prediction about our species' impending future. That chapter focused on the still to be worldwide crude oil and natural gas shortages. Still, in less than a decade, given no future pandemics, these shortages will be sharply felt by the economy of most of the nations of the world and especially those with the greatest population to natural resource imbalances. The serious economic consequences for our world will be direct. To add to future concerns, the impact of this 2020 Covic-19 pandemic will long be felt, unlike previous plagues—the pandemics of the 20th and 21st centuries.

The Corona Virus is the latest "black swan" event hinted at in the above chapters. As predicted, there are already resources mustered to bringing this pandemic under scientific, medical control and then, hopefully, economic resuscitation. Soon, there will be a great effort to rebuild to the model of the current status quo. It will not work. Compounded by the inevitable drying up of fossil fuel resources, and possibly recurring virus attacks, the future of our species is opaque and vulnerable.

Where we are

As argued throughout this book, our species is an unfinished evolutionary work in progress. We are, as the present chaos reveals, reaping the harvest of a runaway set of dynamics in clear violation of the Darwinian adaptive and selective rules of the game. Fundamentally, we are an integral part of organic life's fight against the implications of Carnot's 2nd law of thermodynamics—*entropy*—requiring constant mobilizing of external energy sources to maintain our existence. Our human part in this galactic struggle is to create an adaptive and selective defensive position for the long-run, to try to protect our form of life against evolutionary events still beyond our ken, and perhaps ever beyond our control.

Four Horsemen of the Apocalypse

The theme of human weakness and the appeal for supernatural intervention goes back long in human history. Today we put our hopes in the mysterious powers of our "medicine men." The four horseman theme constitutes the back ground noise of every society, we also, are struggling against ourselves and nature.

New Testament, *Revelations 6.* These four horsemen appear to symbolically portray four disastrous occurrences that will take place before the second coming of Jesus Christ. Historians have argued over whether these events had already happened or whether mankind had yet to experience them. Most evidence points to the fact these had yet to have taken place. The prophetic characters of the "four horsemen" previously appeared in the Old Testament's Book of Zechariah, and in the Book of Ezekiel, where they are named as punishments from God.

We can propose four horsemen of the modern apocalypse, three, human initiated, the monstrous fourth, beyond the hands of human kind:

1. Overpopulation
2. Intellectual capital
3. Ideology/war/terrorism
4. Fear of Nature: resource depletion; radical climate/geological change; desertification/starvation, pandemics.

Overpopulation

We now consider this issue in the light of the Covid-19 pandemic. What was an ignored cloud on the horizon in 2019, now becomes a factor in the survival of Western civilization. For a few years, we will not have to face the looming prospect of demand for fossil fuels out-spanning possible supply; or, even the willingness of fossil energy surplus nations to deliver the last readily available barrel of oil to the wider world.

Some seers estimate the 2100 world population at 11 billion. Today, we have 7.9 billion. The increase is predicted to be centered in the so-called developing world. European and North American indigenous *Hss* population are slated to remain stable and even be reduced. Remember, even while humanity was engaged in the blood baths that characterized the 20th and 21st centuries, human fecundity, scientific medicine, and fossil fuels still exploded the total demographic imprint of *Homo sapiens*.

Immigration into wealthy lands will give rise to their increase. But that increase is estimated on the basis of our former global economic fortunes. It is clear that the quarantines around our planet, the wave of unemployment, and the psychological impact of happy generational dreams now evaporated, else placed on hold, will soon give us a very different set of predictions for economic growth.

We have, in earlier chapters, chronicled the explosive growth of world-wide populations, both of *Hs* and *Hss*, in terms of potential intellectual capital. But it was the *Hss* who created and developed the harnessing of fossil energy, including coal, as well as invented the medical miracles that the scientific mentality created. These two factors were fundamental in lifting up our world demography to what every educated person should long have realized are unsustainable levels. And now we will know this reality.

The shutdowns of the economies of all the nations by the governmental response to the virus has left vast numbers of the least educated in limbo, momentarily supported by printed "dollars" in the trillions. Many predict that the recovery by both the private and public spheres will fail. Already, governors of the states are teetering on the edge of "bankruptcy," as their budgets are deeply in the red and they cannot pay their current workers as well as the pensions of those in retirement.

The corona virus, while having infected the powerful as well as the weak, is by and large an urban catastrophe. The United Nations estimates that there are eighty-one cities in the world with populations over five million, and many more vastly beyond this figure. This is hardly the demographic picture we have described, which created the proto civilization of the Cro-Magnons, as well as the literacy of the urban Sumerians. Here, we tabulate hundreds of thousands of generations in the making. We have had devastating plagues in the recent past, such as the terrible flu epidemic of 1918. They were not global in impact.

We are only on first base in seeing the consequences of this 2020 event. When governments are printing untold monies in the attempt to help the masses, there are still consequences to be discovered. The financial system itself is slated for collapse. Forget the democratic institutions as we have built them since the Enlightenment. Forget the dreams of the libertarians. It is difficult enough management for any nation when coal, crude oil, and natural gas are plentiful and cheap. Uranium, ad the wind that blows over the wheat field, the minerals that create the solar panels, dream on! The governments that print the *Euros* will have to manage (*control*) the masses of humans in panic within these imploding urbs.

The viruses that continually challenge modern human life confront us—a memory from the earliest eons of life itself. They mutate constantly and kill, especially as they penetrate the cancer that is human overpopulation.

Intellectual capital

In the period before the Enlightenment and the beginning of the scientific/industrial period, the world population had many ups and downs. One might deem it a continuum of prosperity and decline. We write here of the progeny of Cro-Magnon, discussed earlier in the context of civilizational advances from the Atlantic, east to the Pacific Ocean. Inevitably, the rise and decline of these *Hss* societies gave impetus to varying skill and educational levels. The example of Rome, the urban center of the empire, is illustrative. Rising up over a period of 500 years, it had attained through conquest and economic unification a cosmopolitan mix of the educated, which engendered games, sometimes brutal, to entertain the 1–2 million inhabitants who resided there, under Augustus, 1st century CE. The continual wars with the invading German tribes over the following centuries reduced the population, so that when Alaric the Visigoth gained control in 410 CE, there were only several hundred thousand inhabitants.

As time went on, even under Papal control, the city became a shell of its greatness. Estimates are that in the period 1000–1200 CE, there may have been only 30 thousand inhabitants in Rome. One wonders about the character of the populations, not only of Rome, but of the other cities and towns of Italy during the supposed "dark ages" of Europe. Where one would describe the culture of imperial Rome, compared to the Hellenistic Greeks' of a similar time frame, leading up to c.100 CE, as populist and brutally low in intellectual elan except for a literary minority, the later burst of Renaissance creativity, now under the aegis of a religious hierarchy, and in the same geographical space and economic and technological advance as imperial Rome, must have arisen from a very different "populist" ecology. Clearly, quiet events in the demographic transition of the populace provided an environment of hardship, struggle and a selective intellectual rise in the character of the people. It could not alone have been the mere addition of those "violent" gothic genes to the population of Italy that created this oncoming Renaissance.

Many view the "Black Death" of 1346–1369, which ravaged Europe as well as the Near East and killed more than 50 million people, as a puzzling event in the history of humanity. Coming at the beginning of the Renaissance in Europe, it seemed to have ricocheted into a turning point in the humanistic perspective on life. The pickup in economic activity was one puzzling element, as was the seeming emancipation from the fatalism of the religious explanations of its horror—it was an invitation to secular thought and inquiry.

The Italian writer and poet, Giovanni Boccaccio (1313–1375 CE), author of *The Decameron*, which recounts the tales of a group of ten citizens trying to escape the plague by seclusion, describes in his introduction the main ways in which people reacted to the pestilence:

> There were some people who thought that living moderately and avoiding any excess might help a great deal in resisting this disease, and so they gathered in small groups and lived entirely apart from everyone else. They shut themselves up in those houses where there were no sick people and where one could live well by eating the most delicate of foods and drinking the finest of wines, allowing no one to speak about or listen to anything said about the sick and dead outside...Others thought the opposite: they believed that drinking excessively, enjoying life, going about singing and celebrating, satisfying in every way the appetites as best one could, laughing, and making light of everything that happened was the best medicine for such a disease...Many others adopted a middle course between the two attitudes just described: They did not shut themselves up, but went around carrying in their hands flowers, or sweet-smelling herbs, or various kinds of spices and they would often put these things to their noses, believing that such smells were wonderful means

> of purifying the brain, for all the air seemed infected with the stench of dead bodies, sickness, and medicines. (Mark 2020, 7–8)

There is a possible message here from our own pandemic, which, if not as devastating in terms of human mortality as the "Black Plague," is sure to leave its economic, intellectual capital, psychological and even ideological mark. Our world being so interconnected and heavily populated, will feel the sharp brand of this Pandemic, as well as the consequences of the spectacular efforts made around the world to protect vulnerable humans.

Note, the new heroes of this battle to protect the civilians of our nations are the medical savants, from the brainiest medical researchers to those assigned to clean and disinfect the hospitals and research labs. In this war, the tanks, missiles, aircraft carriers are in abeyance except as mobile hospitals. The strategists against this mysterious enemy are the medical experts, here garbed in an almost religious mantle of prophetic leadership. The vast amounts of debt nations are issuing are to protect the viability of least specialized members of the economic hierarchy. The expectation in this descent into defensive expediency is that we will have to fight hard to recover a decent working class economy for masses of humans now furloughed or laid off their jobs. Many of these jobs will disappear, as the employer classes attempt to maintain solvency by biting hard on the human and taxation costs. Instead of an advancing, growing financial economy that needs to absorb more hands, a world in fear of a recurrence of viral pandemics will now tread very carefully as they plan for the future. An economy which again begins to use fossil fuels—the blood that flows through the arteries of a 20th and 21st-century prosperity and war—will soon reveal to its users that these resources are not immune from history.

In the short run, the next few decades, things will be rough in terms of the political conflicts over government controls, attempts at egalitarian socialism, the inevitabilities of debt repayment and progressive tax policies. We are likely to witness a universal search for skilled, abstract, intellectual abilities to do the work that a modern robotic economy will now demand. Soon, from the police to the military, even parking lot and cleanup skills, the efficient mind will move to the front. Governments which are not lean in the use of tax revenues, along with the private sector's efficiency in the use natural and human resources, will bankrupt their nation. As the parable notes, "when the tide goes out we will know who is swimming naked."

Every nation that is an ethnicity will be selfishly pulling in its borders for protection in the search for economic and then social survival. The smart ones at the top will be rewarded for creating the most efficient modes of production. Even Vladimir Lenin, taking his lesson from Marx's historical view of economic modernity, realized that the Soviet Union could never compete with its neighbors unless it allowed the free mentality of his people to drive the nation towards modernity. By freeing brain power with all its sociological and economic *hybris*, Lenin, in 1921, was opening the door to national survival.

What we will be witnessing in the decades to come is the continuation of an ancient evolutionary dialogue between the dynamics of plenty and scarcity. For two centuries, the human species has been expanding on the wings of an indulgence of plenty. Science and natural resources have exploded our population within the world of *Hss* humans. They have also extended philanthropy to the outlier *Hs* domains. It is doubtful that this dynamic can continue to maintain a large portion of our nearly 8 billion humans aboard this planet.

From plenty, we slip into scarcity, to witness a new human move. The war for survival between the intellectually competent producers and protectors of our civilization, and those who are the receivers of this diminishing surplus of care for the poor, continues. In the West, will we continue the maintenance of Medicare and Medicaid, government-subsidized health, food stamps, varieties of government assistance to the "needy," including "great society" grants to single parents? Government policy is built on the taxes produced by the free marketplace, else in a totalitarian state, ladled out by fiat. The poor have little philanthropic recourse. In semi-democracies, this war for survival becomes political. How long will the tax payers tolerate such "misuse" of their labor? In any case, scarcity argues for the renewed dominance of the *Hss* mentality, as it adjusts to the now ever-advancing cognitive demands of science and technology for social survival.

Ideology

We have noted in earlier chapters of the historical transition in humanity's need to believe, this from theistic and monarchical obeisance, given the failures to predict and realize their commands, to a more secular search for factual, symbolic methodologies that one could commit to. The issue of hierarchy in human society has always been a contentious element in our social interrelationships. For a few thousand years, priests, with their literate and intellectual skills, along with the charisma of warrior kings, gave these groups authority and privilege.

Even in the democratic and republican political environment exemplified by ancient Greece and Rome, the oligarchic classes held sway in many communities, fighting for privilege with egalitarian and populist persuasions. Indeed, in dynamic expansive times, wealthy privilege was sustained, while the *hoi polloi*, engorged on their crumbs, kept quiet. In our own dynamic centuries, wealth has accumulated outside of the political arena due to the creative thrust of industrial and technological innovation. When this innovation and wealth began to create significant institutional class and political power difference, the masses shifted their allegiance to socialism's claim of populist intellectual equality. And therefore, social class power immorality, *a la* Marx.

As we have previously explained, there came to be one massive, century-long and worldwide ideological commitment of belief that this wealth and seeming power differential was not earned. Rather, it was an illegitimate and grasping moral evil that required extirpation. And thus, the 20th century saw hundreds of millions of talented *Hss* humans destroyed in wars and genocides by their fellow *Hss* compatriots, often because of their ethnicity.

In so-called leftist circles, mostly in "democratic" environments such as the USA, enormous accumulations of private wealth have been created along with a thriving working and middle class. Yet, a very large part of its population is still essentially economic and political wards of the state. What the USA had going for it in the late 20th and 21st centuries was a dynamic which allowed for the government subsidization of a large welfare class, including millions of illegal immigrants magnetized by the "freebies"—medical, educational, and economic benefits not available in their countries of origin.

Hypothetically, our new virus might never obtain classical immunity relief for its recipients. Other viruses have remained beyond medical reach. If this occurs, it could engender a permanent imprint on the world, of fear, scarcity, and morbidity, such that the generosity of political philanthropy both within nations and between nations could create a "me first" nationalistic tonality in our planet of disappointed expectations. It will be fatal for globalization and the wars of ideology will return.

Argument 1: Why are groups of humans within and between nations having such tough luck? Is it not a conspiracy of the well-off, the power makers, to take it out on the poor, especially when they are identified by having different ethnic orientation or skin coloration?

What will the other side counter to this revolutionary accusation?

Argument 2: No, the other side will say, it is a matter of personal lack of discipline, not of studying, working hard enough, or engaging in bad behavior.

In democratic societies, the poor can still vote and agitate, assisted by committed ideological leaders, "socialist or fascist." The issue of human intelligence variability is on the prohibition index. Such ideas cannot be entertained without significant duress. One cannot predict the outcome in a universal environment of blindness and ignorance.

If secular rationality could obtain a place in our discussion, as we have argued in this and in other writings, there could be mustered national and international efforts to gradually complete our "unfinished biological revolution." We could be on our way to create a world wherein all humans have sufficient intellectual and educational capacities to compete economically and socially. This, while the smartest *Hss* are creating ever-heightening, abstract educational and work skills. Remember, we are on the road for a new intellectual direction in economic productivity.

Plague: Athenian preamble

From this time distance, there can only be controversy over the fall of Athens by the Spartans and their allies in 404 BCE. The plague itself hit Athens between 430–429 BCE. In between these two events, a vicious war was fought between Athens, with her Greek allies, and Sparta with their Greek cohorts. This was a cousin's war similar to what took place in the USA between 1861–1865. It is important to relate these historical events and their consequences to the contemporary picture of our world still undergoing the pangs of the Corona Virus. What will be the outcomes of this global event already having a most disruptive impact on the nations of the world, no less the West itself?

Athens was acknowledged as the great symbolic actor on the world stage in 431 BCE, at the outbreak of new conflicts between these two groups of Hellenic powers. Two generations earlier, the Athenians in 480 BCE had abandoned their city to the Persian host who burned down not only the city, but the wooden painted holy Acropolis that hovered over their city. But as with their strategic victory ten years earlier at Marathon (490 BCE), they had a different plan to deal with a much larger Persian invasion, now including Greek forces. The Athenians and Spartans were then, 480 BCE united. Remember the heroic defeat of the small Spartan force at Thermopylae, by the massive Persian forces with Greek assistance. This time, the Athenians had different skills in their plan to repel this Indo-European Asiatic host. Off the coast of Athens, near the refugee island of Salamis, the Athenian naval leader Themistocles ambushed the international Persian fleet and destroyed it. A year later, an international Greek army led by Sparta at Plataea in Boeotia gave the *coup de grace* to Persia which fled back to Asia.

Athens took advantage of this unexpected triumph by emphasizing the democratic elan that the city had been building into an international Hellenic league led by its navy. Long in the making, the various Greek city states subsidized the Athenian fleet, now always at the ready. The gold was to be deposited on the island city of Delos. At first, it was a product of benign imperialism, led by a successful Athenian military leader from the aristocratic classes, Pericles, who had an even greater democratic vision for Athens. Although the contributed wealth of the allies was supposed to be held on Delos, the gold was used by Pericles to build his new holy city on the hill, now with marble.

Greece itself was alive with new secular ideas and practices, partially stimulated by an open political system, even in constant ferment. Athens under Pericles' intellectual vision became a veritable international city in this moment of history. Downhill, in the port of Piraeus, merchants from the world created a ferment of wealth. These were the *perioeci*, "dwellers around," the practice first originating within Sparta. The *perioeci* were non-citizens, but free workers who flooded into Athens much like the craftsmen of the amphorae, designers who began to make Athens wealthy by trading the newly planted products of the vineyards and olive plantation of the 6th century BCE. Now in the fifth century, Athens attracted playwrights, philosophers of science, sophists (lovers of wisdom [usually political or moral thinkers] from throughout the Hellenic world. Among these were talented adventurous women, such as Aspasia, who hailed from the great Ionian city of Miletus [on the western shores of today's Turkey]). Then, Athens became the originating city of natural philosophy, as well as a goad of Persia. Aspasia caught Pericles' intellectual eye, causing a political ruckus as he divorced his current wife, mother of his two sons. Once she, in turn, was remarried, things calmed down. Aspasia gave Pericles a son.

During this period, there was plenty of military conflict, mostly with Sparta **and** its allies, as well as its helots, usually the semi-enslaved, then contracted warriors. The Spartan phalanx had defeated the neighboring Peloponnesian city states. In the great war, 431 BCE, both Corinth on the Peloponnese and Megara, a neighbor of Athens, were significant allies of Sparta. Thus, in opposition to the expansive Athenian Delian league, Sparta became a well-armed, highly militarized, semi-democratic state.

War with Sparta, 431–404 BCE

Pericles ordered all residents of Attica (Athenian homeland) into the city when Sparta invaded Attica in 431 BCE, and began burning the Attic farms. Pericles returned the favor by sending his fleet to attack coastal cities in the Peloponnese. His famous funeral oration, 430 BCE, was written down by the Athenian historian Thucydides. It exemplifies for all time the vision of human freedom and the rights and obligations of the individual in a democratic society. In the summer of 430 BCE, an epidemic broke out in this now overcrowded urb. The exact identity of the disease is uncertain, typhus or typhoid fever are suspected, but this has been the source of much debate.

> The most terrifying aspect of the whole affliction was the despair which resulted when someone realized that he had the disease: people immediately lost hope, and so through their attitude of mind were much more likely to let themselves go and not hold out ... If people were afraid and unwilling to go near to others, they died in isolation, and many houses lost all their occupants through the lack of anyone to care for them. Those who did go near to others died, especially those with any claim to virtue, who from a sense of honor did not spare themselves in going to visit their friends. [Thucydides II.vii.3—54].

It is estimated that over 100 thousand city dwellers out of the extended 300 thousand population of Athens died in the plague.

Pericles witnessed the death of both his legitimate sons, Paralus and Xanthippus, from his first wife, in the epidemic (429 BCE). His morale undermined, he burst into tears and not even Aspasia's companionship could console him. He himself died of the plague later in the year.

The story of the defeat of Athens after this terrible disaster is a long chronicle of democratic chaos and incompetence, without the practical intelligence of Pericles. In moments of truce, the Athenians completed their holy city on high, fought bravely and stupidly as citizen hoplites or sailors, committed war crimes, and were finally crushed by an international fleet sponsored by Sparta, which blocked their food supply from around the Black Sea. The Spartans refused to allow their own allies to burn Athens down. They did hire an orchestra of female flutists to accompany the walls of Athens as they were torn down.

Plato, from a wealthy Athenian family, wrote his dialogues as a testimony both to the philosophical mind which Athens continued to lead, with his

Academy, and the memory of a brilliant, free commoner hoplite, Socrates. A renewed populist Athenian government voted to execute Socrates, who could have, but decided not to run (399 BCE).

Plato's devoted student, Aristotle, was born into a medical family from Stagira, in the north of Greece. Both men had been recruited to serve the Macedonian monarchical leaders. Aristotle returned from the Academy to teach the son of Macedonian King Philip, Alexander. After the battle of Chaeronea (338 BCE) and possibly with Macedonian subsidies, he established his own Lyceum in Athens. Both institutions, the Academy and Lyceum would for centuries maintain Athens' position as a center of secular learning. They would lead Hellenic intelligence and science into the international Hellenistic world of Greek culture, this transformation abetted by Aristotle's student Theophrastus, a botanist.

The Spartan leadership of the Hellenes was short lived. At Leuctra (in Boeotia) in 371 BCE, an upstart Theban army, they once a Persian ally against their fellow Hellenes, now under the leadership of a unique general Epaminandos, destroyed the Spartan army. When the survivor Spartans returned to the Peloponnese, many without their shields or those bearing their dead comrades on them the, women of Sparta are said to have emerged from their homes crying shame at their defeated husbands and sons. These Spartan women, at the core of the Spartan eugenic tradition to create warriors, were free, independent, unique amongst the Greeks. Perhaps there is one other parallel in the ancient world, in role of the Egyptian women. The finale came a few decades later at Chaeronea, (again near Thebes), 338 BCE. Facing the increased strength and military sophistication of Macedonia, led by King Philip, there now were two views about the future of Greece and it civilization. One, led by the thinking of the Athenian, Isocrates, saw a pan-Hellenic unity which included the country folk world of Macedonia, now increasing in wealth and power. (Euripides the Athenian dramatist migrated to Pela in Macedon in 408 BCE amidst the chaos in Athens.)

The other was led by the Athenian political orator Demosthenes (384–322) who saw Philip as the antitheses of the Hellenic vision. Withal his power and status in Athens, Demosthenes went out as a mere hoplite to Chaeronea, Athens now allied with Thebes and in the latter's home territory. It was no contest. Philip and the young Alexander outmaneuvered the allies with innovative technical weapons and the remnant allies fled home. Perhaps it was the teachings of Aristotle to Alexander that rendered the fate of Athens under Macedon, benign.

Lessons of the Plague

Pericles was approximately 65 years old. He could have been elected to leadership office for another ten years. He was a committed democrat from wealthy ancestors. He understood the potentiality of a democratic polis linked to a high civilizational vision, the power of secular knowledge bonded to artistic expression, but with a vigorous citizen's armored defense. The war with Sparta could have been renegotiated. The world within which Sparta had flourished was rapidly changing in the direction of the Athenian Periclean vision. Epaminandos in 371 BCE, revealed the potential weakness of Sparta. But these are hypothetical thoughts

A great new militarism took over the world scene parallel to Athens' political and economic fourth century BCE decline. Demosthenes in 338 BCE was a voice that truly exposed the helpless decline of a once great vision of humanity and the coming of a new if ancient militaristic solution. But something did remain; the vision of a Socrates, Plato, Aristotle. The world had already seen centuries of Hellenic artists, physicists, sophists, medical doctors, and philosophers. These three figures centered in Athens brought together all of these tendrils of art and thought into a unified vision of human kind. Gone was theological servitude. Instead, the theme of intellect serving the investigation our universe of human experience, flourished in this polis without the blockade of societal (Roman) power. Once unleashed by these thinkers, it flowed around the world, first by a military opening of the doors for a new vision of a Hellenistic secular culture for humanity. At the upper ranges of *Hss* intelligence, it could continue on it own impetus for the search for factual knowledge, the esthetic potential of the human soul, even into our own time.

An interesting example of the power and continuity of this new systematic intellectualism which the three great Athenians had set loose is the fact that when Aristotle was teaching Alexander in Pela there was at least one other student in the class. This was Ptolemy 1, who laid the groundwork for the historic Museum and Library in Alexandria after the death of Alexander. Also it noteworthy to recall that Aristotle's grandson, of Aristotle's daughter Pythius, with whom Aristotle sought refuge when he escaped Athens, himself became a medical researcher. This Herophilus was born at Chalcedon in Bithynia and became a famous anatomist and surgeon, considered to have been the first to dissect the human body systematically.

In this way the Greek intellectual genius made its way through time to Rome and then Christianity, ultimately fulfilled in the European Renaissance.

But down below the military unifications, the wealth that international trade brought to the Mediterranean littoral, the undermining of the old civic religions, and the intellectualism at the top, was a growing malaise, the growth of mystery religions.

Early in the 20th century, the English classicist Gilbert Murray wrote an essay, "Four Stages of Greek Religion" (1910). While about Hellenism, he pointed it as an arrow toward the war that Europe was soon to be engaged in, and then possibly destroying itself. It was Gilbert Murray's thesis that in the period between 340 BCE and 200 CE, the so-called Hellenistic Age, there was a "failure of nerve," a retreat from the growth of human intellectual development. In that period there is a change in the whole relation of people to the world about them. There "is a rise of mysticism, a loss of self-confidence, of hope in this life and of faith in normal human effort." There is also "a cry for infallible revelation."

Today

What is the message for us now having experienced a global pandemic that surely will have left a mark on the minds of these 7.7 billion humans on planet Earth. As we write, the residents of this planet have an extra few years before the eventual peaking of oil/natural gas production and experiencing the gradual repeat of a shrinking economy and the social distress which it will engender.

The only thing that one can predict, is the relentless millennial filtering of our *Homo sapiens* population upward into *Homo sapiens sapiens*' compatibility with the cognitive demands of scarcity. As we have experienced in the 20th century genocidal struggles of *Hss* against *Hss*, there may be sharp and possibly tragic historical events in this century-by-century biologically selective process.

The rational solution would be for the elite scientific minds of the day to engage in an educational process to teach humankind how it could complete this biological transition in social peace and moral stability.

Index

D

I

www.ingramcontent.com/pod-product-compliance
Lightning Source LLC
LaVergne TN
LVHW052214200926
848481LV00001B/103
* 9 7 8 1 6 8 0 5 3 9 2 5 7 *